The Lithium Air Battery: Fundamentals

Nobuyuki Imanishi • Alan C. Luntz • Peter Bruce
Editors

The Lithium Air Battery: Fundamentals

Springer

Editors
Nobuyuki Imanishi
Faculty of Engineering
Department of Chemistry
Mie University
Tsu, Japan

Peter Bruce
School of Chemistry
University of St. Andrews
St. Andrews, UK

Alan C. Luntz
IBM Research, Almaden Research Center
San Jose, CA, USA

SUNCAT, SLAC National Accelerator Laboratory
Menlo Park, CA, USA

ISBN 978-1-4939-4380-7 ISBN 978-1-4899-8062-5 (eBook)
DOI 10.1007/978-1-4899-8062-5
Springer New York Heidelberg Dordrecht London

Printed on acid-free paper

Springer is part of Springer Science+Business Media (www.springer.com)

Preface

Energy storage has emerged as one of the greatest challenges facing society in the twenty-first century. With the world population increasing and many parts of the world becoming more industrialized, the demand for usable energy is predicted to rise inexorably (by 56 % from now to 2040). As a result, coal, oil, and gas will be joined increasingly by nuclear and renewable fuels for electricity generation. Deploying renewable energy will certainly become necessary to mitigate CO_2 emissions and their effects on climate change as energy demand grows. The use of renewables transforms the demand for energy storage. As this component grows relative to traditional sources, it will ultimately be necessary to store the unpredictable supply of electricity from renewable generation and provide it to consumers when required. While the time scale for employing electrochemical energy storage on the grid is not immediate, the electrification of transport is driven by the same imperatives and is beginning now. The lack of cost-effective, safe, long-lasting electrical energy storage with sufficient energy densities is right now the only significant bottleneck preventing mass market conversion to electric vehicles.

Meeting the needs for energy storage will require a range of solutions, including lithium-ion batteries. Due to its relatively high energy density, the lithium-ion battery—which Sony introduced in 1991—led to the portable electronics revolution of the last two decades. The Lithium-ion battery will continue to evolve slowly providing increasing energy density; it is currently the technology of choice for electric vehicles and will be critical for many years to come. However, the general consensus today is that for true mass market acceptance of electric vehicles, a battery chemistry beyond Li-ion with a higher energy density (and hence range-cost tradeoff) must ultimately be employed. Therefore, it is important to look beyond the horizon of lithium-ion and explore alternative rechargeable batteries that might exceed what lithium-ion batteries could deliver. The options are limited and include zinc air and lithium sulfur; however the battery with the highest theoretical specific energy is obtained by combining a lithium anode with an O_2 cathode, i.e., the lithium air battery.

Using O_2 as a fuel at the cathode of a battery is not new. For example, primary zinc air batteries have been used for many decades. The lithium air battery has been explored since the 1970s; especially noteworthy is the pioneering work by Abraham in 1996. Today, the need for better energy is driving intense interest in the rechargeable lithium air battery. As with any technology that has the potential to be transformational, significant barriers need to be overcome. Prior to the introduction of the lithium-ion battery, many believed rechargeable lithium batteries would never become a commercial reality. Work on the early incarnations of the rechargeable lithium air battery revealed that the reactions taking place in such cells were not those expected or desired. The clear lesson from this work was that fundamental understanding of the chemistry and electrochemistry underpinning the operation of the rechargeable lithium air battery was essential and that only by acquiring such knowledge would it be possible to address the barriers preventing commercial realization of lithium air. The contributors to this book are actively engaged in research at the cutting edge of lithium air batteries and focus on understanding the processes taking place in the cell and overcoming the hurdles found therein. The role of electrolyte and electrode stability is discussed, as are the mechanisms of the electrode reactions and the morphologies of the products. The use of solid electrolytes, protected lithium anodes, and the issues of air handling at the cathode are all examined. The state of development in the field encouraged us that this was an appropriate time for this book. Although progress has been significant, much still remains to be done to explore the science behind the lithium air battery so that informed and evidence-based decisions can be made concerning the ultimate viability of this technology.

Tsu, Japan
CA, USA
St. Andrews, UK
November 2013

Nobuyuki Imanishi
Alan C. Luntz
Peter Bruce

Contents

Contributors

Paul Albertus Bosch Research and Technology Center, Palo Alto, CA, USA

Fanny Bardé Advanced Technology 1, Toyota Motor Europe NV/SA, Zaventem, Belgium

Peter G. Bruce School of Chemistry, University of St. Andrews, St. Andrews, Fife, UK

Thomas J. Carney Department of Materials Science and Engineering, Massachusetts Institute of Technology, Cambridge, MA, USA

Maria K.Y. Chan Center for Nanoscale Materials, Argonne National Laboratory, Lemont, IL, USA

Yuhui Chen School of Chemistry, University of St. Andrews, St. Andrews, Fife, UK

Jake Christensen Bosch Research and Technology Center, Palo Alto, CA, USA

Larry A. Curtiss Material Science Division, Argonne National Laboratory, Lemont, IL, USA

Center for Nanoscale Materials, Argonne National Laboratory, Lemont, IL, USA

Stefan A. Freunberger Christian Doppler Laboratory for Lithium Batteries, Institute for Chemistry and Technology of Materials, Graz University of Technology, Graz, Austria

Betar M. Gallant Department of Mechanical Engineering, Massachusetts Institute of Technology, Cambridge, MA, USA

Nikolay Goncharenko PolyPlus Battery Company, Berkeley, CA, USA

S. Gowda IBM Research, Almaden Research Center, San Jose, CA, USA

Jeffrey P. Greeley School of Chemical Engineering, Purdue University, West Lafayette, IN, USA

Mary Hendrickson US Army CERDEC, Aberdeen, MD, USA

H. Horn IBM Research, Almaden Research Center, San Jose, CA, USA

Nobuyuki Imanishi Department of Chemistry, Mie University, Tsu, Japan

Lutgard De Jonghe PolyPlus Battery Company, Berkeley, CA, USA

B. Kumar Electrochemical Power Group, Energy Technology and Materials Division, University of Dayton Research Institute, Dayton, OH, USA

Mechanical and Aerospace Engineering, University of Dayton, Dayton, OH, USA

J. Kumar Electrochemical Power Group, Energy Technology and Materials Division, University of Dayton Research Institute, Dayton, OH, USA

David G. Kwabi Department of Mechanical Engineering, Massachusetts Institute of Technology, Cambridge, MA, USA

Kah Chun Lau Material Science Division, Argonne National Laboratory, Lemont, IL, USA

Timm Lohmann Bosch Research and Technology Center, Palo Alto, CA, USA

Yi-Chun Lu Department of Mechanical Engineering, Massachusetts Institute of Technology, Cambridge, MA, USA

A.C. Luntz IBM Research, Almaden Research Center, San Jose, CA, USA

SUNCAT, SLAC National Accelerator Laboratory, Menlo Park, CA, USA

B.D. McCloskey IBM Research, Almaden Research Center, San Jose, CA, USA

Robert R. Mitchell Department of Materials Science and Engineering, Massachusetts Institute of Technology, Cambridge, MA, USA

Fuminori Mizuno Battery Research Division, Toyota Motor Corporation, Susono, Shizuoka, Japan

Materials Research Department, Toyota Research Institute of North America, Ann Arbor, MI, USA

Vitaliy Nimon PolyPlus Battery Company, Berkeley, CA, USA

Eugene Nimon PolyPlus Battery Company, Berkeley, CA, USA

Alexei Petrov PolyPlus Battery Company, Berkeley, CA, USA

Edward Plichta US Army CERDEC, Aberdeen, MD, USA

Kirill Pridatko PolyPlus Battery Company, Berkeley, CA, USA

Yang Shao-Horn Department of Mechanical Engineering, Massachusetts Institute of Technology, Cambridge, MA, USA

Department of Materials Science and Engineering, Massachusetts Institute of Technology, Cambridge, MA, USA

Philippe Stevens R&D Division, Les Renardières, Electricité de France (EDF), Cedex, France

Kensuke Takechi Advanced Battery Laboratory, Toyota Central R&D Laboratories, Inc., Nagakute, Aichi, Japan

Carl V. Thompson Department of Materials Science and Engineering, Massachusetts Institute of Technology, Cambridge, MA, USA

Gwenaëlle Toussaint R&D Division, Les Renardières, Electricité de France (EDF), Cedex, France

Steven J. Visco PolyPlus Battery Company, Berkeley, CA, USA

V. Viswanathan SUNCAT, SLAC National Accelerator Laboratory, Menlo Park, CA, USA

Department of Mechanical Engineering, Stanford University, Stanford, CA, USA

Jie Xiao Pacific Northwest National Laboratory, Richland, WA, USA

Wu Xu Pacific Northwest National Laboratory, Richland, WA, USA

Osamu Yamamoto Faculty of Engineering, Department of Chemistry, Mie University, Tsu, Mie, Japan

Ji-Guang Zhang Pacific Northwest National Laboratory, Richland, WA, USA

Chapter 1
Introduction

Osamu Yamamoto

1.1 Energy Density of Lithium Air Batteries

Lithium air batteries have far higher theoretical energy density than lithium-ion batteries, which have the highest specific energy density of conventional battery systems at present. The calculated specific energy density of the lithium air battery, excluding oxygen, is as high as 11,430 Wh kg^{-1}, which is comparable to the energy density of fossil fuels. Lithium air batteries are now attracting increasing attention and R&D effort as possible power sources for electric vehicles (EVs). However, several factors reduce the theoretical maximum energy density. For one, the weight of oxygen must be included, because part of the discharge product is stored in the battery. In this introduction, the practical achievable energy density of lithium air rechargeable batteries will be compared with those of conventional battery systems.

Since the discovery of practical use lead–acid batteries by Planté in 1859, many types of battery systems have been developed. The energy density of the lead–acid battery calculated from the mass of the active material and the open-circuit voltage (OCV; 2.05 V) is 171 Wh kg^{-1}. The cell performance of the first practical rechargeable batteries has improved drastically within the last half century; however, the energy density is not yet completely sufficient for portable electric devices and EVs. The driving range of an EV with lead–acid batteries was estimated to be less than 100 km with a full charge. The Agency of Industrial Science and Technology (AIST), under the jurisdiction of the Ministry of International Trade and Industry (MITI) of Japan, instituted a battery research program for EVs in 1971. Improved lead–acid, iron–air, iron–nickel, zinc air, and sodium–sulfur batteries were studied for 6 years under this program. The specific energy density of the improved lead–acid battery (140 Ah pack) was reported to be 64.5 Wh kg^{-1}

O. Yamamoto (✉)
Faculty of Engineering, Department of Chemistry, Mie University,
Tsu, Mie 514-8507, Japan
e-mail: Yamamoto@chem.mie-u.ac.jp

N. Imanishi et al. (eds.), *The Lithium Air Battery: Fundamentals*,
DOI 10.1007/978-1-4899-8062-5_1,

(115 Wh L^{-1}) at 0.2 C rate (12 W kg^{-1}) with a 200 cycle lifetime until a capacity loss of 55 % [1]. The highest specific energy density of 131.5 kW kg^{-1} was developed for a 2.46 kWh zinc air system unit. The specific energy density was two times higher than that of the lead–acid battery, but the columbic efficiency for the charge and discharge processes was as low as 37.0 %. Sodium–sulfur batteries (108 Ah packed unit) showed an energy density of 89.4 Wh kg^{-1} (265.5 Wh L^{-1}) at 0.2 C rate. The cycle life (20 % capacity decrease) was 150 cycles. However, there was no battery developed among these systems that was acceptable for EV applications

In 1989, an advanced nickel metal hydride rechargeable battery was commercialized. This battery consists of a hydrogen-absorbing metal alloy anode, a nickel oxyhydroxide cathode, and a potassium hydroxide electrolyte. The energy density of a 1 kWh module was around 60 Wh kg (150 Wh L^{-1}). The volumetric energy density is higher than that of lead–acid batteries and this system showed excellent cycle life. This battery was used for EV by Toyota Motors. The driving range with a full charge was 178.5 km at a constant speed of 72 km h^{-1}; the battery weight was 451 kg at approximately 30 % of the total car weight. The nickel metal hydride battery has been widely used for hybrid vehicles at present. In 1991, Sony commercialized lithium-ion batteries. This type of battery is quite unique, because lithium intercalation compounds were used as positive and negative electrode materials. The positive electrode material is typically a metal oxide with a layered structure, such as $LiCoO_2$, and the negative electrode typically employs graphitic carbon. The specific energy density of the initial lithium-ion battery was less than 100 Wh kg^{-1} for a small size cell, while that for small size batteries is now around 200 Wh kg^{-1}. In 1992, the United States and Japan started a project to develop high energy density batteries for EV applications. At the first stage of the US Project (USABC project), several types of batteries such as nickel–metal hydride, lithium sulfur, lithium–iron sulfide, lithium-ion, and lithium polymer batteries, were investigated as candidates, and finally lithium polymer batteries were made the focus. The long-term goal of the USABC project was to achieve specific energy densities of 200 Wh kg^{-1} and 300 Wh L^{-1} at 1/3 C rate, and a power density of 400 W kg^{-1} [2]. The energy densities of the lithium polymer batteries developed by this project were 155 Wh kg^{-1} (220 Wh L^{-1}) for a 2.2 kWh module. This polymer battery consisted of a lithium metal anode, a vanadium oxide cathode, and a polyethylene oxide (PEO)-based polymer electrolyte. In Japan, lithium-ion battery were the main focus for EV applications in the New Sunshine Program. The specific energy density targets of the 3 kW model were 150 Wh kg^{-1} and 300 Wh L^{-1}, and a power density of 400 W kg^{-1}. After 10 years development, the specific mass energy density target was achieved and a specific volumetric energy density of 252 Wh L^{-1} was obtained, which was slightly lower than the target [3]. The battery materials for the EV battery are almost same as those used for conventional lithium-ion batteries. In the third stage of EV battery development, a challengeable target of 700 Wh kg^{-1} and 2,000 Wh L^{-1} was proposed by the New Energy and Industrial Technology Development Organization of Japan (NEDO) in 2006. The energy density is comparable to that of an internal

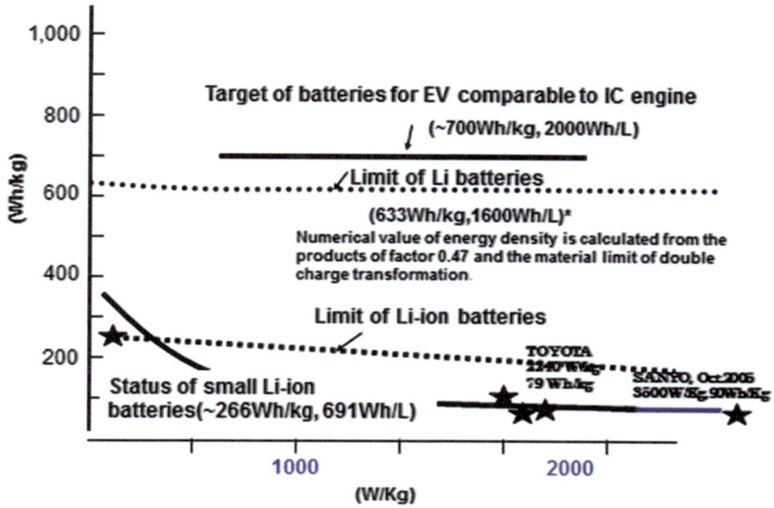

Fig. 1.1 Status and targets of advanced batteries

conversion engine, i.e., the driving range with a full charge is the same as a vehicle with an internal conversion engine. However, such high energy density batteries, including primary batteries, have yet to be developed.

The energy density of batteries is dominated by the electrode materials. There are many electrode couples for lithium-ion cells. A high energy density couple is selected from the high capacity anode with a low potential and the high capacity cathode with a high potential. There are many choices for anode materials with high capacity, but no high capacity cathodes with high potential. Thus, at present, the capacity of lithium-ion cells is limited by the cathode material, and the energy density limit of lithium-ion batteries is recognized to be around 250 Wh kg^{-1}. To obtain high energy density lithium-ion batteries requires the search for a new type of cathode material with a multivalent M species in a Li–M–O system. If the valence of M in Li_xMO_2 is changed from M^{4+} to M^{2+}, then there is a possibility to obtain an energy density of 600 Wh kg^{-1} by coupling with a high capacity anode. $Li_xCr_3O_8$ is an example, where the valence of Cr^{6+} is reduced to Cr^{3+}, and the capacity is approximately 400 mAh g^{-1}, which is three times higher than that of $LiCoO_2$; however, this is not an environmentally friendly compound [4]. At present, the highest energy density of mass produced small size lithium-ion cells for mobile phones is 266 Wh kg^{-1}. The specific energy calculated according to the active material weights of these lithium-ion batteries is around 400 Wh kg^{-1}. The specific energy density is generally dependent on the specific power density. Figure 1.1 shows the specific energy density of commercialized lithium-ion batteries as a function of the specific power density. The specific energy density is significantly decreased with increasing power density, and that of large size lithium-ion batteries for EVs is approximately 100 Wh kg^{-1}. Nissan Motors has

Table 1.1 Specific energy densities for various battery systems

System	Mass-specific energy density ($Wh\ kg^{-1}$) Calculated	Practical	Conversion yield[a]	Volume-specific energy density ($Wh\ L^{-1}$) Calculated	Practical
Li/O_2	3,458	?		6,170	?
Li/S	2,566	350[1]	0.128	4,260	350[1]
Zn/air	1,086	180[2]	0.166	6,091	208[2]
$C/LiCoO_2$	387	100[3]	0.258	1,015	150[3]
		245[4]	0.633		585[4]
M-H/NiOOH	180	63[5]	0.35		142[5]
		58[6]	0.32		84[6]
Pb/PbO_2	171	60.6	0.35	370	108

[a]Practical/calculated energy density, [1]ca. 5 Wh, [2]120 Wh, [3]375 Wh, [4]12 Wh, [5]1.14 Wh, [6]336 Wh

commercialized an EV that employs lithium-ion batteries, of which the driving range was announced to be 200 km with a full charge, although the driving range may be reduced when the air conditioning system is used. The battery capacity is 24 kWh and the weight of the batteries is approximately 15 % of the total car weight. The driving range is too short compared to that of internal combustion engine vehicles. An energy density for a battery that is comparable to that of the internal combustion engine is approximately 700 $Wh\ kg^{-1}$. However, the energy density gap between that of conventional lithium-ion batteries and the target is too large. Therefore, a new battery system should be developed for EV applications with high energy density and power density beyond that of lithium-ion batteries.

High energy density batteries can be obtained using a combination of high capacity anode and cathode materials with a high potential difference. The highest calculated specific energy density of 6,347 $Wh\ kg^{-1}$ is obtained for the fluorine–lithium couple. However, this couple has not been studied due to the difficulty of handling fluorine gas. A lithium carbon fluoride primary cell was commercialized by Panasonic in 1971, of which the specific energy density was approximately 360 $Wh\ L^{-1}$. However, a rechargeable battery of this system has not yet been developed. The second highest calculated energy density battery is the lithium air couple. The specific energy densities calculated from reaction (Eq. 1.1) and the OCV of 2.96 V are 3,458 Wh kg and 4,170 $Wh\ L^{-1}$ for the discharged state:

$$2Li + O_2 = Li_2O_2 \tag{1.1}$$

The other possible high specific energy couple is lithium sulfur, of which the calculated specific energy densities are 2,566 $Wh\ kg^{-1}$ and 4,260 $Wh\ L^{-1}$ for reaction (Eq. 1.2) at a cell voltage of 2.2 V:

$$2Li + S = Li_2S \tag{1.2}$$

Table 1.1 summarizes the calculated and measured (practical) specific energy densities of various rechargeable batteries that have been developed and are presently under development. Lithium sulfur batteries have been extensively studied for more than 40 years [5, 6]. Sion Power announced a high energy density lithium sulfur battery of 350 Wh kg^{-1} [7]. The calculated to practical energy density conversion yield for the lithium sulfur battery is 0.13, which is much lower than that of 0.26–0.63 for lithium-ion batteries. The low conversion yield of lithium sulfur batteries may be due to the high resistance of sulfur and the Li_2S reaction product because of the addition of a large amount of a conductive additive into the cathode. However, lithium sulfur batteries are quite attractive batteries in view point of the energy density, although there are many serious problems to solve for their practical use, such as high self-discharge rate and poor cyclic performance. Zinc air batteries also show a high calculated specific energy density, and especially, a high specific volumetric energy density. This type battery has been extensively studied in Japan as a power source for EVs. The energy density of the zinc air battery has been improved to 180 Wh kg^{-1} at 10 W kg^{-1} [8]. The conversion yield is as low as 0.16; therefore, higher energy density batteries could be developed with further study. However, this battery exhibits high overpotential for the charge and discharge processes and the energy efficiency is less than 50 %. The energy loss should therefore be improved for EV applications.

The highest specific energy density battery can be expected for the lithium air system, because the calculated specific energy density is the highest among the possible anode and cathode couples. The concept of lithium air rechargeable batteries was proposed by Abraham and Jang in 1996 [9]. Practical lithium air batteries have yet to be developed, so that the practical-specific energy density has not been estimated. However, it is important to estimate the calculated energy density of realizable lithium air cells using the mass and volume of the active material and the cell voltage. From the calculated energy density, the practical energy density can be estimated using a conversion yield of >0.13, which was observed for lithium sulfur and zinc air batteries. Calculation of the specific energy density of lithium air batteries is sometimes given as the specific energy density based on the mass of lithium atoms in the charged state (11,430 Wh kg^{-1}); however, as the cell discharges, oxygen from the atmosphere enters the cathode, which results in an increase in mass (discharged state), so that the mass of oxygen should be included in the determination of specific energy.

Two types of lithium air batteries have been developed; nonaqueous and aqueous. In the nonaqueous lithium air system, two cell reactions are considered; the formation of Li_2O_2 and Li_2O, given in reactions (Eqs. 1.1 and 1.3), respectively.

$$4Li + O_2 = 2Li_2O \tag{1.3}$$

The standard reversible cell potentials of reactions (Eqs. 1.1 and 1.3) are 2.96 and 2.91 V, respectively [10]. Some authors have reported that the formation of Li_2O is possible, but may be difficult to reverse upon charging [10]. The calculated energy density of reaction (Eq. 1.3) is higher than that of Eq. (1.1); however, the energy density calculation for nonaqueous lithium air batteries should be based on reaction (Eq. 1.1). In aqueous lithium air systems, water molecules are involved in the redox reaction at the air cathode:

$$4Li + 6H_2O + O_2 = 4(LiOH \cdot H_2O) \tag{1.4}$$

The cell voltage is dependent on the concentration of OH^- in the solution and is 3.86 V in a neutral solution (see Chap. 10). However, the concentration of OH^- is increased with discharge. Saturation of LiOH is reached at 5.3 M L^{-1}, which corresponds to discharge depth of only ca. 5 %. The reaction product is deposited in the electrolyte and on the air electrode. The observed OCV for aqueous lithium air batteries with a neutral solution was 3.75 V, which is comparable to the calculated OCV [11], and the OCV with a LiCl and LiOH saturated aqueous solution was around 3.0 V, where LiCl is saturated to prevent the reaction of LiOH and the lithium-protective layer of a water-stable lithium ion conductor, as discussed later (see Chap. 8). Calculation of the specific energy density for the aqueous system should be based on the OCV of 3.0 V and the formation of $LiOH{\cdot}H_2O$, because LiOH deposits with crystalline water. The specific energy density calculated from an OCV of 3.75 V (Li/neutral aqueous solution/air) and the reaction product of LiOH is 4,200 Wh kg^{-1}, which is higher than that of the nonaqueous system. Table 1.2 shows the calculated energy densities of the lithium air batteries along with that for the conventional lithium-ion battery, where the mass and volume of only the active materials are accounted for. The mass and volume-specific energy densities of the nonaqueous system are respectively 8 and 2.3 times higher than that for a conventional lithium-ion battery with a carbon anode and $LiCoO_2$ cathode. The mass-specific energy density of the aqueous system is around five times higher than that of the lithium-ion battery. The aqueous system also has an advantage of mass-specific energy density over the lithium-ion battery, but the volume-specific energy density is only twice as high as the lithium-ion battery. As shown in Chap. 11, the volume-specific energy density is lower than that of the Li/MO_2 cell, where MO_2 has a specific capacity of 275 mAh g^{-1}, a density of 4.25 g cm^{-3}, and an average potential 3.75 V, while such a materials is not in widespread use today. Some authors have claimed that because the lithium cycling efficiency is less than 100 %, the lithium anode should be charged more than threefold [12]. The calculated energy density, and in particular, the energy density per volume, is decreased by the excess lithium charge. The most attractive application of high energy density batteries is for EVs, where the volumetric energy density is important, in addition to the mass energy density. The volumetric energy densities of the nonaqueous and aqueous lithium air batteries with threefold charge lithium anodes are comparable to that of the lithium-ion battery. Thus, improvement of the cycling efficiency of the lithium electrode or replacement with an alternative is an important challenge to maximize the specific energy density of lithium air batteries [13]. Metallic lithium is a particularly good anode candidate for lithium air batteries, because it has a high theoretical specific capacity (3,860 mAh g^{-1}) and a high negative potential (−3.05 V vs. NHE). Nevertheless, the use of a metal lithium anode is limited due to the occurrence of dendrite growth

Table 1.2 Specific energy densities for lithium air batteries

System	Anode	OCV (V)	Energy density Wh kg^{-1}	Energy density Wh L^{-1}	Cell reaction
Li/nonaqueous/air	Li	3.0	3,458	6,170	$2Li + O_2 = Li_2O_2$
	Li threefold excess	3.0	2,170	1,245	
	Si	3.0	2,738	3,302	$2/4.4Li_{4.4}Si + O_2 = Li_2O_2 + Si$
Li/aqueous/air	Li	3.0	1,910	2,004	$4Li + O_2 + 6H_2O = 4(LiOH{\cdot}H_2O)$
	Li threefold excess	3.0	1,400	1,210	
	Li	3.69	1,319		$2Li + 1/2O_2 + 2CH_3COOH = 2LiCH_3COO + H_2O$
	Si	3.0	1,663	2,605	$4/4.4Li_{4.4}Si + 6H_2O + O_2 = 4(LiOH{\cdot}H_2O) + 4/4.4Si$
$C/LiCoO_2$	carbon	3.8	387	1,016	$LiCoO_2 + 3C = Li_{0.5}CoO_2 + 0.5C_6Li$

during the lithium deposition process [14]. One alternative for the lithium anode is $Li_{4.4}Si$, which has a high specific capacity of 2,012 mAh g^{-1} [15]. The calculated volumetric energy density of lithium air batteries is improved using a silicon anode; that of the nonaqueous system is three times higher than that of lithium-ion batteries, while that of the aqueous system is two and half times higher. These high mass and volume-specific energy densities for lithium air batteries have promoted much research activity into these systems. However, there are still many problems to be solved for the practical application of a silicon anode, such as the cyclic performance and the first cycle irreversible reaction.

Unfortunately, the calculated to practical energy density conversion yield may be lower for lithium air batteries than that for lithium-ion batteries, which is 0.25 for large and 0.633 for small scale batteries, because the reaction products are insulative, as for the lithium sulfur batteries. The cathode in the lithium air cell must be a porous-conducting matrix within which Li_2O_2 or $LiOH{\cdot}H_2O$ can form. The conversion yield of the well-established lead–acid battery is 0.35, as shown in Table 1.2. Accordingly, the practical-specific energy densities for the nonaqueous lithium air (>800 Wh kg^{-1}) and for aqueous lithium air (>500 Wh kg^{-1}) have been predicted for the full developed system, assuming a conversion yield of 0.33.

1.2 History of Lithium Air Batteries

The concept of lithium air batteries was introduced by Littauer and Tsai [16, 17] in 1974, where the use of an aqueous alkaline solution as the electrolyte was proposed. Lithium air batteries with water reduce the introduction of oxygen from the environment at the major portions of the electrode surface to prevent the formation of an oxide. The anode oxide film is of such a form that it permits the inert cathode structure to be pressed against it without short-circuiting. Typical OCVs of 2.9–3.0 V, and cell voltages of 2.0 V are achieved at current densities of approximately 200 mA cm^{-2}. This is only a primary or mechanically rechargeable system. Hydrogen evolution at the lithium surface always creates pores within the film according to the following reaction:

$$2Li + 2H_2O = 2LiOH + H_2 \tag{1.5}$$

Unlike conventional batteries, the current efficiency of lithium air cell is governed by the ratio of two competing reactions; the oxygen reduction, reaction (Eq. 1.1), at the cathode and corrosion, reaction (Eq. 1.5), at the anode [18]. The corrosion reaction represents a loss in current efficiency. The stability results from the in situ formation of a surface oxide/hydroxide film on the lithium metal anode that retards the rapid corrosion effect after its initial formation. During discharge, lithium is polarized at approximately 300 mV from the initial OCV, so that columbic efficiency for the anodic reaction closely approaches 100 %. However, at OCV and low current density, self-discharge of the lithium anode is rapid,

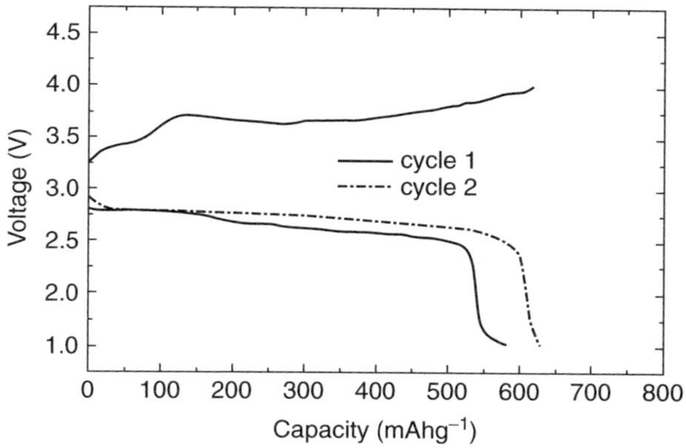

Fig. 1.2 Cycling performance of a Li/PAN-based polymer electrolyte/carbon with cobalt phthalocyamine/O_2 cell at room temperature. The cell was discharged at 0.1 mA cm^{-2} and charge at 0.05 mA cm^{-2} (from [9])

which results in low electrochemical efficiencies. To suppress the lithium corrosion reaction, an alternative to the aqueous electrolyte should be used that is stable with lithium metal, or the lithium metal electrode should be protected by a water-stable lithium ion-conducting electrolyte. In 1988, Batalov [19] proposed a lithium air rechargeable cell of Li/$LiBe_2O_3$/LiCl–$LiCO_3$/$Li_{0.1}NiO$ (or $La_{0.7}Sr_{0.3}MnO_3$)/O_2. $LiBe_2O_2$/LiCl–$LiCO_3$ (1:1) is a molten lithium-conducting electrolyte with a $LiBe_2O_2$ matrix. This cell was operated at 580–600 °C and showed a cell voltage of 2 V at 100 mA cm^{-2}, where charging was performed at a voltage of 3 V. However, the detailed cell performance was not clear. This type of lithium air cell was based on a molten salt-type fuel cell, where the anode is nickel metal and is active toward hydrogen.

In 1996, Abraham and Jang [9] reported a lithium air battery using a nonaqueous electrolyte. The cell consisted of a lithium metal anode, a gel-type polymer electrolyte and a carbon air electrode with a catalyst. The polymer electrolyte was a gel-type lithium-conducting polymer electrolyte of $LiPF_6$ in polyacrylonitrile (PAN) with ethylene carbonate and propylene carbonate, which is stable in contact with lithium metal and exhibits a conductivity of around 10^{-3} S cm^{-1} at room temperature. Cobalt phthalocyanine was used as a catalyst for the air electrode. The observed OCV was ca. 3.0 V at room temperature, which is comparable to that calculated from reaction (Eq. 1.1). The formation of Li_2O_2 in the discharged carbon electrode was observed in the Raman spectrum. This cell was rechargeable and had good columbic efficiency with the cobalt phthalocyanine-catalyzed carbon electrode. Figure 1.2 shows the cycling performance of the cell. The cell was first discharged at 0.1 mA cm^{-2} to a 1.5 V cut off to yield a capacity of 580 mAh g-carbon^{-1}. In the following charge at 0.05 mA cm^{-2} to 4.0 V, a capacity of 630 mA g-carbon^{-1} was obtained. In the second discharge, the capacity was 630 mA g-carbon^{-1}. The energy density of the test cell was estimated to be 559 Wh kg^{-1}, using the weights of lithium metal, carbon, and oxygen at a discharge voltage of 2.7 V, which is higher than that for a typical lithium-ion cell (388 Wh kg) using the weights of the carbon anode and the $LiCoO_2$ cathode at a discharge voltage of 3.7 V. The capacity of the nonaqueous lithium air battery is dependent on the

Fig. 1.3 Cyclic performance of a $Li/PC\text{-}DME\text{-}LiPF_6/carbon\text{-}\lambda\text{-}MnO_2\text{-}PVDF$ cell at 0.1 mA cm^{-2} (from ref. [20])

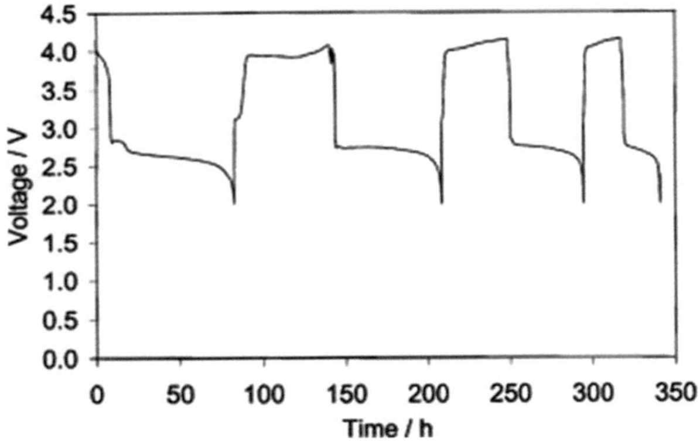

weight of the carbon air electrode and the surface area of carbon, because the Li_2O_2 reaction product is insoluble in the nonaqueous electrolyte and is deposited on the carbon air electrode. Therefore, to improve the energy density of lithium air batteries, higher capacity carbon air electrodes should be used. Read [20] reported a high capacity carbon air electrode after investigating various carbon materials and nonaqueous electrolytes; Super P carbon black in propylene carbonate (PC) and diethyl carbonate (DME) with $LiPF_6$ showed a high capacity. Discharge capacities of 1,934 and 1,069 mAh g-carbon^{-1} were observed at 0.05 and 0.2 mA cm^{-1}, respectively. The discharge capacity is more than two times higher than that of the cell with a polymer electrolyte reported by Abraham et al. The specific energy density of the cell with the high capacity carbon air electrode was calculated to be 2,081 Wh kg^{-1}, which is approximately five times higher than that of typical lithium-ion batteries. However, the cyclic performance of the $Li/PC\text{-}DME\text{-}LiPF_6$/Super P carbon black with $\lambda\text{-}MnO_2/O_2$ cell cycled at 0.1 mA cm^{-1} was poor, as shown in Fig. 1.3. The charge and discharge capacity is significantly decreased with cycling. The details of the carbon electrode for oxygen reduction will be discussed in Chap. 10.

In 2006, Bruce and coworkers reported the possibility of improved cycling for $Li/PC\text{-}LiPF_6$/Super P carbon black with an electrolytic manganese dioxide (EMD) cell [21]. Figure 1.4 shows the cycling performance of the cell. Although the capacity does fade, the rate of fail is slow, and a discharge capacity of ca. 600 mAh g-carbon^{-1} (441 mAh g^{-1} including oxygen) is still obtained after 50 cycles at a rate of 70 mA g^{-1}. Although, the capacity is lower than that calculated based on $2Li + O_2 = Li_2O_2$ (1,170 mAh g^{-1}), it is four times higher than that of the typical cathode $LiCoO_2$ of lithium-ion batteries. Similar high capacity retention was observed for the $Li/PC\text{-}Li(CF_3SO_2)_2N$/Super P carbon black with an EMD cell by Mizuno et al. [22], where a capacity of 600 mA g-carbon^{-1} was obtained after 100 cycles at 0.02 mA cm^{-2}. However, the expected reaction product of Li_2O_2 was not detected, but lithium carbonate and lithium alkylcarbonate (RO–(C=O)–OLi) were observed using Fourier transform infrared (FTIR) spectroscopy analysis. Significant studies have also shown that there is little or no evidence of Li_2O_2 formation occurring in parallel with electrolyte degradation in alkyl carbonate

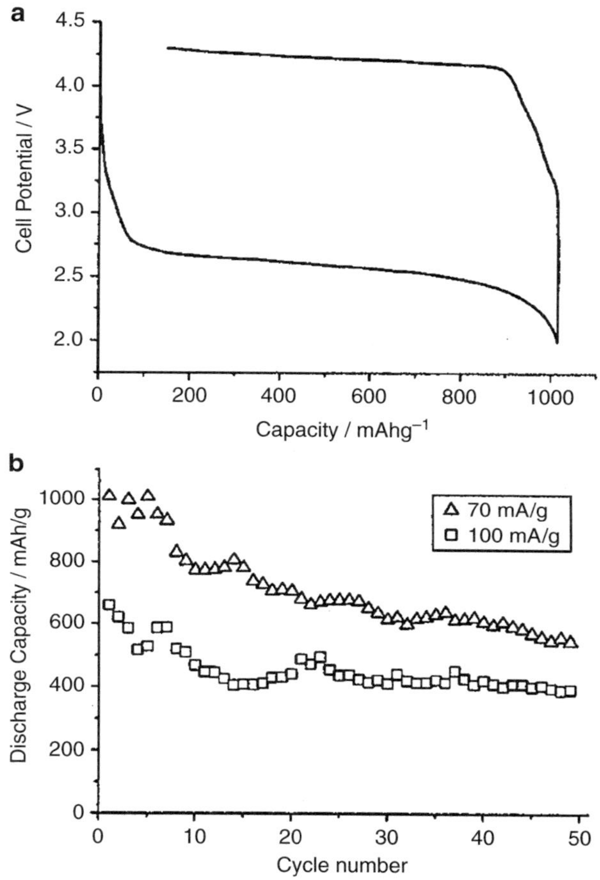

Fig. 1.4 (**a**) Charge and discharge curves for a Li/PC-LiPF$_6$/Super P carbon-EMD-PVDF cell at 50 mA g^{-1}. (**b**) Variation of discharge capacity with cycle number; rate = 70 and 50 mA g^{-1}. Capacities are expressed per gram of carbon in the electrode (from [21])

electrolytes [23] (see Chap. 2). This is important, because it rules out the possibility of Li_2O_2 formation for Li O cells with organic carbonate electrolytes. Recently, attention has been turned to other electrolytes, such as ether, ionic liquids, and polymer electrolytes, to avoid electrolyte decomposition during the charging process. More recently, Scrosati and coworkers have presented attractive results using an electrolyte of tetra(ethylene) glycol dimethyl ether (TEGDME)-$LiCF_3SO_3$ [24]. The test cell with a lithium metal foil anode and Super P carbon black cathode showed no discharge capacity degradation after 100 cycles at the high specific discharge rate of 1,000 mA g-carbon^{-1} (1 mA cm^{-2}) and a discharge capacity of 1,000 mAh g-carbon^{-1} (358 mAh g^{-1} including oxygen). At a low rate of 500 mA g^{-1} (0.5 mA cm^{-2}), a capacity of 3,000 mAh g-carbon^{-1} (1,075 mAh g^{-1} including oxygen) was obtained for 30 cycles without degradation. However, as the carbon electrode packing density was as low as 0.125 g cm^{-3}, the calculated volumetric energy density using lithium and the carbon volume was around 1,000 Wh L^{-1}, which is comparable to that of lithium-ion batteries. If these results

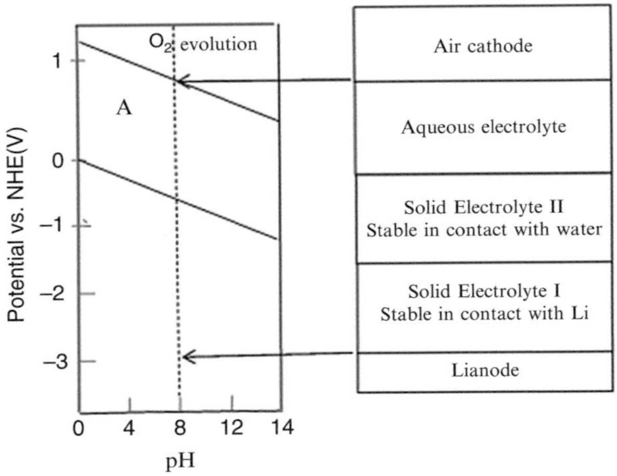

Fig. 1.5 Pourbaix diagram and schematic diagram of an aqueous lithium air battery

are confirmed by further studies, then the implication is that ethers are not a final solution as a suitable nonaqueous electrolyte for lithium air batteries [13] (see Chaps. 2, 3, and 4). Kuboki et al. reported that a cell with large surface area air electrode consisting of conducting carbon with a hydrophobic ionic liquid of 1-methyl-3-octylimidazolium bis(trifluoromethylsulfonyl)imide operated in air for 56 days and showed a high discharge capacity of 5,360 mAh g-carbon^{-1} at 0.01 mA cm^{-1} [25]. Lithium oxygen cells with Super P carbon air electrodes and polyethylene-based polymer electrolytes can be charged at relative low voltage [26]. However it will be important in the future to investigate the nature of the discharge product and the reaction mechanisms in ionic liquids and polymer electrolytes.

Lithium reacts with water to produce LiOH and hydrogen gas; therefore, to avoid this parasitic corrosion reaction, most research on rechargeable lithium air batteries have been focused on using aprotic solvents as electrolytes. However, nonaqueous lithium air batteries have some severe problems that must be still addressed, such as lithium corrosion by water and CO_2 when operated in air, decomposition of the electrolyte during the discharge process, and high polarization of the air electrode. Some of these problems could be removed with the use of an aqueous electrolyte and a water-stable lithium electrode (see Chaps. 6, 7, and 8). Thus, the prerequisite to obtain practical aqueous lithium air rechargeable batteries is the development of a water-stable lithium electrode that can survive lithium stripping/deposition for a long lifespan in an aqueous electrolyte. The electrochemical window of aqueous electrolytes can be analyzed using the Pourbaix diagram shown in Fig. 1.5. Water is usually regarded to be stable in area *A*, which is well known as the electrochemical window of water based on the two following electrode reactions:

$$2H^+ + 2e^- = H_2, \quad E^\circ = 0.00\,V \tag{1.6}$$

$$1/2O_2 + 2H^+ + 2e = H_2O, \quad E^\circ = 1.229\,V \tag{1.7}$$

If a stable electrode such as Pt, Au or carbon is inserted into the aqueous solution, the theoretical electrochemical window can be easily identified in area *A*. However, if a negative electrode such as lithium metal is placed into the aqueous electrolyte, the following reaction occurs:

$$2Li^+ + 2e^- = 2Li, \quad E^\circ = -3.040\,V \tag{1.8}$$

The Li/aqueous electrolyte/oxygen cell potential with a neutral solution is 3.856 V vs. the standard oxygen electrode, and water decomposes spontaneously. If lithium metal contacts with a lithium-conducting electrolyte that is stable at a potential lower than −3.040 V, then water is not decomposed. The lithium-protective layer should be stable in water. A schematic of the aqueous lithium air battery is also shown in Fig. 1.5. The concept of a water-stable lithium electrode was proposed by Visco et al. in 2004 [27] (see Chap. 6). This electrode concept adopts a water-stable NASICON (Na^+ super ionic conductor)-type lithium ion-conducting solid electrolyte as a protective layer that covers and isolates the lithium metal from direct contact with the aqueous electrolyte. The NASICON-type lithium ion-conducting solid electrolyte, $Li_{1+x}M_xTi_{2-x}(PO_4)_3$ (M = Al, Sc, Y), was reported by Aono et al. [28]. The maximum conductivity of 10^{-3} S cm^{-1} was observed for $Li_{1+x}Al_xTi_{2-x}(PO_4)_3$ (LATP; $x = 0.3$) at room temperature. The water stability of this compound was examined by Imanishi and coworkers [29, 30] (see Chap. 8), where LATP was unstable in an aqueous solution of 0.1 M HCl and 1 M LiOH, but stable in an aqueous solution saturated with LiOH and LiCl. The lithium ion conductivity of LATP did not change after immersion in the saturated LiOH/LiCl aqueous solution at 50 °C for 1 week. These results suggest that LATP could be used as the protective layer for the water-stable lithium metal electrode in saturated LiOH/LiCl aqueous solution, because the reaction product of the aqueous lithium air battery is LiOH. However, LATP is unstable in direct contact with lithium metal by the reduction of Ti^{4+} with lithium. Therefore, an interface layer between lithium metal and LATP should be used; high lithium ion conductivity and stability in contact with lithium metal are required. Visco et al. have used a thin Li_3N layer, which was prepared by a sputtering technique on LATP. Imanishi et al. proposed a PEO-based polymer electrolyte that has low lithium ion conductivity at room temperature, but ca. 10^{-3} S cm^{-1} at 60 °C. The change in impedance of the water-stable lithium electrode in 1 M LiCl aqueous solution with storage time is shown in Fig. 1.6, where a composite polymer electrolyte of PEO_{18}-lithium bis(trifluoromethanesulfonyl)imide (LiTFSI) and $BaTiO_3$ was used as the protective layer between lithium metal and LATP. No significant change of the impedance profiles was evident after 30 days, which suggests that this type of water-stable lithium electrode is acceptable for the lithium electrode of aqueous lithium air batteries. The OCV of the cell was 3.70 V at 60 °C and this was maintained for 2 months. The OCV was comparable to that calculated from the cell reaction of Li + H_2O + $1/2O_2$ = 2LiOH (ca. 3.8 V at 25 °C).

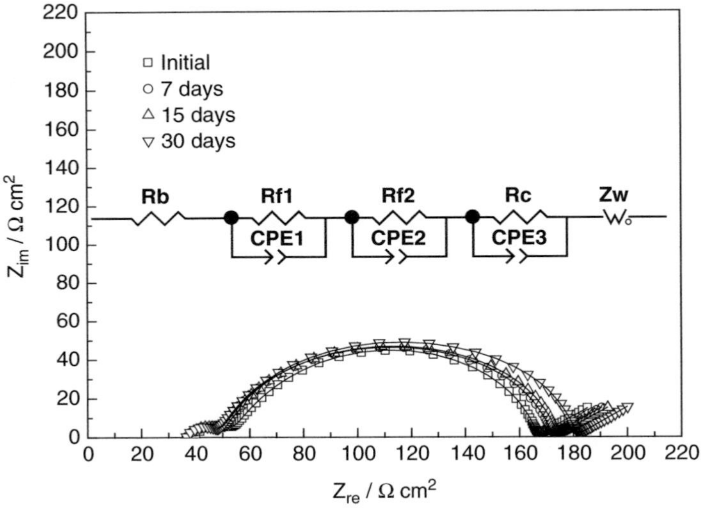

Fig. 1.6 Time dependence of impedance profiles for a Li/PEO_{18}LiTFSI-10 wt% $BaTiO_3$/LATP/ aqueous 1 M LiCl/Pt cell at 60 °C (from [36])

Recently, Weppner and coworkers presented an interesting garnet-type lithium ion-conducting solid electrolyte of $Li_7La_3Zr_2O_{12}$ (LLZ) [31]. This new compound exhibits a high lithium ion conductivity of 2.3×10^{-4} at room temperature and is stable in contact with lithium. The stability of this solid electrolyte in water was examined by Imanishi et al. and it was found to be stable in an aqueous solution with LiCl [32], but unstable in LiOH aqueous solution. Therefore, this lithium ion-conducting solid electrolyte could be used as the protective layer for the water-stable lithium metal electrode without the second solid electrolyte II, shown in Fig. 1.5, but the interface phenomena between lithium and LLZ should be studied in more detail, especially lithium dendrite formation during the lithium deposition process.

The reduction and evolution of oxygen in aqueous media has been extensively studied for many years. Bifunctional oxygen electrodes, which consume oxygen during discharge and evolve oxygen during charge, were developed by Siemens Corp. in the early 1970s for rechargeable zinc air batteries. The electrode is composed of two layers, one layer of which is a hydrophilic porous nickel sheet adjacent to the electrolyte, while the other hydrophobic carbon layer borders the gas phase [33]. During the oxygen reduction reaction, the carbon layer acts as an active layer, while the nickel layer catalyzes the oxygen evolution reaction. Research on bifunctional catalysts has, until now, been focused mainly on perovskite, spinel, and pyrochlore catalysts. The perovskite oxide $LaA_{1-x}M_xO_3$ (A = Sr, Ca, M = Fe, Co, Mn) exhibits a low oxygen reduction overpotential of 0.1 V and an oxygen evolution overpotential of 0.2 V at 100 mA cm^{-2} [34]. These overpotentials are

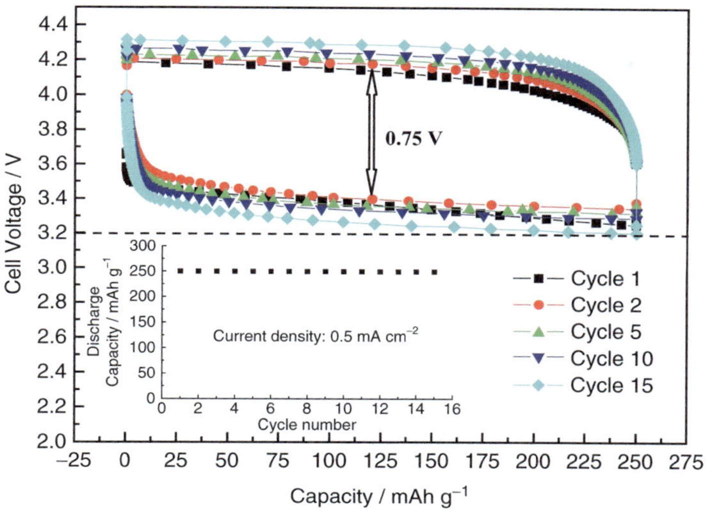

Fig. 1.7 Charge–discharge performance of the prototype acidic lithium air cell (from [35])

significantly lower than those of the nonaqueous system, which gives the aqueous lithium air system an advantage over the nonaqueous system.

The cell performance of rechargeable aqueous lithium air batteries has not yet been presented to allow estimation of the energy density. The rechargeability of aqueous lithium air batteries was demonstrated by Imanishi and coworkers [35]. They reported successful charge and discharge performance using an aqueous electrolyte solution of acetic acid and lithium acetate as shown in Fig. 1.7. The calculated specific energy density is 1,319 Wh kg^{-1} and an energy density of 779 Wh kg^{-1} (including the weight of lithium anode and the electrolyte) was demonstrated, which is much higher than that of conventional lithium-ion batteries using graphite and $LiCoO_2$. The potential difference between charge and discharge was ca. 0.75 V at 0.5 mA cm^{-1}. The estimated energy density of this cell was approximately two times higher than that of the lithium-ion battery. Table 1.2 shows that higher energy density aqueous lithium air batteries can be expected using water as the active material. The charge and discharge cell performance for a cell with excess water was reported [36]; however, the cell performance with a limited amount of water has not been demonstrated yet to estimate the energy density, due to the difficulty of the oxygen evolution reaction in the small amount of electrolyte at high current density. Although the basic mechanisms for the oxygen reduction and evolution reactions in an aqueous electrolyte are well known, these processes in an electrolyte containing a large amount of lithium salt have received much less attention and thus merit further investigation in view of the present interest in aqueous lithium air cells [37].

1.3 Key Issues and Further Prospects for Lithium Air Batteries

The energy density of a battery is regulated by the chemistry of the electrodes. The specific energy density of the present lithium-ion batteries is limited mainly by the specific capacity of the cathode materials. Therefore, air is the most attractive cathode material to improve the cell-specific energy density, because it is supplied from environment. The primary zinc air cell has been commercialized as a battery for hearing aids, due to its high mass and volumetric energy densities (440 Wh kg^{-1} and 700 Wh L^{-1}). Rechargeable zinc air batteries for EV applications have been extensively studied for many years; however, an acceptable system with a high specific energy density and long cycle life has not yet been successfully produced. The main problems of the zinc air battery are zinc dendrite formation during the charging process (low cycle life), contamination by CO_2 from the air (low storage life), and high overpotentials for the charge and discharge processes (low energy conversion efficiency). The energy densities of the practical rechargeable zinc air battery are 180 Wh kg^{-1} and 208 Wh L^{-1}, as shown in Table 1.1, and the conversion yield of this battery is only 0.166.

The theoretical energy density of the lithium air battery, excluding oxygen (11,430 Wh kg^{-1}), is comparable with the specific energy density of fossil fuel. Therefore, the lithium air system is expected to provide a possible energy density of 700 Wh kg^{-1}, which could deliver a driving range comparable to that for internal engine vehicles. In this decade, especially in the last few years, the chemistry and cell performance of lithium air cells have been extensively studied, and many interesting and important results have been presented. However, the highest practical energy densities of the small test cells are only moderately higher than that of the lithium-ion battery. Thus, to achieve batteries capable of discharge rates and energy densities appropriate for EV applications requires further breakthroughs. The deep cycling capability of the lithium air battery is yet unproven. The prospect of lithium air batteries with adequate safety for EV applications is clouded by unresolved issues of lithium dendrite formation.

The main issues for lithium air batteries with nonaqueous electrolytes that require improvement to achieve the requirements for EV applications are:

1. A nonaqueous electrolyte that is stable to peroxide radical attachment with high oxygen solubility
2. An excellent catalyst for the reduction of Li_2O_2
3. The suppression of lithium dendrite formation
4. Protection from carbon dioxide and water in the ambient air

Issue (1) is the key point to be resolved for an acceptable battery system. Recently, Scrosati and coworkers [24] presented a stable electrolyte of TEGDME-$LiCF_3SO_3$. The Li/TEGDME-$LiCF_3SO_3$/Super P carbon black/O_2 cell has excellent cyclability at a current density of 1 A g-$carbon^{-1}$, which corresponds to approximately 1 mA cm^{-2}. A reversible capacity of 1,000 mAh g^{-1} was achieved for 100 cycles. These results should encourage further research into stable nonaqueous electrolytes for lithium air batteries. The energy density of this cell calculated using the weights of the lithium anode, carbon cathode, and oxygen is 1,449 Wh kg^{-1}, which is approximately four times higher than that of the lithium-ion cell with a carbon anode and $LiCoO_2$ cathode. This result indicates the potential for the development of lithium air batteries with high energy density and excellent cyclability. However, the excellent performance was established for a low packing density carbon electrode, and the volumetric energy density is comparable to that of lithium-ion batteries. Therefore, a more compact air electrode with high capacity at high current drain should be developed. And also, the stability of TEGEME is still questionable. The other problem of the nonaqueous system is the high electrode polarization, issue (2). The energy loss during the charge and discharge processes at 1 mA cm^{-2} in the cells reported by Scrosati and coworkers [24] was approximately 50 %, which is much higher than the several percent energy loss of lithium-ion batteries. Therefore, an excellent catalyst for Li_2O_2 reduction should be developed. The third issue is also important for practical batteries aimed at EV applications. In the 1980s, lithium rechargeable batteries with a lithium metal anode and Mo_2S cathode were commercialized for cellar phones; however, these batteries had many accidents, due to lithium dendrite formation on the lithium metal electrode during the charging process. To reduce such dendrite formation, lithium polymer batteries that consist of a dry polymer, lithium metal anode, and VO_x cathode have been developed for EV applications. Lithium dendrite formation was suppressed using a polymer electrolyte, but the safety problem of the lithium metal anode was not completely resolved. The carbon anode (LiC_6) has now been widely used for lithium-ion batteries instead of a lithium metal anode. The same problem of dendrite formation is present in nonaqueous lithium air batteries. Thus, an alternative for lithium metal is also required for lithium air batteries. The calculated specific energy density of a lithium air cell with the LiC_6 anode is 845 Wh kg^{-1}, which is not attractive for EV applications. The silicon lithium alloy $Li_{4.4}Si$ is another candidate for the anode material. This alloy has a high reversible capacity that is as high as 2,000 mAh g^{-1} [38], which is five times higher than that of LiC_6 and is free from lithium dendrite formation in nonaqueous electrolytes. The fourth issue appears unresolved at the present (see Chap 11). The previously reported cell performance has been measured using pure oxygen or dry air. However, preliminary results have shown a rapid decrease in cell capacity by cycling using air, where a hydrophobic membrane was used to prevent the ingress of water from the air [39]. The removal of water and carbon dioxide from air remains a significant challenge that must be overcome to allow lithium air battery operation with repeated cycling and long-term use in the atmosphere. An oxygen-selective membrane that selectively permeates oxygen while retaining other gas from air has been proposed [40]. However, the incorporation of such a membrane remains a critical challenge for high rate and long-term applications, such as the energy source for EV applications.

Aqueous electrolyte-based lithium air batteries also have many problems to be solved before practical EV applications can be realized. The reduction of oxygen and the evolution of oxygen in concentrated alkaline solutions for zinc air secondary batteries and alkaline fuel cells have been studied extensively for many years. Although the basic mechanisms of the oxygen reduction and oxygen evolution reactions are well known, the specific process in a weak alkaline solution with a high concentration of lithium ions has received much attention and merits further study in view of the interest in the aqueous lithium air system [13]. Significant attention has been focused on protection of the lithium metal anode from the aqueous electrolyte, because without a water-stable lithium electrode, the aqueous lithium air cell could not function. The water-stable lithium electrode consists of lithium metal protected with a NASICON-type lithium ion-conducting LATP solid electrolyte, which is stable with lithium ions in aqueous solution. However, this electrolyte is unstable in contact with lithium metal, so a protective layer is required between the lithium metal and LATP. Lithium-conducting polymer electrolytes have been used for this protective layer. The cell resistance of the Li/PEO_{18}LiTFSI-PP13TFSI/LATP/1 M LiCl aqueous solution/Pt, air cell was reported to be 130 Ω cm^2 at 60 °C, where PP13TFSI is an ionic liquid [41]. The limiting current density of this type electrode was less than several milliamps per square centimeter. The cell short-circuit of Li/PEO_{18}LiTFSI-PP13TFSI/Li was observed after 35 h polarization at 1.0 mA cm^{-2} and 60 °C and the dendrite formation onset time was 17 h [42]. The specific capacity of a lithium anode with a 10 μm thick copper current collector is 1,270 mAh g^{-1}, which is significantly higher than that of nonaqueous liquid electrolyte 22 mAh g^{-1} [43]. However, the dendrite formation onset time is significantly decreased with increasing current density, and the lithium deposition and stripping limiting current densities are too low for EV applications. In addition, the weight of a lithium-conducting solid electrolyte should be reduced to obtain a high total specific energy density of the cell. At present, the thickness of the available LATP plate without water penetration is around 150–250 μm. Assuming the thickness of a carbon electrode with a capacity of 1,000 mAh g^{-1} is 100 μm, then the weight of LATP is comparable to the air electrode. Higher capacity air electrodes and an anion exchange membrane as a separator to deposit the LiOH reaction product into the aqueous solution should also be developed [44] (see Chap. 7). The aqueous lithium air system has the potential to be developed into high power density batteries, because of the high solubility of LiOH in aqueous solution and the higher solubility of oxygen gas in the aqueous solution than that in nonaqueous solutions. However, the ionic conductivity and mechanical properties of the protective layer between lithium metal and water should also be improved and a system that deposits the LiOH reaction product into the aqueous solution should be established to realize high power and energy density lithium air rechargeable batteries.

The lithium air rechargeable batteries are the most promising battery system with respect to energy density for the development of EVs with driving ranges comparable to conventional vehicles with internal combustion engines. Candidate materials for lithium air battery systems have been reported by many researchers, especially the nonaqueous electrolytes, water-stable solid electrolytes, and catalysts for Li_2O_2 reduction. However, the technological basis for high power density and

Table 1.3 Theoretical and expected energy densities for lithium air batteries

Cell reaction	Theoretical maximum energy density Wh kg^{-1}	Theoretical maximum energy density Wh L^{-1}	Expected anode capacity (mAh g^{-1})	Expected cathode capacity (mAh g^{-1})	Expected total cell energy density Wh kg^{-1}	Expected total cell energy density Wh L^{-1}
$2/4.4Li_{4.4}Si + O_2 = Li_2O_2 + Si$	2,738	3,302	2,000	2,000	900	1,300
$4/4.4Li_{4.4}Si + 6H_2O + O_2 = 4(LiOH{\cdot}H_2O) + 4/4.4Si$	1,663	2,605	2,000	900	560	740

extended deep cycling has yet to be demonstrated. Lithium air battery research and technology is still in the first stages. Some researchers are not optimistic for the future of lithium air batteries, especially with respect to the volumetric energy density and power density. The capacity of the air electrode is dependent on the specific area of the carbon electrode [45, 46]. High capacities of the carbon oxygen electrode of more than 5,000 mA g-carbon^{-1} have been reported for nonaqueous lithium oxygen cells, where the current density was as low as 0.01 mA g-carbon [25] or the packing density as low as 0.125 g mL^{-1} [24]. An air electrode capacity of 2,000 mAh g^{-1} could be expected at moderate current and packing densities [47]. Lithium metal electrodes could not be used for EV battery applications, due to safety reasons. Therefore, candidates such as $Li_{4.4}Si$ with a capacity of 2,000 mAh g^{-1} could be expected [38]. For aqueous lithium air batteries, the air electrode capacity is not dependent on the specific area of the carbon, but on the amount of aqueous electrolyte, assuming the reaction product of $LiOH{\cdot}H_2O$ is deposited in the electrolyte [44]. Table 1.3 shows the expected specific energies with anode and cathode capacities of 2,000 mA g^{-1} for the nonaqueous and aqueous systems, where the conversion yield is assumed to be 0.3. The expected energy density could be acceptable for EV applications; however, such batteries must be capable of being discharged and charged at the rate required for EV applications, which may be the most challenging target yet.

References

1. Takahashi S, Higuchi T, Ashimura S et al (1979) Report of the Government Industrial Research Institute, Osaka No. 354
2. Kalhammer FR (1999) Batteries for electric and hybrid vehicles recent development progress. Report for State of California Air Resources Board, Sacramento, CA
3. Terada N (2000) In: 11th electric vehicle symposium
4. Takeda Y, Kanno R, Tsuji Y et al (1984) Rechargeable lithium/chromium oxide cells. J Electrochem Soc 131:2006–2010
5. Coleman JR, Bates MW (1968) Power sources 2. In: Collins DH (ed) Proceedings, 6th international symposium, Brighton. Pergamon, New York, p 289
6. Rauh RD, Abraham KM, Pearson GF et al (1979) Lithium/dissolved sulfur battery with organic electrolyte. J Electrochem Soc 126:523–527
7. Sion Power. www.sionpower.com

8. Linden D, Reddy TB (eds) (2002) Handbook of batteries, 3rd edn. McGraw-Hill, New York
9. Abraham KM, Jang Z (1996) A polymer electrolyte based rechargeable lithium oxygen battery. J Electrochem Soc 143:1–5
10. Lu YC, Gasteiger HA, Parent MC et al (2010) The Influence of catalysts on discharge and charge voltages of rechargeable Li–oxygen batteries. Electrochem Solid-State Lett 13:A69–A72
11. Zhang T, Imanishi N, Hasegawa S et al (2008) Li/polymer electrolyte/water stable lithium-conducting glass ceramics composite for lithium–air secondary batteries with an aqueous electrolyte. J Electrochem Soc 155:A965–A969
12. Dahn J (2009) Scalable energy storage: beyond Li-ion. Almaden, San Jose, CA
13. Bruce PG, Freunberger SA, Hardwick LJ et al (2012) Li-O_2 and Li-S batteries with high energy storage. Nat Mater 11:19–29
14. Peled F (1979) The electrochemical behavior of alkali and alkaline earth metals in nonaqueous battery systems—the solid electrolyte interphase model. J Electrochem Soc 126:2047–2051
15. Choi NS, Yao Y, Coi Y et al (2011) One dimensional Si/Sn-based nanowires and nonotubes for lithium-ion energy storage materials. J Mater Chem 21:9825–9840
16. Littauer EL, Tsai KC (1976) Anodic behavior of lithium in aqueous electrolytes: I. Transient passivation. J Electrochem Soc 123:771–776
17. Littauer EL, Tsai KC (1977) Corrosion of lithium in aqueous electrolytes. J Electrochem Soc 124:850–855
18. Bennion DN, Littauer EL (1976) Mathematical model of a lithium-water electrochemical power cell. J Electrochem Soc 123:1462–1469
19. Batalov NN, Arkhigov GG (1988) Power sources 16, Abstract, p 30
20. Read J (2002) Characterization of the lithium/oxygen organic electrolyte battery. J Electrochem Soc 149:A1190–A1195
21. Ogasawara C, Depart A, Holzaplel M et al (2006) Rechargeable Li_2O_2 electrode for lithium batteries. J Am Chem Soc 128:1390–1393
22. Mizuno F, Nakanishi S, Koani Y et al (2010) Rechargeable Li-air batteries with carbonate-based liquid electrolytes. Electrochemistry 78:403–405
23. Freunberger SA, Chen Y, Peng Z et al (2011) Reaction in the rechargeable lithium-O_2 battery with alkali carbonate electrolytes. J Am Chem Soc 133:8043–8047
24. Jung H-G, Hassonu J, Park J-B et al (2012) An improved high-performance lithium-air battery. Nat Chem 4:579–582
25. Kuboki T, Okuyama T, Ohsaki T et al (2005) Lithium-air batteries using hydrophobic room temperature ionic liquid electrolyte. J Power Sources 146:766–769
26. Hassoun J, Croce F, Armond M et al (2011) Investigation of the O_2 electrochemistry in a polymer electrolyte solid state cell. Angew Chem Int Ed 50:2999–3002
27. Visco SJ, Nimon E, Katz B, et al (2004) In: 12th international meeting on lithium batteries, Abstract # 53, Nara, Japan
28. Aono H, Subimoto E, Sadaoka Y et al (1989) Ionic conductivity of the lithium titanium phosphate ($Li_{1+X}M_XTi_{2-X}(PO_4)_3$, M = Al, Sc, Y, and La) systems. J Electrochem Soc 136:590–591
29. Hasegawa S, Imanishi N, Zhang T et al (2009) Study on lithium/air batteries—stability of NASICON type lithium ion conducting glass-ceramics with water. J Power Sources 189:371–377
30. Shimonishi Y, Zhang T, Imanishi N et al (2011) A study on lithium/air secondary batteries—stability of the NASICON-type lithium ion conducting solid electrolyte in alkaline aqueous solution. J Power Sources 196:5128–5132
31. Murgan R, Thangadurai V, Weppner W (2007) Fast lithium ion conduction in garnet-type $Li_7La_3Zr_2O_{12}$. Angew Chem Int Ed 46:7778–7781
32. Shimonishi Y, Toda A, Zhang T et al (2011) Synthesis of garnet-type $Li_{7-x}La_3Zr_2O_{12}$ and its stability in aqueous solutions. Solid State Ion 183:48–53

33. Carlson L, Ojefors L (1980) Bifunctional air electrode for metal-air batteries. J Electrochem Soc 127:525–528
34. Shimizu Y, Nemoto A, Kyodo T et al (1993) Gas diffusion-type oxygen electrode using perovskite-type oxides for rechargeable metal-air batteries. Denki Kagaku 61:1458–1458
35. Zhang T, Imanishi N, Shimonishi Y et al (2010) A novel high energy density rechargeable lithium/air battery. Chem Commun 48:1661–1663
36. Zhang T, Imanishi N, Hasegawa S et al (2009) Water-stable lithium anode with three-layer construction for aqueous lithium-air secondary batteries. Electrochem Solid-State Lett 12:A132–A135
37. Suntvick J, Gasteiger HA, Yabuuchi N et al (2011) Design principle for oxygen reduction activity on perovskite oxide catalysts for fuel cells and metal-air batteries. Nat Chem 3:546–550
38. Cho J (2010) Porous Si anode materials for lithium rechargeable batteries. J Mater Chem 26:4009–4014
39. Younsi SQ, Cioseka K, Elstrom K (2008) In: 214th meeting of ECS, Abstract #465, Honolulu, HI
40. Zhang D, Li R, Hung T et al (2010) Novel composite polymer electrolyte for lithium air batteries. J Power Sources 195:1202–1206
41. Zhang T, Imanishi N, Hirano A et al (2011) Stability of Li/polymer electrolyte-ionic liquid composite lithium conducting glass ceramics in an aqueous electrolyte. Electrochem Solid-State Lett 14:A45–A48
42. Liu S, Imanishi N, Zhang T et al (2010) Lithium dendrite formation in Li/poly(ethylene oxide)–lithium bis(trifluoromethanesulfonyl)imide and *N*-methyl-*N*-propylpiperidinium bis (trifluoromethanesulfonyl)imide/Li cells. J Electrochem Soc 157:A1092–A1098
43. Park HE, Hong CH, Yoon WY (2008) The effect of internal resistance on dendritic growth on lithium metal electrodes in the lithium secondary batteries. J Power Sources 178:765–768
44. Puech L, Cantau C, Vinatier P et al (2012) Elaboration and characterization of a free standing LiSICON membrane for aqueous lithium–air battery. J Power Sources 214:330–336
45. Andrei P, Zheng PJ, Handrickson M et al (2010) Some possible approaches for improving the energy density of Li-air batteries. J Electrochem Soc 157:A1287–A1295
46. Zhang JF, Liang BY, Hendricken M et al (2008) Bifunctional air electrode for metal-air batteries. J Electrochem Soc 155:A432–A437
47. Bealttle SD, Manoleseu DM, Blair SL (2009) High-capacity lithium–air cathodes. J Electrochem Soc 156:A44–A47

Chapter 2
Nonaqueous Electrolytes

Stefan A. Freunberger, Yuhui Chen, Fanny Bardé, Kensuke Takechi, Fuminori Mizuno, and Peter G. Bruce

2.1 Introduction

The electrolyte is what distinguishes the nonaqueous (aprotic) lithium air battery from its aqueous sibling. However, its significance is far greater than mere definition. The electrolyte has a profound influence on the reactions that occur at the anode and cathode and hence the overall operation of the nonaqueous lithium air cell on discharge/charge. It must possess a wide range of attributes, as listed in Table 2.1. The problems of cycling efficiency and dendrite formation at the Li metal/nonaqueous electrolyte interface are well known. Although the presence of O_2 adds to the challenges at the anode, it is the reactions at the O_2 cathode/electrolyte

S.A. Freunberger
Christian Doppler Laboratory for Lithium Batteries, Institute for Chemistry and Technology of Materials, Graz University of Technology, Stremayrgasse 9, 8010 Graz, Austria

Y. Chen • P.G. Bruce (✉)
School of Chemistry, University of St Andrews, Purdie Building, St Andrews KY16 9ST, Fife, UK
e-mail: p.g.bruce@st-andrews.ac.uk

F. Bardé
Advanced Technology 1, Toyota Motor Europe NV/SA, Hoge Wei 33A, B-1930 Zaventem, Belgium

K. Takechi
Advanced Battery Laboratory, Toyota Central R&D Laboratories, Inc., Nagakute 480-1192, Aichi, Japan

F. Mizuno (✉)
Battery Research Division, Toyota Motor Corporation, 1200 Mishuku, Susono 410-1193, Shizuoka, Japan

Materials Research Department, Toyota Research Institute of North America, 1555 Woodridge Avenue, Ann Arbor, MI 48105, USA
e-mail: fuminori.mizuno@tema.toyota.com

N. Imanishi et al. (eds.), *The Lithium Air Battery: Fundamentals*,
DOI 10.1007/978-1-4899-8062-5_2,

Table 2.1 Requirements on electrolytes for the nonaqueous Li–O_2 battery

Conductivity	• Sufficiently high for the anticipated rate capability
Stability	• Within potential window used on discharge and charge
	• In contact with O_2 and its reduced species on discharge
	• In contact with Li_2O_2 and its intermediates on charge
	• In contact with the anode or a stable SEI formed on the anode
Low volatility	• To minimise evaporation at the porous O_2 cathode
O_2 solubility and diffusivity	• To ensure adequate rate of mass transport to the cathode
Able to wet the electrode surfaces	
Promotes some solubility of Li_2O_2	Interaction with intermediates for high rate and packing density of Li_2O_2
Safety, low cost and toxicity	

interface that are new and less well understood, yet critical to the operation of the nonaqueous Li–O_2 battery. In particular achieving the reversible formation/decomposition of Li_2O_2 at the cathode on discharge/charge presents a major challenge for the electrolyte. Many aprotic electrolytes exhibit decomposition at the cathode during discharge and charge affecting the discharge and charge potential, the kinetics of discharge and charge, cyclability and calendar life. Identifying electrolytes that permit the efficient and highly reversible formation of Li_2O_2 at the cathode is one of the key challenges for the nonaqueous lithium air battery at the present time. As we shall see, the nature of the cathode itself can also influence the stability of the electrolyte during operation of the cell.

Of course no electrode reaction can proceed with 100 % purity, and side reactions are inevitable in chemistry and electrochemistry; however, they must be reduced to a sufficiently low level to support the cycle and calendar life required for a given application. We shall begin by reviewing the early studies, which employed organic carbonate electrolytes; they serve as a useful warning that electrochemical measurements alone can sometimes be misleading. This will be followed by considering the methods best suited to study stability and exploration of the more stable electrolytes. The chapter ends with a Sect. 2.4. Much of the current focus concerning electrolytes from nonaqueous lithium air batteries concerns stability and this will be reflected in the present chapter.

2.2 Early Electrolytes

Organic carbonates were used widely in the early lithium air cells. Their ubiquity in lithium ion cells was undoubtedly a factor in their use; however, they also, in the case of propylene carbonate (PC), offered low volatility, important to avoid evaporation of electrolyte at the O_2 electrode, and a high oxidation stability (up to 4.5 V) considered to be necessary for charging cells in early studies. However, more recent work has shown that the organic carbonates are unstable in Li–O_2 cells, even to the extent of showing little or no evidence for Li_2O_2 formation on discharge.

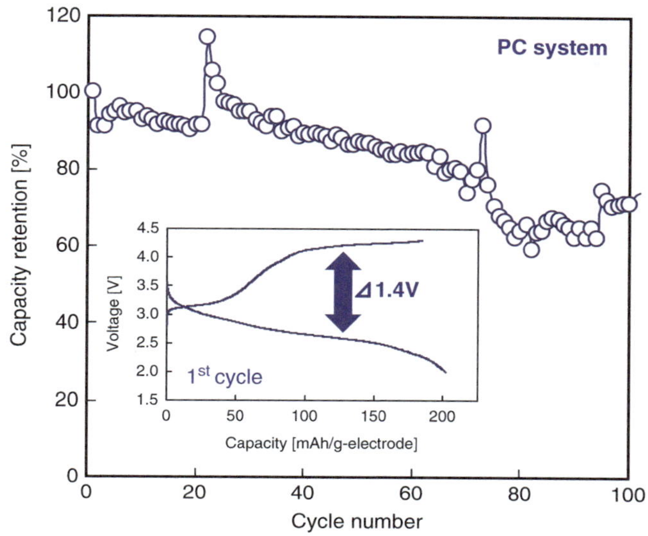

Fig. 2.1 Capacity retention on cycling a Li–O_2 cell composed of lithium metal anode, a 1 M solution of LiTFSI in propylene carbonate as the electrolyte and a porous cathode (Super P carbon black/α-MnO_2 catalyst/Kynar binder). The *inset* shows the variation of voltage during the first discharge then charge [1]. Reprinted with permission from [1]. Copyright (2010) The Electrochemical Society of Japan

Li–O_2 cells containing organic carbonate electrolytes such as propylene carbonate (PC), ethylene carbonate (EC), dimethyl carbonate (DMC), ethylmethyl carbonate (EMC) and mixtures thereof do demonstrate rechargeability as evidenced in Fig. 2.1 [1]. However, despite the ability to recharge such cells, it is now understood that the process is dominated by electrolyte composition. Some degree of decomposition is not unexpected, based on important studies several years ago of the reactivity of superoxide with carbonyl compounds including organic carbonates [2, 3]. Recent investigations by several groups have highlighted just how serious the problem of stability can be [1, 4–10]. FTIR spectra collected from cathodes at the end of discharge to 2 V versus Li/Li^+, in Li–O_2 cells containing organic carbonate-based electrolytes, exhibit many peaks that may be attributed to a variety of decomposition products (Fig. 2.2) [1, 4, 6]. A substantial amount of Li_2CO_3 is present along with other compounds containing C=O, C–O, C–C and C–H groups. Washing the cathode with D_2O then analysing the resulting solution with ^{1}H NMR (Fig. 2.3) identified products that indicated the presence of lithium propylene dicarbonate, lithium formate and lithium acetate, in addition to Li_2CO_3, in agreement with solid-state ^{13}C and ^{1}H MAS-NMR studies of the cathode [4]. Mass spectrometric analysis of the gas phase after discharge revealed that H_2, H_2O and CO_2 were also evolved.

Mechanisms have been proposed previously for the reaction between reduced O_2 species and PC; however, none of the mechanisms are able to reproduce the

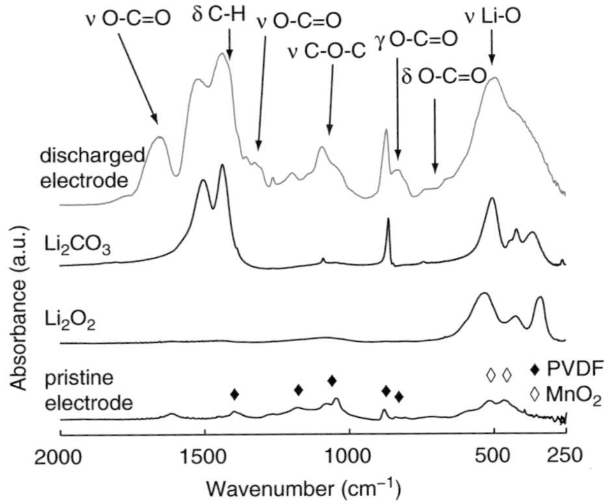

Fig. 2.2 FTIR spectrum of a composite electrode (Super P/α-MnO_2/Kynar) discharged in 1 M $LiPF_6$/PC to 2 V under 1 atm. O_2. The reference spectra for Li_2O_2, Li_2CO_3 and the electrode before discharge are also shown [4]. Reprinted with permission from [4]. Copyright (2011) American Chemical Society

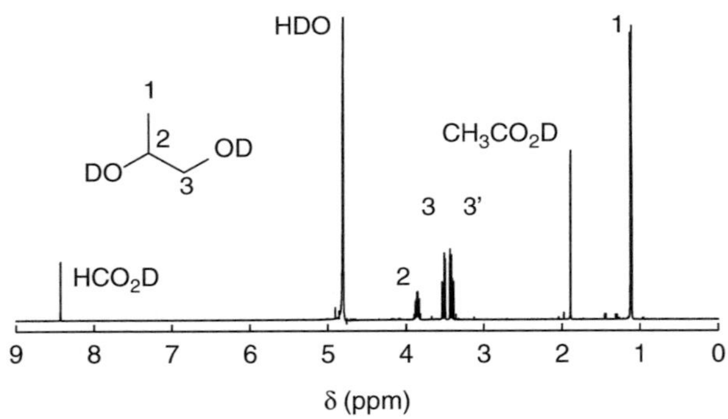

Fig. 2.3 ^{1}H solution NMR of the same electrode as in Fig. 2.2, extracted with D_2O. Reprinted with permission from [4]. Copyright (2011) American Chemical Society

products observed here [3, 11]. We therefore proposed the mechanism shown in Scheme 2.1 [4]. The reactions on discharge commence with O_2 reduction to O_2^-. This then attacks the CH_2 group, rather than the carbonyl C, in accord with DFT studies [8], via an S_N2 mechanism, reactions 2 and 3, to yield the ring-opened species **2**. The latter reacts via loss of O_2 and uptake of CO_2 (formed via the reactions below) resulting in formation of lithium propyl dicarbonate **4**, reactions 4 and 5. Intermediate **2** is known, in the presence of O_2, to easily undergo oxidative decomposition reactions to form H_2O and CO_2, analogous with combustion reactions [12]. Incomplete decomposition leads to the formation of formic and acetic acid or in the environment of the Li^+ electrolyte to Li formate and Li acetate. Finally, Li_2CO_3 can form by the reaction between O_2^- and CO_2, reactions 6 and 7 [13].

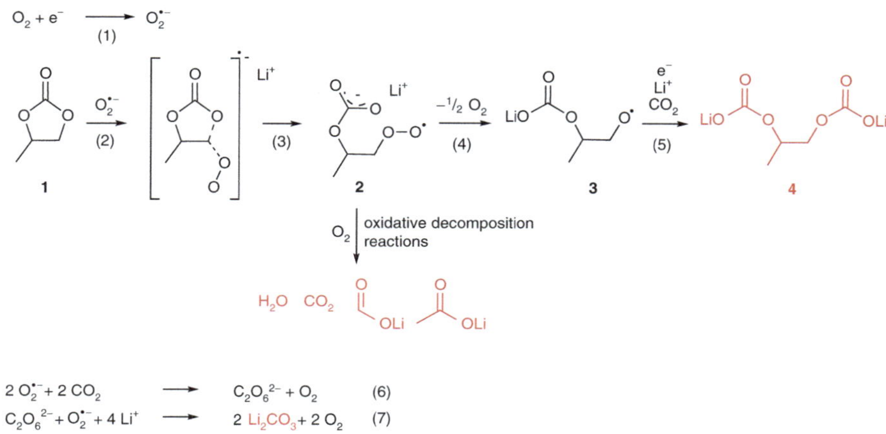

Scheme 2.1 Proposed reaction scheme on discharge to explain formation of the compounds: Li propyl dicarbonate, Li formate, Li acetate, Li_2CO_3, CO_2 and H_2O. The O_2^- generated in the first step could combine rapidly with Li^+, forming LiO_2, which may act as a nucleophile instead of, or as well as, O_2^-. It is expected to be reactive by analogy with HO_2. The subsequent reactions in the scheme would be the same. Reprinted with permission from [4]. Copyright (2011) American Chemical Society

Despite the presence of electrolyte decomposition, the Li–O_2 cell with organic carbonate electrolytes can, as we saw in Fig. 2.1, sustain rechargeability/cycling. Could this observation indicate that reversible Li_2O_2 formation/decomposition does occur, explaining the rechargeability, in parallel with electrolyte degradation, which would, in turn, explain the capacity fading? To investigate whether Li_2O_2 formation/decomposition is significant, in situ surface-enhanced Raman spectroscopy (SERS) was carried out at a roughened Au electrode. The in situ nature of the experiment ensures no contamination from air. The SERS data (Fig. 2.4a) show no evidence of any Li_2O_2, but a band associated with Li_2CO_3 is apparent. XPS spectra of a discharged carbon composite electrode confirm the absence of Li_2O_2 in the O 1*s* spectrum and suggest instead carboxylate and carbonate groups (Fig. 2.4b) [14]. Equally the C 1*s* spectra (Fig. 2.4c) confirm the presence of Li_2CO_3, lithium propyl dicarbonate and lithium carboxylates [14].

How then do Li–O_2 cells with organic carbonate electrolytes exhibit rechargeability? To answer this question, studies were carried out in which each of the solid discharge products, Li_2CO_3, Li propyl dicarbonate, HCO_2Li and CH_3CO_2Li, was subjected to charging in a Li–O_2 cell and the process monitored by a combination of in situ differential electrochemical mass spectrometry (DEMS), MS, FTIR and PXRD. The results showed that each of the identified products can be decomposed separately on charging up to 4.25 V (Fig. 2.5). Quantitative MS analysis of the gases evolved on charging the individual compounds allowed determination of the decomposition mechanisms. Generally, after removal of an electron and concurrent delithiation, the resulting radicals rearrange,

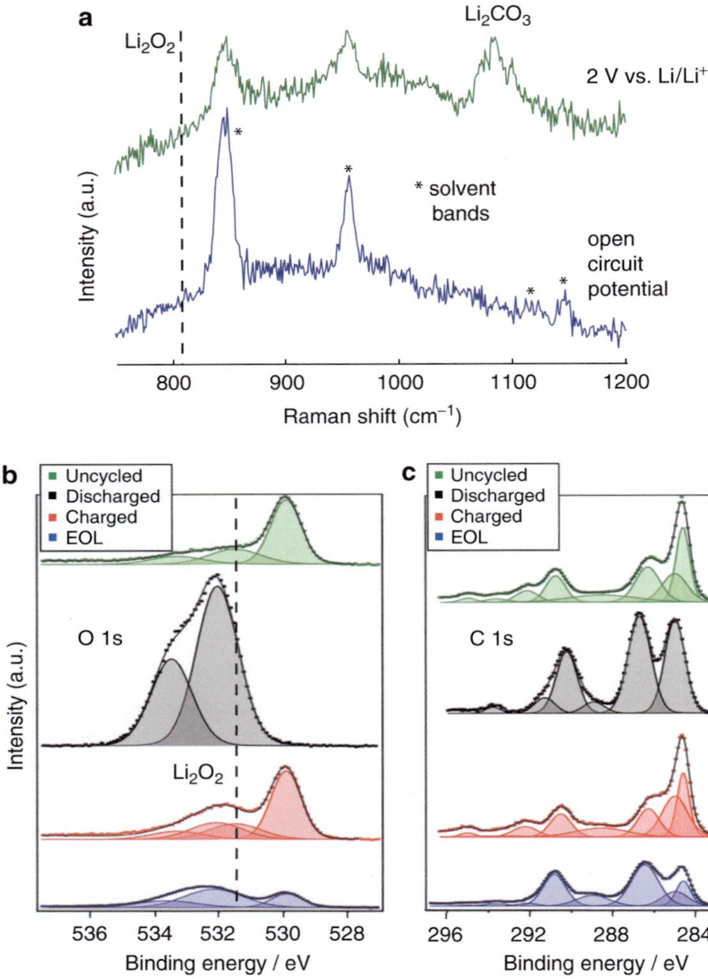

Fig. 2.4 (**a**) SERS on a Au electrode in 1 M $LiPF_6$/PC under O_2 at open circuit potential and after discharge to 2 V [4]. (**b**) O 1*s* and (**c**) C 1*s* XPS spectra of Super P/α-MnO_2/Kynar cathodes at the indicated state of cycling in 1 M $LiPF_6$/PC [14]. The *dashed lines* in (**a**) and (**b**) mark the position where Li_2O_2, if present, would occur. Reprinted (adapted) with permission from [4, 14]. Copyright (2011) American Chemical Society

involving loss of CO_2, and undergo further oxidative decomposition in the presence of O_2 to yield CO_2 and H_2O. No O_2 is evolved as exemplified by the charging of Li_2CO_3 under Ar (Fig. 2.5e). Therefore if a Li–O_2 cell containing an organic carbonate electrolyte is cycled between 2 and ~4 V, then the compounds formed by electrolyte decomposition on discharge are expected to be oxidised on charge. This is confirmed in Fig. 2.5 [4]. The FTIR spectra from the pristine electrode and the electrode after one cycle are identical, indicating that the products formed on

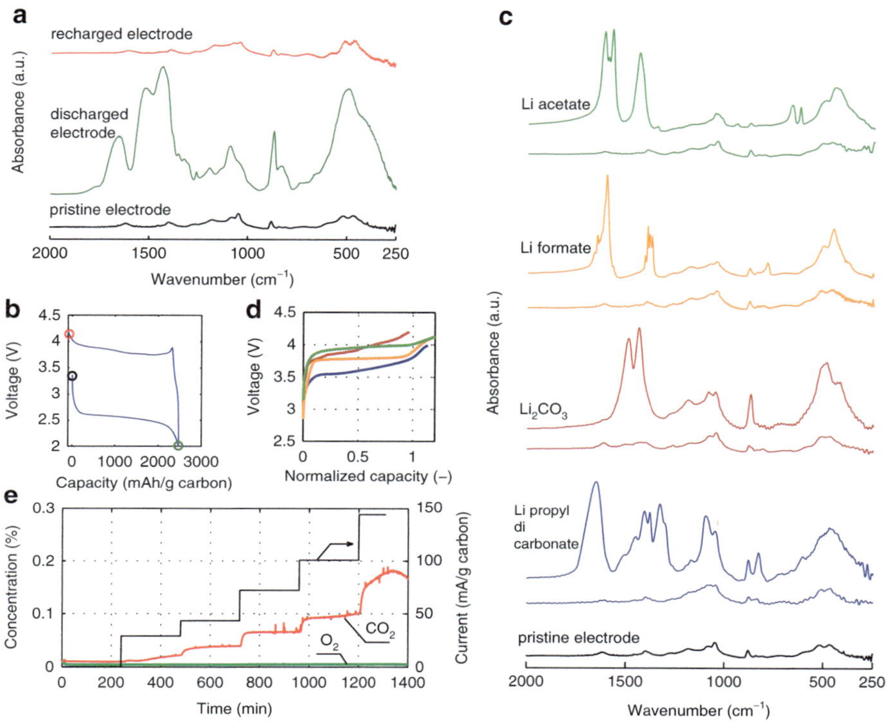

Fig. 2.5 (**a**) FTIR spectra of a pristine electrode (Super P/α-MnO_2/Kynar) and after the first discharge and then charge in 1 M $LiPF_6$ in PC under O_2. (**b**) Variation of voltage as cell is discharged and then charged. (**c**) Composite electrodes that contain the discharge products individually were subjected to charging in 1 M $LiPF_6$ in PC under O_2. FTIR spectra of the as-prepared electrodes and the charged electrodes for each of the compounds, together with the spectrum of a pristine electrode. (**d**) The corresponding charging curves at 70 m Ag^{-1} (based on mass of carbon). (**e**) Gas evolution measured by DEMS on oxidation of a composite electrode containing Li_2CO_3 in response to a stepwise increased current under Ar [4]. Reprinted (adapted) with permission from [4]. Copyright (2011) American Chemical Society

discharge are decomposed on charging. FTIR spectra of composite electrodes containing each of the discharge products separately before and after charging confirm their decomposition on charging (Fig. 2.5c). Likewise XPS spectra of the recharged electrodes confirm oxidation of the discharge products (Fig. 2.4b, c). These results are in good accord with those of other authors, for example, the DEMS studies of various carbonate-based electrolytes [9].

By studying the processes that occur at the cathode in Li–O_2 cells with organic carbonate-based electrolytes, we learn that the ability to cycle cells with such electrolytes has nothing to do with reversible Li_2O_2 formation but is entirely sustained by back-to-back irreversible reactions involving decomposition of the electrolyte on discharge and then decomposition of the solid discharge products on charging. This has a number of important consequences. At best, cycling will continue only until a significant proportion of the electrolyte has been consumed, then the cell will fail, explaining the often observed drastic and sudden termination of cycling (cell death). However capacity fading generally occurs even before electrolyte starvation, because not all the compounds formed on discharge are entirely oxidised on charge. For example, Li_2CO_3 is, as can be seen from Fig. 2.5d, the most difficult to oxidise requiring the highest voltage (usually >4 V) to remove it completely. Accumulation of insulating Li_2CO_3 at the electrode results in passivation and capacity fading, especially if the cells are not charged to potentials >4 V or high rates are used. The fact that the degree of accumulation of products like Li_2CO_3 and the degree of electrolyte starvation will depend on the cell construction, voltage range, rate, etc., explains why capacity fading and cell failure are different for different cells reported in the literature. Finally, we learn an important lesson from the work on organic carbonate electrolyte. In a new and relatively unexplored field, such as Li–O_2 cells, it is necessary to conduct fundamental investigations of the processes occurring in the cells, and using a range of techniques, electrochemical measurements alone and the methods familiar in the study of Li-ion cells, is not sufficient.

2.3 The Search for Stable Electrolytes

As we shall see, unlike organic carbonates, several electrolytes do support formation/decomposition of Li_2O_2 at the cathode, but in most cases this is accompanied by significant electrolyte decomposition. This emphasises the difficulty in finding electrolytes stable towards the strong nucleophiles O_2^-/LiO_2 that are expected to be intermediates in Li_2O_2 formation/decomposition, as well as towards Li_2O_2 itself [15–17]; certainly it is known that fresh surfaces of Li_2O_2 particles react with many electrolytes [1, 6, 9, 11, 18]. LiO_2 is somewhat analogous to HO_2 and as such may be even more reactive than O_2^-, since as well as O_2^- being a strong nucleophile, the associated Li^+ can react together in a concerted process, like H^+. Following extensive investigation of aprotic electrolytes for Li–O_2 cells, electrolytes based on ethers, amides, ionic liquids (IL), sulfones and dimethyl sulfoxide (DMSO) have been identified to form predominantly Li_2O_2 on a first discharge and to allow its removal on recharge [19–29]. To sustain Li_2O_2 formation/decomposition with high purity and high reversibility on cycling depends on the synergy between electrolyte and electrode [27, 29–31]. The stability of the electrolyte cannot be considered in isolation. Since only a couple of electrolytes have been investigated in this respect, further advances towards better electrolytes for the aprotic Li–O_2 cell are still required. Following consideration of the methods used to study electrolyte stability, the results of such investigations of electrolytes based on ethers, ionic liquids, amides, sulfones and DMSO will be discussed.

Three approaches to seeking a stable electrolyte for a nonaqueous Li–O_2 battery have been taken: modelling of the reactions between O_2^-, LiO_2 and the solvent;

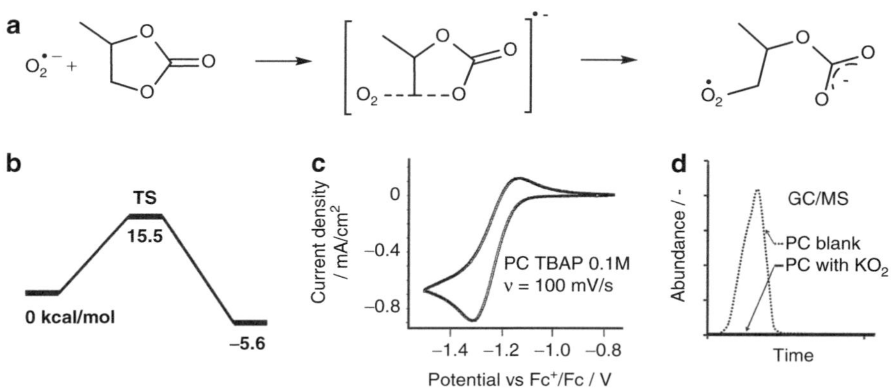

Fig. 2.6 The combined screening approach for electrolyte stability using computational modelling, cyclic voltammetry and chemical reaction with KO_2, exemplified with PC. (*top* (**a**)) Reaction for superoxide addition to PC with ring opening. (*left* (**b**)) Reaction-free energy profile for the above reaction obtained from DFT calculations. (*centre* (**c**)) Cyclic voltammogram on a glassy carbon electrode in O_2 saturated PC/$TBAClO_4$ electrolyte showing side reactions of O_2^- formed. (*right* (**d**)) GC-MS analysis of a dilute solution of the solvent to be screened (here PC) in CH_3CN with and without addition of excess KO_2 showing complete reaction of PC with KO_2 [34, 38]. Reprinted with permission from [23]. Copyright (2011) American Chemical Society

chemical screening using KO_2; and direct electrochemical measurements combined with spectroscopic methods. All three are valuable and interrelated:

1. Increasingly, computational methods are used to gain insight into solvent stability and to identify promising directions for new developments [8, 32–36]. Various levels of quantum chemical calculations have been employed to predict the likelihood of solvent attack. Calculated Mulliken charges, i.e. polarity of the solvent molecules, have revealed a trend between the presence of positively charged carbons and susceptibility of the solvent to attack by O_2 reduction products [32]. A wide variety of solvents have been screened for stability using DFT methods [8, 33–37]. Generally reaction-free energy profiles are calculated for various possible reaction pathways involving O_2^- or LiO_2 attack on the solvent by, e.g. involving nucleophilic reaction or proton abstraction (Fig. 2.6a). Both species turn out to be highly reactive. The activation energies thus obtained are in excellent agreement with electrochemical and chemical screening studies [34]. The mechanistic insight obtained from such calculations provides valuable guidance for the design of more stable electrolytes.
2. KO_2 is a readily available source of superoxide, and it has been used extensively in organic synthetic studies and is frequently used for screening of solvent stability in combination with various analytical methods [3, 17, 19, 32, 34, 39–42]. The relatively poor solubility and therefore slow reaction rate are overcome by complexation with crown ethers, e.g. 18-crown-6. Application of the method to a PC electrolyte is shown in Fig. 2.6.

3. Direct electrochemical measurement of the reversibility of the O_2/O_2^- couple in nonaqueous electrolytes has been employed to investigate the stability of solvents [3, 19, 32, 34, 39, 40, 43–45]. Typically, salts with large cations, e.g. tetrabutylammonium salts, are used for the supporting electrolyte since they leave the O_2^- in its most unperturbed state [15, 17, 46]. Cyclic voltammograms (CVs) at moderate rate (100 mV s^{-1} or below) may show a diminished or absent anodic current, indicative of reaction between O_2^- and the electrolyte. In order to obtain quantitative values, rotating ring disc electrode measurements have been used [44]. Figure 2.6 summarises the combination of these three screening approaches for the case of PC [34].

Although the above methods are useful for initial screening, ultimately assessing stability requires the construction of Li–O_2 cells and analysis of the reactions occurring at the cathode during sustained cycling. As we have seen, electrochemical rechargeability is a necessary but not sufficient proof of stable and reversible Li_2O_2 formation/decomposition, yet only if the cathode reaction is dominated by stable and reversible Li_2O_2 formation/decomposition can a truly rechargeable Li–O_2 cell exist. It is also the case that the technique of first choice for studying Li-ion cells, powder diffraction, may, if used alone, be misleading, as it can emphasise the presence of crystalline solids, such as Li_2O_2, over less crystalline compounds, such as Li carboxylates, that may be present indicating decomposition. Only combining electrochemical measurements with several spectroscopic/analytical techniques can one determine true stability. Several of such methods are described below.

FTIR

Examination of the cathode by FTIR is useful as it captures solid products that form in the cathode whether or not they are crystalline. Using a spectrometer with a low cut-off of 250 cm^{-1} permits identification of Li_2O_2.

In Situ Differential Electrochemical Mass Spectrometry

To confirm that the reaction at the cathode is overwhelmingly Li_2O_2 formation/decomposition, it is important to establish charge/mass balance, i.e. that the charge passed on discharge and charge corresponds to the amount of Li_2O_2 formed/decomposed and in turn O_2 consumed on discharge and evolved on charge. The charge/mass balance on discharge/charge should be $2e^-/O_2$ with no further gases evolved. This is most readily measured using an in situ DEMS system such as is shown in Fig. 2.7a; the system fulfils the requirement of quantifying all the gases consumed and evolved during discharge and charge [9, 10, 27, 29, 47]. Typically, the continuous or intermittent purge gas flow through the cell headspace is analysed by the mass spectrometer, after it exits the cell. An alternative approach to the measurement of integral gas consumption or evolution is to follow the pressure change in the sealed headspace on cycling, although this does not permit identification of the gases consumed/evolved [9, 48].

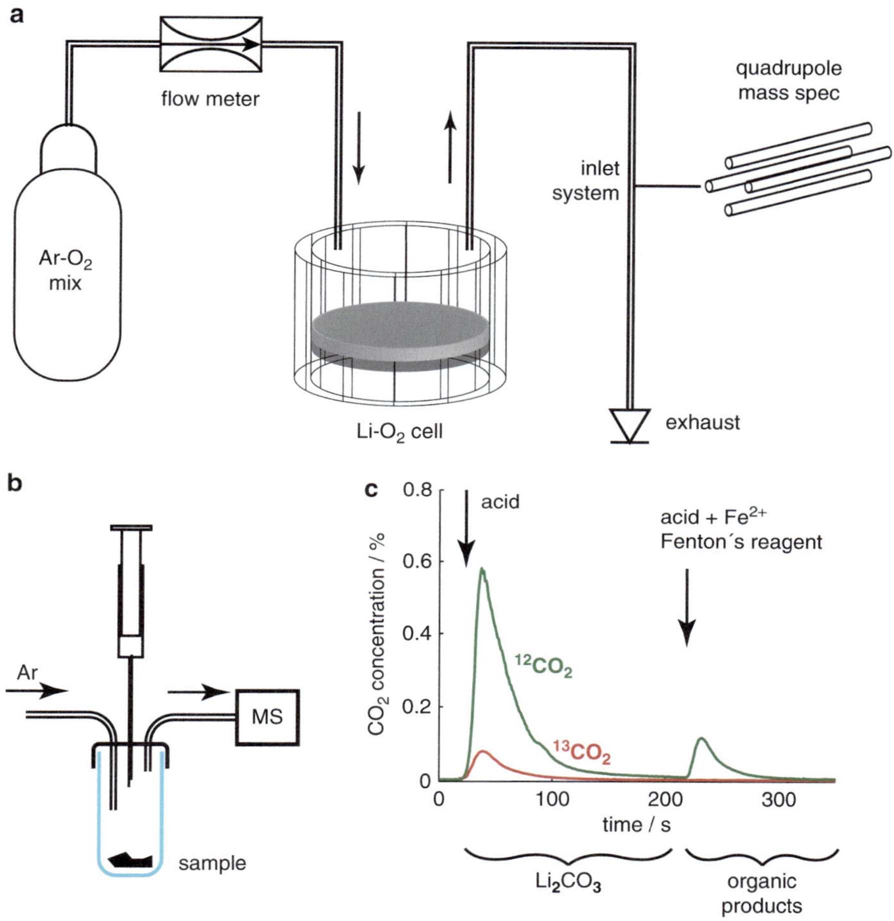

Fig. 2.7 (**a**) Schematic of an in situ differential electrochemical mass spectrometry (DEMS) [47]. The cell headspace above the cathode is continuously purged with a mixture of ca. 5 % Ar in O_2, which is analysed by the mass spectrometer after it exits the cell. Ar acts as an inert tracer with known invariant flux. The measured concentrations allow calculation of the flux of each component. (**b**) Setup for chemical analysis using the same system as for DEMS. (**c**) Typical measurement using the apparatus in (**b**), which is able to distinguish inorganic and organic side products and their source through isotope labelling [31]

NMR

FTIR alone can only suggest the presence of functional groups but cannot unambiguously identify particular organic compounds. Extracting the electrodes with D_2O and subjecting the solution to ^{1}H NMR allows the determination of the compounds, such as Li carboxylates, at trace sensitivity (<0.1 %) [4, 49–51]. Furthermore ^{19}F and ^{11}B NMR can be used to detect salt decomposition products in the electrolyte and discharge product [51, 52]. Solid-state NMR (^{1}H, ^{13}C, ^{7}Li, ^{17}O) on discharged cathodes is useful to identify the nature of main compounds formed, including Li_2O_2 and Li_2CO_3 [4, 53, 54].

Chemical Analysis

Although spectroscopic methods are invaluable in order to identify the compounds that form, they are often less successful at quantifying their amounts. To address this problem, a procedure for direct chemical analysis has been developed (Fig. 2.7b, c) [31]. It involves extracting the cathodes from the Li–O_2 cells, treating them with acid to decompose the Li_2CO_3 present in the electrode and following Fenton's reagent to oxidise organic compounds. The CO_2 evolved in each case is determined by mass spectrometry. In this way, inorganic and organic decomposition products can be quantified with very high sensitivity, including compounds that cannot be oxidised and hence would not be detectable with. The mass spectrometric detection allows assignment of the source of the products using isotope labelling.

2.3.1 *Ether-Based Electrolytes*

Following the realisation that carbonates are unsuitable for Li–O_2 cells, much attention shifted to electrolytes based on ethers [9, 20–23, 55–61]. Ethers are attractive for the Li–O_2 battery because they are one of the few solvents that combine the following attributes: most of them are capable of operating with a lithium metal anode, stable to oxidation potentials up to 4 V versus Li/Li^+, safe, of low cost and, in the case of higher molecular weights, such as tetraglyme, of low volatility. Crucially, they are also anticipated to show greater stability towards reduced O_2 species compared with organic carbonates. A variety of ether-based electrolytes both linear (stemming from the glyme family, e.g. $CH_3O(CH_2CH_2O)_4CH_3$) and cyclic (e.g. 1,3-dioxolane—$C_4O_2H_8$ and 2-methyl tetrahydrofuran—$C_5H_{10}O$) have been studied. They are not expected to be susceptible to nucleophilic attack by O_2 [9, 20], something which is supported by theoretical studies [34, 60, 61]. However, despite the anticipated advantages of ethers, there is growing experimental and theoretical evidence that ether-based electrolytes are not suitable for Li–O_2 cells [9, 24, 37, 62–66]. On the first discharge to 2 V, Li–O_2 cells with ether-based electrolytes form as the major product Li_2O_2, as identified by many groups (Fig. 2.8a). However, careful analysis with a variety of the techniques described above revealed that, in addition, electrolyte decomposition takes place forming HCO_2Li, CH_3CO_2Li, polyethers/esters and Li_2CO_3 (Fig. 2.8b–d) [9, 24, 37, 62–66]. On subsequent charging to 4.7 V, only 60 % of the expected O_2 was evolved (identified by DEMS) (Fig. 2.9a), and this was accompanied by evolution of CO_2 [9, 64]. An average of $3.2e^-/O_2$ was also observed, which conflicts with the ratio of $2e^-/O_2$ expected for pure Li_2O_2 oxidation. The stability of ethers may however be enhanced to some extent by siloxane end groups or optimised salt concentration, although decomposition persists [60, 65, 67]. The extent of

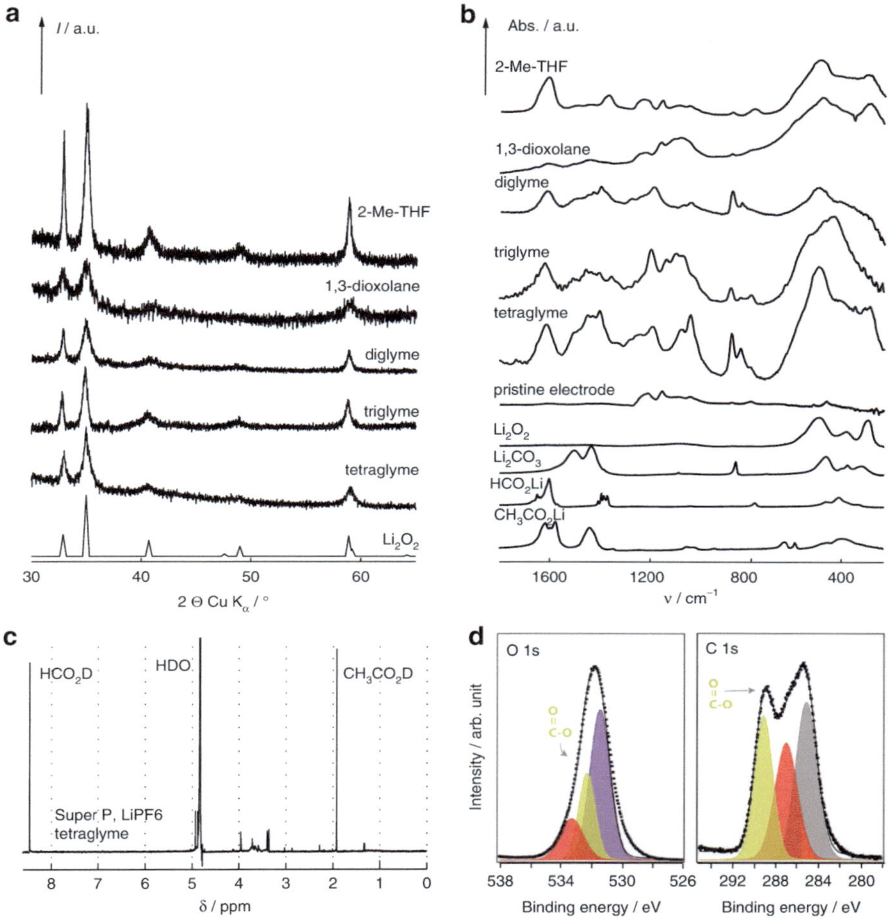

Fig. 2.8 (**a**) Powder X-ray diffraction patterns of composite cathodes (Super P/PTFE for diglyme and cyclic ethers) discharged in 1 M $LiPF_6$ in the indicated solvent to 2 V versus Li/Li^+. (**b**) FTIR spectra of the same cathodes and reference spectra. (**c**) A representative NMR spectrum of the D_2O extract from the discharged cathodes, in this case a tetraglyme electrolyte was used [24]. (**d**) O 1*s* and C 1*s* spectra of a Li_2O_2 sample after being in contact with 1 M $LiClO_4$ in tetraglyme for 48 h [66]. Reprinted (adapted) with permission from [24]. Copyright (2011) John Wiley and Sons. Reprinted with permission from [66]. Copyright (2012) American Chemical Society

electrolyte degradation compared with Li_2O_2 formation on discharge appears to increase rapidly with cycling, such that after only five cycles there is little or no evidence of Li_2O_2 in powder X-ray diffraction (PXRD) patterns at the end of discharge (Fig. 2.8a). The corresponding load curves are shown in Fig. 2.9b [24, 68].

Dimethoxyethane (DME) has been used widely in Li–O_2 batteries since it was recognised that the principal product on the first discharge is Li_2O_2, based on

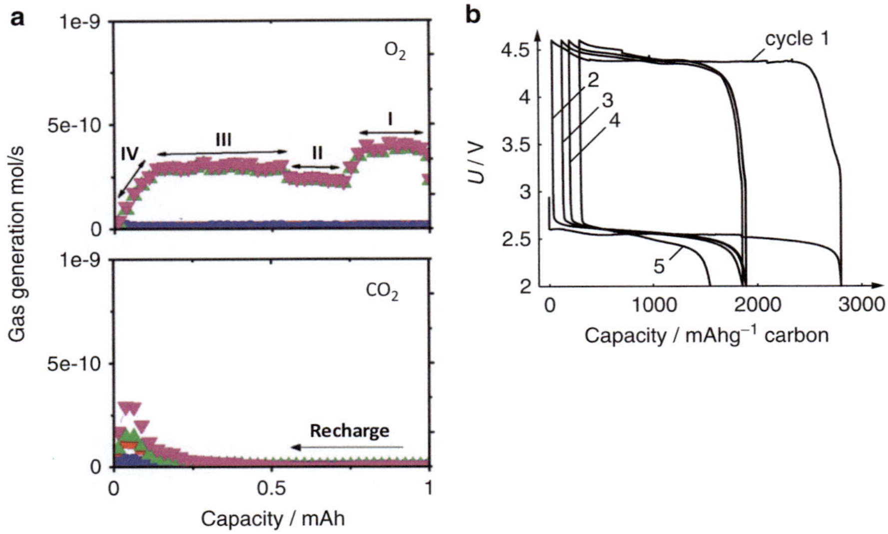

Fig. 2.9 (**a**) O_2 and CO_2 evolution during galvanostatic recharging of a P50 carbon paper cathode to 4.7 V versus Li/Li^+ in 1 M LiTFSI–DME, previously discharged to 1 mAh under O_2 [9]. (**b**) Load curves for Super P cathode cycled in 1 M $LiPF_6$ in tetraglyme at 70 mA g^{-1} carbon between 2 and 4.6 V versus Li/Li^+ [24]. Reprinted (adapted) with permission from [9]. Copyright (2011) American Chemical Society. Reprinted (adapted) with permission from [24]. Copyright (2011) John Wiley and Sons

powder XRD [3, 9]. However, although it may be more stable than the longer chain ethers, it still decomposes in a similar manner to that described above. Indeed the data in Fig. 2.9 are for DME.

Overall ether-based electrolytes, including the most widely used tetraglyme and DME, are unstable in the rechargeable Li–O_2 cell, which results in significant fading after a few cycles (Fig. 2.9b). The fading is likely due to the accumulation of Li_2CO_3 on the electrode surface, which, as noted above (Fig. 2.5), is more difficult to oxidise, requiring a higher oxidation potential (>4 V) for the same current density compared to other decomposition products to fully decompose. As a result, it is easy to accumulate a net amount of Li_2CO_3 on the electrode on cycling, resulting in passivation. A relatively small quantity can lead to a disproportionate amount of capacity fading. Ethers are known to form peroxides under ambient conditions by auto-oxidation reactions [69, 70]. This instability towards auto-oxidation has been identified as a possible pathway of initial attack [37]. Moreover, some ethers will not cycle efficiently with lithium metal, particularly in the presence of O_2 [71, 72]. The latter study suggests that long-term cycling of lithium metal anodes will require a protective layer against ingress of O_2, as is the case with other possible electrolytes discussed below.

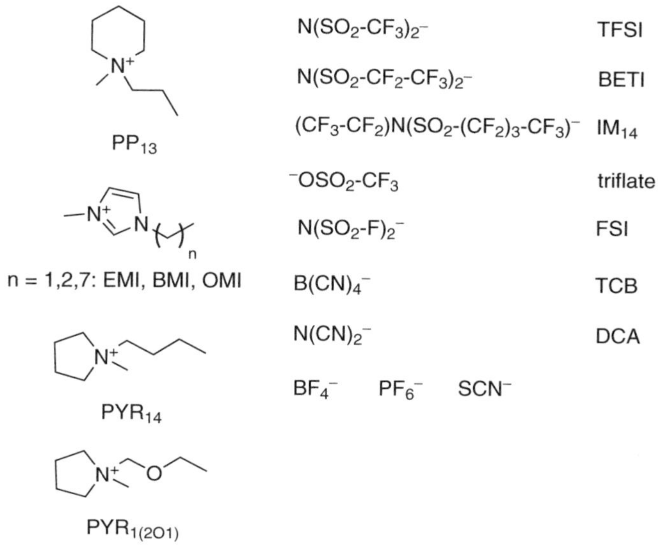

Scheme 2.2 Overview of ionic liquids studied so far for Li–O_2 battery applications. Various combinations of the depicted cations and anions were used

A detailed study of carbon cathodes in Li–O_2 cells has shown that carbon promotes electrolyte decomposition on discharge, leading to side products derived from the electrolyte [31]. Carbon itself decomposes, mainly on charging at voltages >3.5 V and associated with the oxidation of Li_2O_2. Carbon is also thermodynamically unstable in contact with Li_2O_2 [62]. In the case of the tetraglyme-based electrolyte, the quantity of Li_2CO_3 originating from carbon degradation after the first discharge/charge cycle is equal to the amount resulting from electrolyte decomposition after the first discharge. On repeated cycling the proportion and amount of side products derived from carbon increases. The issue of carbon stability is analysed in Chap. 3, which discusses in detail electrodes for nonaqueous Li–O_2 cells. As the vast majority of Li–O_2 studies to date have employed carbon-based cathodes, the results in this chapter are drawn predominantly from measurements involving carbon. However, to ensure that the conclusions presented reflect the instability of the electrolytes due to reaction with reduced oxygen species and not the effect of the carbon electrode, separate experiments, in which carbon was replaced by nanoporous gold (NPG), were carried out. These experiments confirmed that while carbon electrodes exacerbate the electrolyte decomposition, the trends described above persist with NPG, i.e. ethers are unstable in aprotic Li–O_2 cells.

2.3.2 *Ionic Liquids*

Ionic liquids (IL) have been widely used for O_2 reduction studies [73–75]. Early work on the use of room temperature ionic liquids for Li–O_2 applications evoked wide interest in their use [25, 39, 41, 42, 44, 76–80]. Accordingly a wide range of the many possible candidates have been studied (Scheme 2.2). Cations mainly comprise asymmetrically substituted piperidinium (e.g. methyl-propyl, PP_{13}), imidazolium (e.g. ethyl-methyl, EMI) and pyrrolidinium (e.g. butyl-methyl, PYR_{14}, or ethoxymethyl-methyl, $PYR_{1(2O1)}$) and the anion mostly being $TFSI^-$, but also drawn from a large variety of others. Screening by reaction with KO_2 or voltammetric methods suggests largely improved stability over organic carbonate solvents [32, 42, 44]. Only relatively little effort was devoted to the spectroscopic characterisation of ILs in Li–O_2 cells. Raman spectra obtained from the precipitate of the reaction of KO_2 with PP13TFSI/LiTFSI are shown in Fig. 2.10a in comparison to those in other Li^+ electrolytes based on a variety of nonaqueous solvents. Li_2O_2 is the main discharge product with a small fraction of Li_2CO_3 present. A FTIR study of the discharge product in the IL $PP_{13}TFSI$ is shown in Fig. 2.10b. While Li_2O_2 is positively identified, Li_2CO_3 is additionally abundant. This result is in accord with electron energy loss spectroscopy (EELS) (Fig. 2.10c). In contrast to carbonate- or ether-based electrolytes, the FTIR spectra show evidence of only a small fraction of lithium organic compounds.

Cycling curves for a cell comprising a vertically aligned carbon nanotube (VACNT) cathode, a Li metal anode and a 0.4 M $LiCiO_4$ in $PP_{13}TFSI$ are shown in Fig. 2.11a. The capacity retention on cycling is shown in Fig. 2.11c. Powder XRD analysis after various discharge and charge cycles confirms formation of Li_2O_2 during the first discharge and its disappearance on charging (Fig. 2.11b). This is in accord with XPS spectra (Fig. 2.11d) which in addition show Li_2CO_3 to appear on charging. Subsequent cycles show diminishing amount of Li_2O_2 in the XRD with no further evidence after 21 cycles. XPS confirms that Li_2CO_3 formation dominates the discharge reaction. It is not yet clear whether the IL is decomposed or the used carbon substrate, which has been shown to be unstable in the Li–O_2 cell, particularly so on charging Li_2O_2 above 3.5 V [31]. Results on IL-based electrolytes presented so far suggest potential as a base for electrolytes in the nonaqueous Li–O_2 battery due to nonvolatility and potentially high stability. Limited cyclability of cells using IL-based electrolytes and carbon electrodes suggests that side reactions are an issue that requires deeper investigation.

2.3.3 *Amide-Based Electrolytes*

Amides are a major class of solvent that are known to be highly stable against bases and nucleophilic attack and have been extensively used for O_2 reduction studies [81–83]. N-alkyl-substituted amides, e.g. dimethylformamide (DMF)

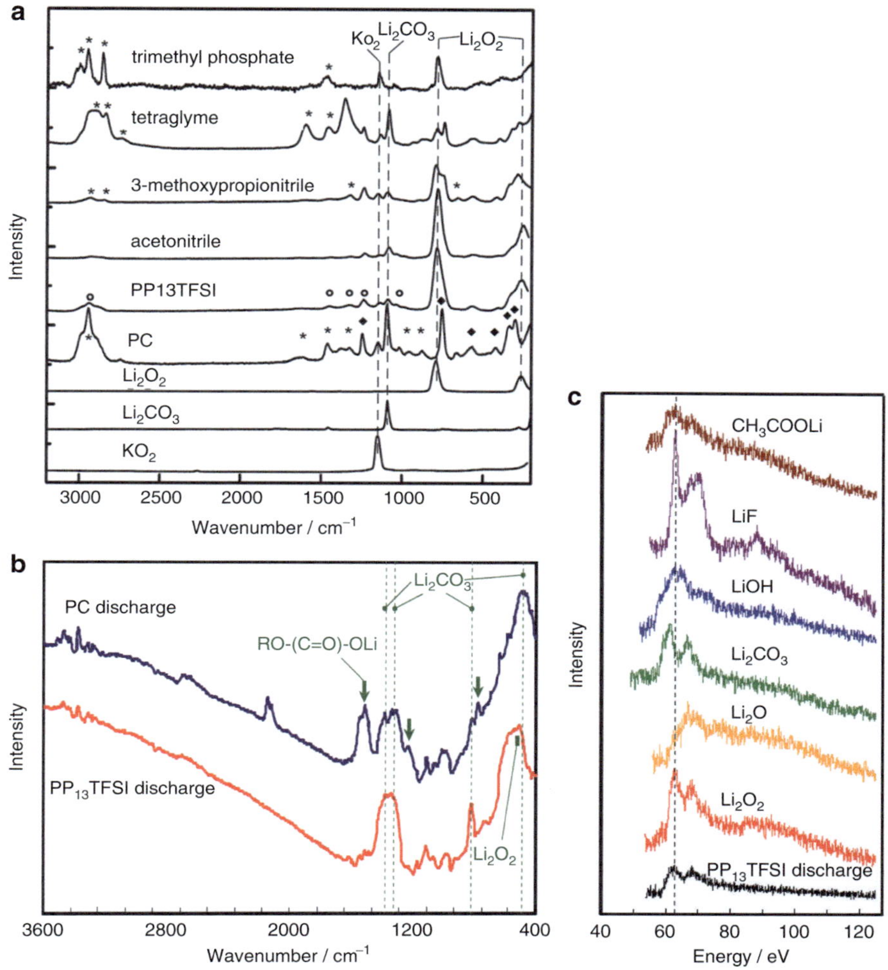

Fig. 2.10 (**a**) Raman spectra of precipitates obtained from reactivity screening with KO_2 in various solvents including the IL $PP_{13}TFSI/LiTFSI$ [42]. Reproduced by permission of The Electrochemical Society. (**b**) FTIR spectra of composite cathodes (Super P/PTFE) in either 1 M $LiPF_6$ in PC or 1 M LiTFSI in $PP_{13}TFSI$, 1 atm. O_2, discharged to 2 V [25]. (**c**) EELS spectra of the same cathode discharged in $PP_{13}TFSI/LiTFSI$ and of various model compounds [25]. Reprinted from [25], Copyright (2013), with permission from Elsevier

being a prominent member, have been commonly used for such studies. Theoretical calculations and KO_2 screening studies suggest high energy barriers against nucleophilic substitution on the carbons, due to unfavourable stability of the reaction products, as exemplified by *N*-methyl-2-pyrrolidone in Fig. 2.12 [34]. Further DFT studies revealed, as for ethers discussed above, auto-oxidation at α-protons as a possible pathway of degradation, which may be further

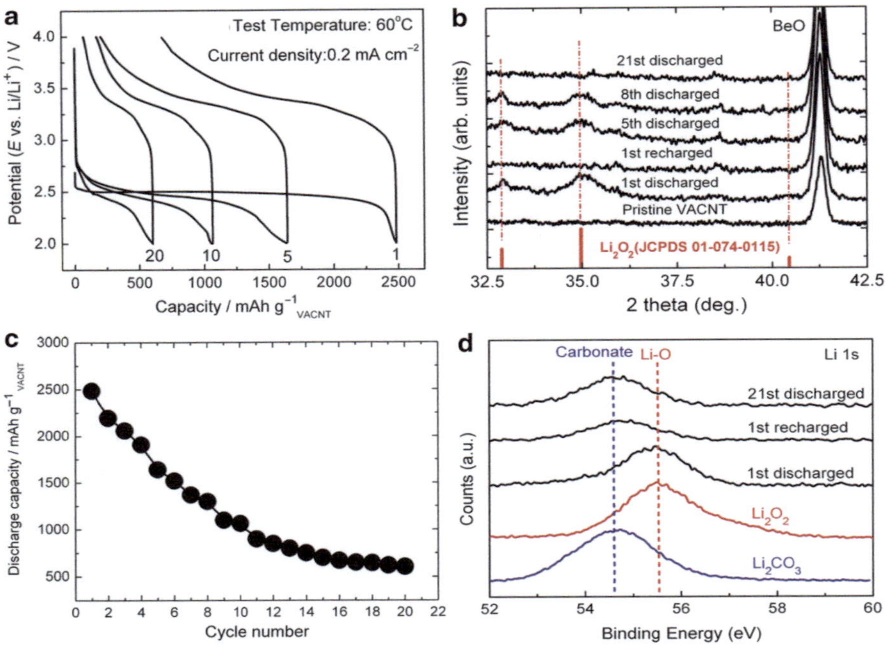

Fig. 2.11 (**a**) Variation of cell voltage in a cell employing a vertically aligned carbon nanotube (VACNT) cathode, a Li metal anode and a 0.4 M $LiCiO_4$ in $PP_{13}TFSI$. (**b**) PXRD after the indicated discharge and charge cycles. (**c**) Retention of discharge capacity with cycle number. (**d**) XPS spectra after the first cycle and after the 21st discharge [79]. Reprinted [79], with permission from Elsevier

accelerated under conditions found in the Li–O_2 battery [37]. This is because hydroperoxides formed upon auto-oxidation undergo superoxide-mediated oxidation to yield finally the highly reactive hydroxyl radical, which could serve as an efficient initiator of further auto-oxidation. While both ethers and amides require a high activation energy for proton abstraction by O_2^-, amides are much more stable against auto-oxidation than ethers. Taken together, the computational and experimental screening studies suggest that N,N-dialkylamides are amongst the most stable solvents towards nucleophilic substitution, H-abstraction and in terms of auto-oxidation stability [38].

Several amides (DMF, dimethylacetamide—DMA and *N*-methyl-2-pyrrolidone—NMP) have been investigated as the basis of electrolytes for Li–O_2 batteries [47]. Load curves of a cell composed of a porous cathode (Super P/PTFE), an electrolyte, comprising 0.1 M $LiClO_4$ in DMF and a $LiFePO_4$ counter electrode are given in Fig. 2.13. $LiFePO_4$ was used as a counter electrode since lithium metal is not stable in DMF. While initially the recharge capacity is nearly identical to the previous discharge, significant capacity fading is observed on cycling with the charge plateaus shifting gradually to higher voltages.

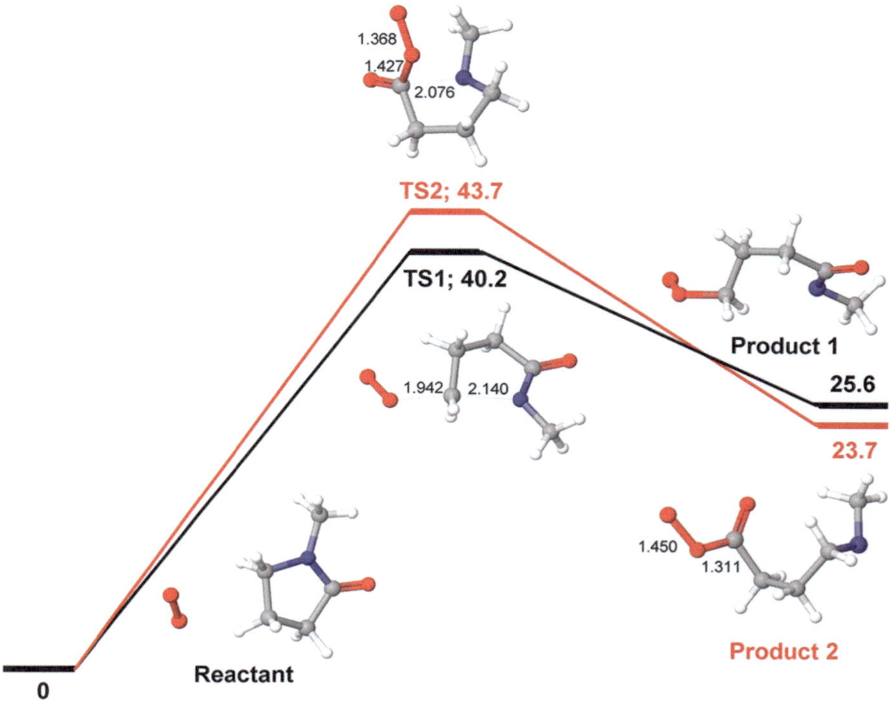

Fig. 2.12 Reaction-free energy profile for nucleophilic attack of O_2^- at the ethereal (*grey*) and carbonyl (*red*) carbon atoms of *N*-methyl-2-pyrrolidone calculated in the field of continuum solvent (kcal mol^{-1}). B3LYP optimised bond lengths are shown [34]. Reprinted with permission from [34]. Copyright (2011), American Chemical Society

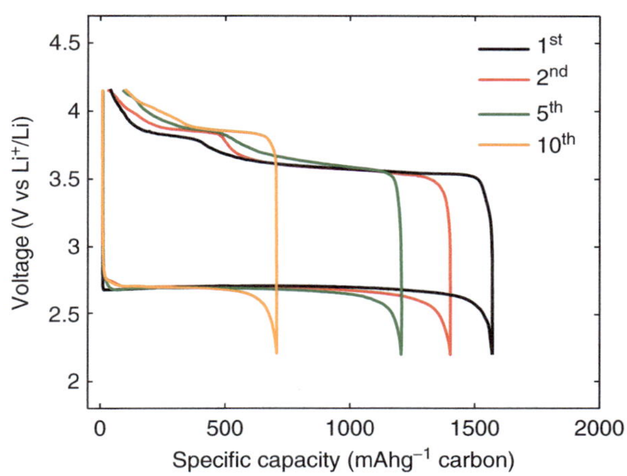

Fig. 2.13 Load curves for a cell containing 0.1 M $LiClO_4$ in DMF and a composite cathode, Super P/PTFE, rate 70 m Ag^{-1}. Reprinted with permission from [47]. Copyright (2012) American Chemical Society

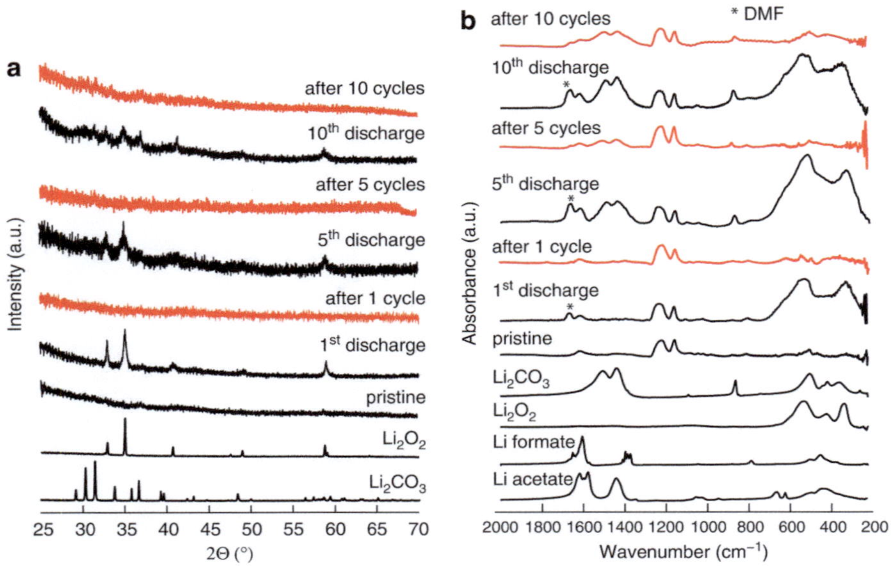

Fig. 2.14 (**a**) PXRD patterns (Cu Kα) of Super P/PTFE composite electrodes cycled in 0.1 M $LiClO_4$ in DMF, rate 70 m Ag^{-1}. Diffraction patterns in *red* correspond to the pristine electrode and electrodes at the end of charge after the indicated number of cycles. (**b**) FTIR spectra of the same cathodes. Again spectra in *red* correspond to recharged electrodes. Reprinted with permission from [47]. Copyright (2012) American Chemical Society

Analysis using FTIR, PXRD (Fig. 2.14), DEMS (Fig. 2.15) and NMR revealed that Li_2O_2 is the dominant product on the first discharge, accompanied by a small degree of electrolyte decomposition, as demonstrated by a very small amount of NO gas (<0.1 % of O_2) evolution on discharge (Fig. 2.15a) and detection of small quantities of HCO_2Li and CH_3CO_2Li in ^1H NMR. As shown previously, this method permits identification of compounds that may not be assigned from FTIR alone.

On the first charge, all discharge products were removed (Fig. 2.14). The DEMS data on the first charge showed only O_2 evolution (Fig. 2.15b) and an e^-/O_2 ratio of 1.96. The lack of any detectable CO_2 and H_2O evolution expected from oxidation of HCO_2Li and CH_3CO_2Li confirms that the amount of electrolyte decomposition occurring on the first discharge is very small. Together, the PXRD, FTIR, NMR and DEMS data demonstrate that the first discharge/charge cycle in DMF is dominated by reversible Li_2O_2 formation with a very small degree of electrolyte degradation. While early cycles show mainly reversible Li_2O_2 formation/decomposition, from the fifth cycle onwards, clear evidence is seen for significant side reactions (Fig. 2.14b) involving formation of HCO_2Li, CH_3CO_2Li and Li_2CO_3. Li_2CO_3 is seen to be more difficult to oxidise, as noted above, and is therefore only partly removed and accumulates as the dominant side product on cycling (Fig. 2.14b) in accord with the observed capacity fading (Fig. 2.13).

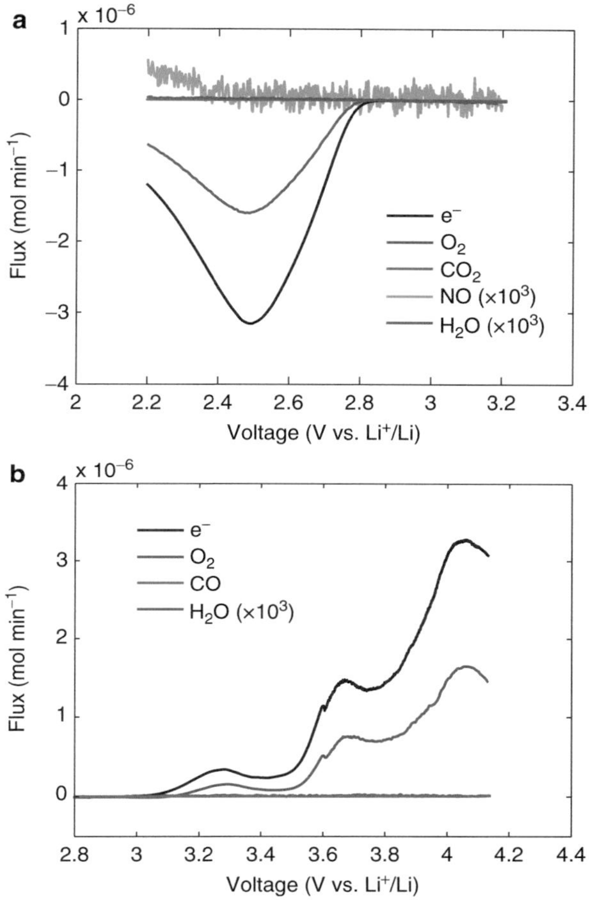

Fig. 2.15 In situ DEMS during a linear potential scan of a cell containing 0.1 M $LiClO_4$ in DMF and a Super P/PTFE composite electrode: (**a**) first discharge, (**b**) first recharge, scan rate 0.05 mV s^{-1}. Note the NO and H_2O fluxes are magnified by 10^3 to make them visible on the same plot. Current decrease during discharge is due to increasing filling of the electrode with product. Reprinted with permission from [47]. Copyright (2012) American Chemical Society

To verify that the side reactions are principally due to the DMF solvent and not the electrode, the Super P cathode was replaced with NPG. To demonstrate that the instability is not confined to DMF, the related amides DMA and NMP were also investigated. In both cases very similar results were obtained by DEMS, FTIR and NMR for the first cycle, confirming no significant difference in nature and extent of decomposition for the different amides.

Most studies that report extended cycling achieve this by restricting the capacity range on each cycle. Generally, the quantity of side reaction products is a function of the amount of charge passed; therefore, restricting the capacity range results in less side products after a given number of cycles and less polarisation. Of course this merely postpones, and does not eliminate, the inevitable accumulation of side products [19, 23, 27, 28, 84–86]; another benefit of shallow cycling is less mechanical stress on the electrode, which aids capacity retention. Load curves for a cell with DMA electrolyte and a Super P/PTFE cathode cycled at a restricted discharge

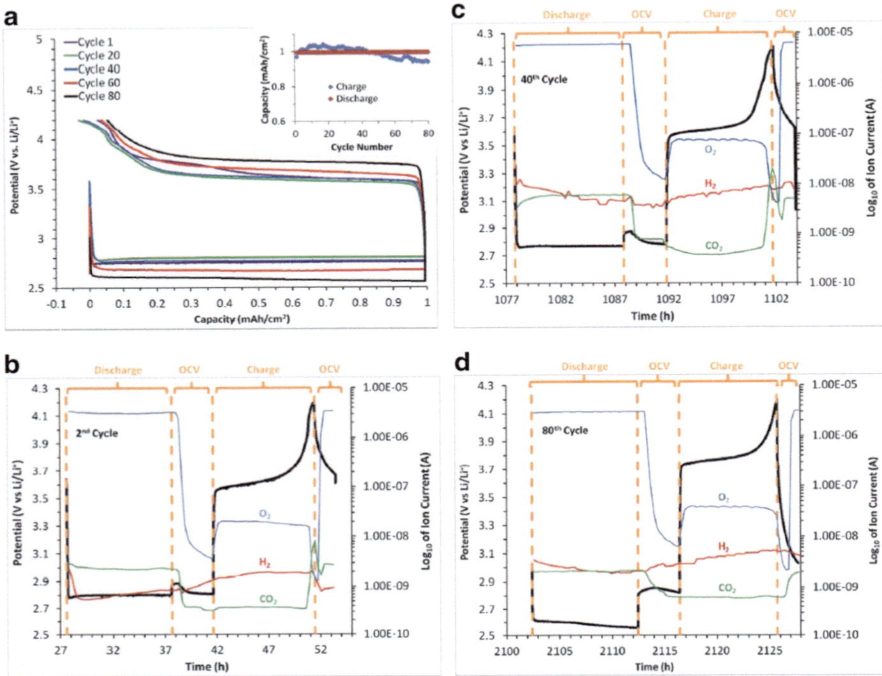

Fig. 2.16 (**a**) Voltage profiles for cycles 1, 2, 20, 40, 60 and 80 of a cell comprising a Li anode, a 1 M $LiNO_3$/DMA electrolyte and a Super P/PTFE porous cathode. The rate is 0.1 mA cm^{-2}, discharge limited to 1 mAh cm^{-2} and charge limited to 4.2 V. The *inset* shows discharge and charge capacity versus cycle number. (**b–d**) Potential and mass spectrometer signal of the same cell during cycles 2, 40 and 80. Discharge was performed under O_2, charge under Ar [27]. Reprinted (adapted) with permission from [20]. Copyright (2013) American Chemical Society

capacity of 1 mAh cm^{-2} (approximately 10 % of the full capacity) are shown in Fig. 2.16a [27]. Up to cycle 40 the voltage profile remains very reproducible, except for a feature at ~3.8 V during first charge which is associated with initial SEI formation. At higher cycle numbers the polarisation on discharge and charge increases, which is consistent with the slower accumulation of sufficient side products to passivate the electrode, in the case of restricted cycling. DEMS results (Fig. 2.16b–d) show that no significant H_2 or CO_2 evolution was observed on discharge and only O_2 on charge, with CO_2 appearing above 4 V. The CO_2 evolution may stem from both oxidation of decomposition products formed on discharge and freshly oxidised electrolyte or carbon which is promoted by the presence of Li_2O_2 [27, 31, 62]. We cannot be sure that the better cyclability and lower amounts of side products observed for DMA during restricted cycling, compared to our observations over a wider capacity range, are in fact due to the different capacity ranges. However, it does rationalise the two sets of results. These battery studies together with experimental and computational screening confirm that the amide functionality is amongst the most stable at the conditions of a Li–O_2 battery.

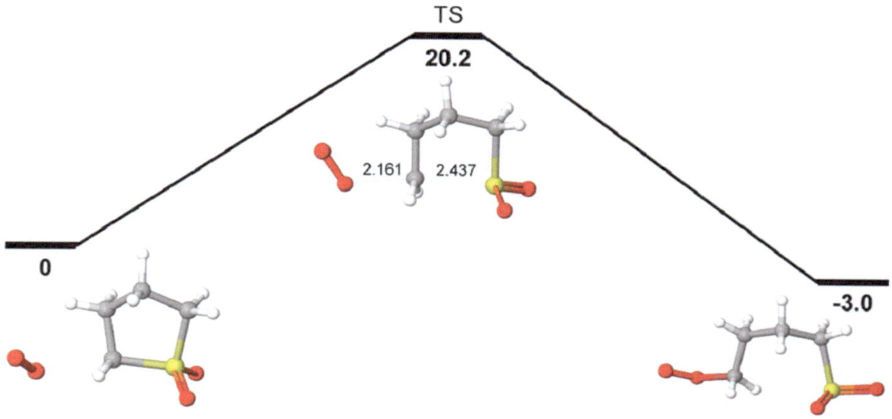

Fig. 2.17 Reaction-free energy profile for nucleophilic attack of O_2^- at the C_α atom of sulfolane (TMS) calculated by the continuum solvent approach (kcal mol^{-1}). B3LYP optimised C–O bond lengths in the transition state are shown. For EMS $\Delta G_{act} = 22.8$ kcal mol^{-1}. The most plausible pathway for the reaction is a nucleophilic attack at the C_α atom, followed by C–S bond cleavage to give a sulfinate anion [34]. Reprinted with permission from [23]. Copyright (2011) American Chemical Society

2.3.4 Sulfone-Based Electrolytes

In the search for electrolytes compatible with high-voltage intercalation cathodes operating beyond the stability limit of carbonate electrolytes (~4.2 V), sulfone-based electrolytes, with an anodic (oxidation) stability exceeding 5 V versus Li/Li^+, have been explored [87–89]. Recently, sulfones have been considered as the basis of electrolytes for nonaqueous Li–O_2 batteries [26, 85]. The melting points of many sulfones are above room temperature, e.g. ethylmethylsulfone (EMS, 27 °C) and tetramethylenesulfone (TMS or sulfolane, 36 °C), precluding room temperature operation. Alternatively, the room temperature liquid ethyl vinyl sulfone (EVS) and oligoether-substituted sulfones have been proposed [88, 89]. DFT calculations on the nucleophilic attack of sulfones by superoxide indicate an activation energy slightly above 20 kcal mol^{-1} for EMS and TMS (Fig. 2.17); values <20 kcal mol^{-1} represent solvents that are likely to be unstable, whereas those with $\Delta G_{act} > 24$ kcal mol^{-1} are expected to show much slower reactivity with O_2^- [34]. The screening approach using cyclic voltammetry and reaction with KO_2 (Fig. 2.6) confirms possible nucleophilic attack by O_2^-, albeit at slow rate. It has also been suggested that sulfones do not form a stable SEI in contact with low-voltage anodes such as Li or graphite [89].

Several sulfones (TMS, EMS, EVS) have been tested experimentally as the basis of electrolytes in nonaqueous Li–O_2 cells. Results for a cell composed of a porous cathode (Super P/PTFE), an electrolyte comprising 1 M LiTFSI in various solvents (TMS, EMS or EVS) and a $LiFePO_4$ counter electrode are given in Fig. 2.18 [26].

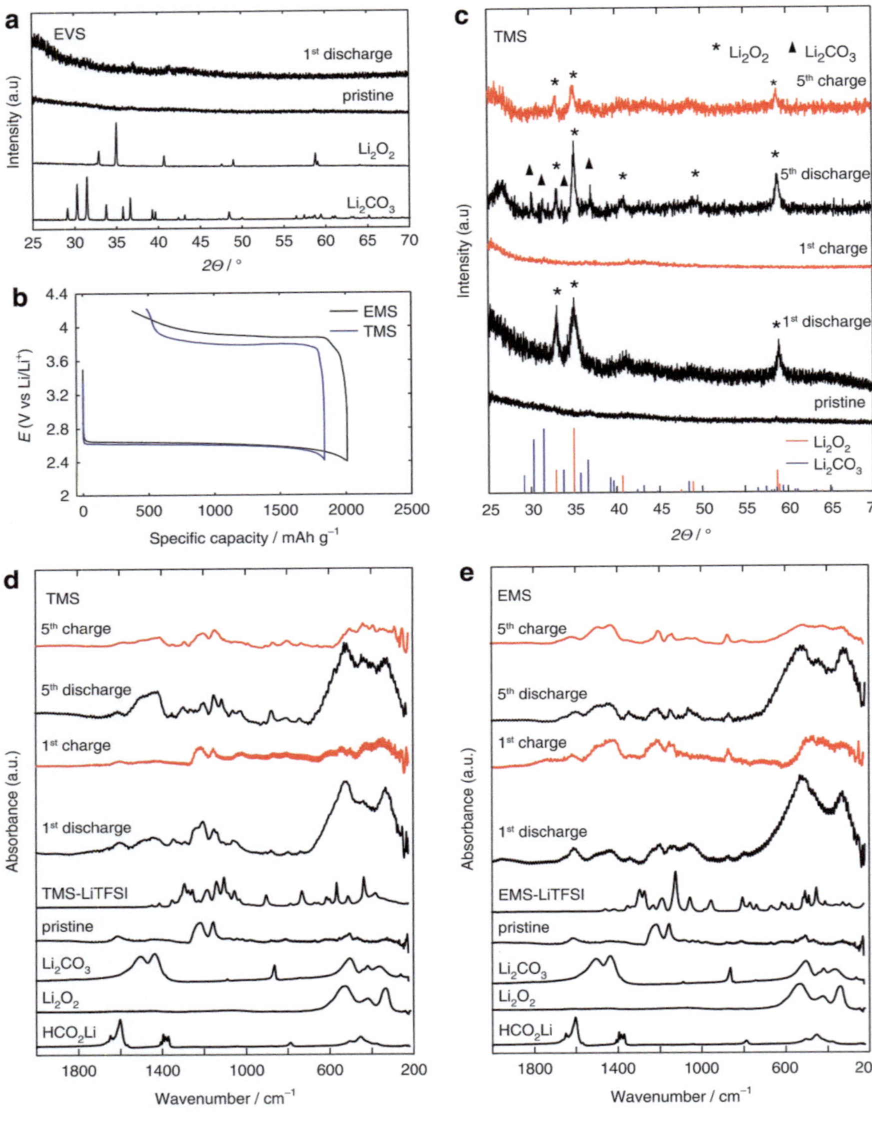

Fig. 2.18 (**a**) XRD analysis of discharge product in ethyl vinyl sulfone (EVS). (**b**) Load curves of first cycle of Super P/PTFE composite cathodes in 1 M LiTFSI/sulfolane (TMS) and in 1 M LiTFSI/ethyl methyl sulfone (EMS). (**c**) XRD analysis of Super P/PTFE cathodes in 1 M LiTFSI/TMS electrolyte at the indicated cycle number and state of charge, FTIR analysis of Super P/PTFE cathodes in (**d**) 1 M LiTFSI/TMS and (**e**) 1 M LiTFSI/EMS electrolyte at the indicated cycle number and state of charge [26]

The vinyl group on EVS rendered it unstable to attack by superoxide, even during the first discharge [26]. Li_2O_2 is absent in the PXRD pattern of the electrode at the end of first discharge (Fig. 2.18a). After discharge to 2.4 V in both TMS and EMS, recharge is incomplete at 4.2 V (Fig. 2.18b). XRD analysis of such cathodes at the end of first discharge/charge and fifth discharge/charge in TMS is shown in Fig. 2.18c. It shows Li_2O_2 as discharge product on first discharge, and it was removed upon charging. At the end of fifth discharge, Li_2O_2 is still the dominant product, accompanied by some Li_2CO_3. Some Li_2O_2 remains upon fifth charge, which means its oxidation on charge is incomplete. The FTIR spectra of the cathodes cycled in TMS were in accord with the XRD results (Fig. 2.18d). Li_2O_2 is the dominant product, on the first and fifth discharge. However, it is accompanied by growing fractions of Li_2CO_3, Li formate and Li acetate, the latter two confirmed by ^{1}H NMR. LiOH was observed neither in XRD nor FTIR. After recharging to 4.2 V, the FTIR spectra show residues of all the discharge products. Cathodes in EMS show very similar FTIR spectra (Fig. 2.18e) [26]. DEMS analysis reveals an e^-/O_2 ratio of 2.05 and 2.33 on discharge and charge, respectively, for TMS and 2.15/2.3 for EMS. This is consistent with considerable charge being consumed, both on discharge and charge, in side reactions, and with the observed capacity fading on cycling [26].

Overall the sulfone-based electrolytes investigated to date confirm the theoretical predictions of nucleophilic attack by O_2^-. Nevertheless, Li_2O_2 is the dominant product during the first cycles. Changing the carbon substrate for NPG reduced the amount of side products but not their nature, rendering the results valid as the inherent behaviour of the sulfone solvents.

2.3.5 *Dimethyl Sulfoxide Electrolyte*

Cells based on DMSO electrolyte have shown to sustain cycling with the reversible formation/decomposition of Li_2O_2 at the cathode. In the past, it was used to studying O_2 reduction in nonaqueous media [2, 90, 91]. It is only very weakly acidic and has a high Gutmann donor number of ~30 kcal mol^{-1}, which is the highest of all solvents discussed above. The basic electrochemistry of O_2 in DMSO solutions containing Li^+ ions has been studied using cyclic voltammetry on planar glassy carbon electrodes [40]. In contrast to the same experiment in tetraglyme electrolyte, consecutive cycles in DMSO electrolyte show virtually the same reduction current indicating efficient removal of the reduction products during the anodic scan (Fig. 2.19) [28]. Galvanostatic discharge/charge cycling using a porous carbon electrode (Ketjen Black/PVdF) in 1 $LiPF_6$/DMSO exhibits capacity fading (Fig. 2.20a). In a different study, FTIR analysis of the discharge product for a Super P/PTFE cathode in 1 M $LiClO_4$ in DMSO confirms dominant formation of Li_2O_2, along with some Li_2CO_3 and HCOOLi [29, 31]. The proportion of the latter may be estimated to be ~15 % after the first discharge. It has been shown that carbon promotes electrolyte decomposition on discharge and is unstable on charge, as noted above in the section on

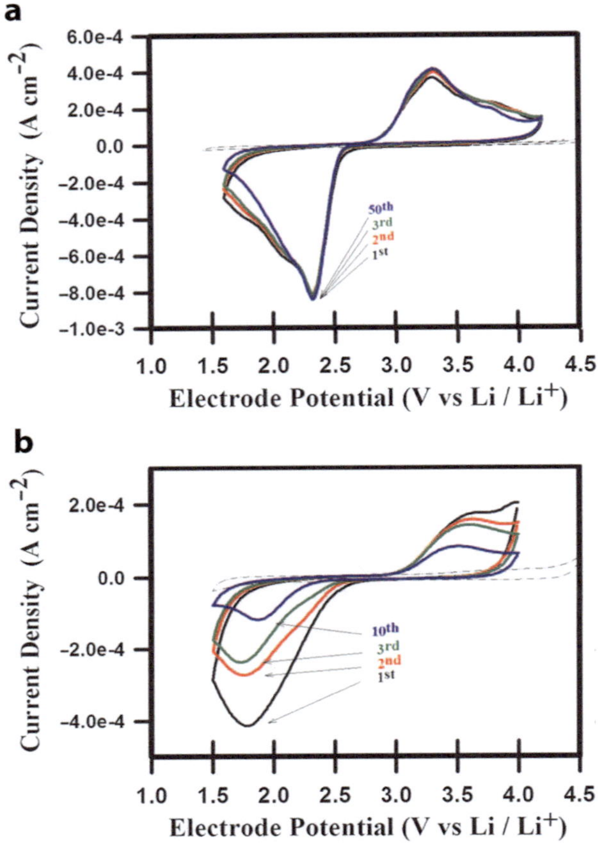

Fig. 2.19 (**a**) Cyclic voltammogram on a glassy carbon disc electrode in O_2-saturated 0.1 M $LiPF_6$/DMSO, 50 consecutive cycles at scan rate 0.1 V s^{-1}. (**b**) 10 cycles at the same conditions as in (**a**) in 0.1 M $LiPF_6$/tetraglyme [28]. Reprinted with permission from J. Electrochem. Soc., 160, A259 (2013). Copyright 2012, The Electrochemical Society

ethers [31]. The significant proportion of side reaction products formed is consistent with poor reversibility of the carbon/DMSO cells.

Replacing carbon by a NPG electrode significantly alleviates this problem when in contact with 0.1 M $LiClO_4$/DMSO electrolyte (Fig. 2.21) [29]. Discharge/charge curves for cycles 1, 5, 10 and 100 are shown in Fig. 2.21a; 95 % of the initial capacity is retained after 100 cycles. As is now recognised in the field (exemplified by the apparent cyclability of the Li–O_2 cell with carbonate electrolytes), electrochemical cycling alone is not a sufficient proof of reversible Li_2O_2 formation/decomposition [1, 4, 6, 7, 9, 24, 63]. Additional spectroscopic methods must be combined with electrochemistry. FTIR and in situ SERS carried out at the end of discharge and charge as a function of cycle number (1, 5, 10 and 100) are shown in Fig. 2.22. Both analyses confirm that Li_2O_2 is the main product in the electrode at the end of each discharge. Small additional peaks in the FTIR are apparent, which can be assigned to Li_2CO_3 and HCOOLi. The latter was confirmed by ^{1}H NMR. Together Li_2CO_3 and HCOOLi accounted for <1 % of the Li_2O_2, i.e. >99 % of the product in the electrode on discharge is Li_2O_2 and there is no evidence of this

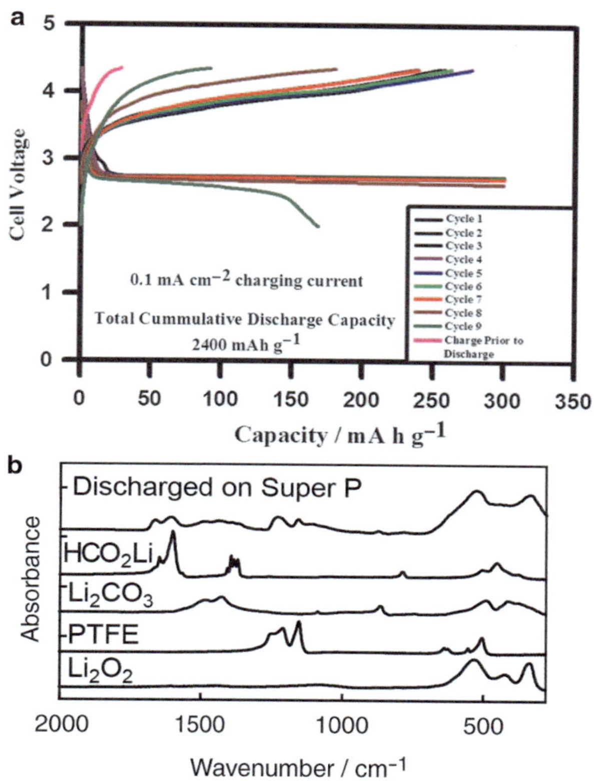

Fig. 2.20 (**a**) Galvanostatic cycling of a Ketjen300/PVdF composite cathode versus a Li metal anode in 1 M $LiPF_6$/DMSO electrolyte at 0.1 mA cm^{-2}.10 % depth of discharge (300 mAh g^{-1} carbon) and charge to 4.35 V [28]. (**b**) FTIR analysis of a Super P/PTFE cathode in 1 M $LiClO_4$/DMSO after the first discharge [29]. Reprinted with permission from J. Electrochem., A259 (2013). Copyright 2012, The Electrochemical Society. From Science 337 (6094):563-566 (2012). Reprinted with permission from AAAS

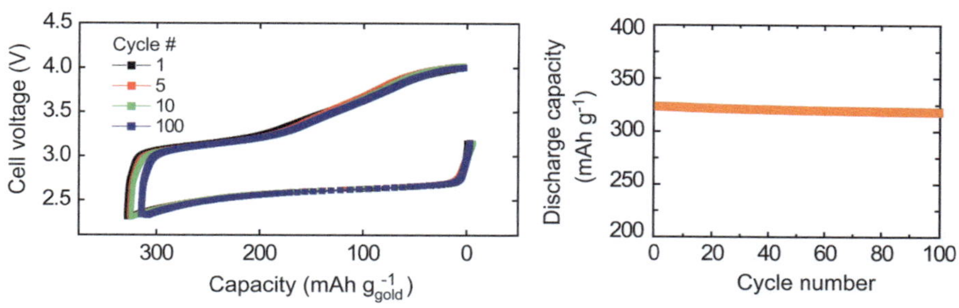

Fig. 2.21 Discharge/charge curves (*left*) and cycling profile (*right*) for a Li–O_2 cell with a 0.1 M $LiClO_4$/DMSO electrolyte and a nanoporous gold cathode, at a current density of 500 mA g^{-1} (based on the mass of Au) [29]. From Science 337(6094):563-566 (2012). Reprinted with permission from AAAS

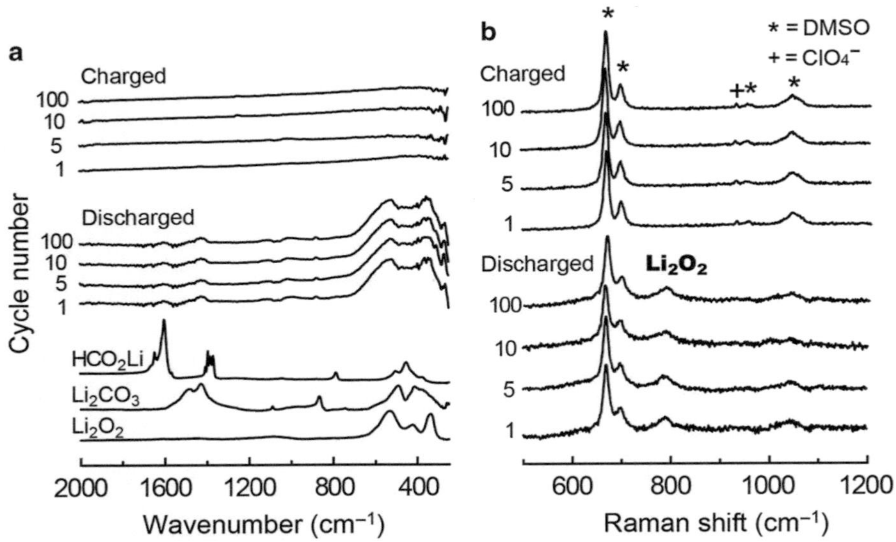

Fig. 2.22 Vibrational spectra of a nanoporous gold cathode at the end of discharge and charge in 0.1 M $LiClO_4$/DMSO. (**a**) FTIR and (**b**) in situ surface-enhanced Raman spectroscopy (SERS) [29]. From Science 337(6094):563-566 (2012). Reprinted with permission from AAAS

Table 2.2 Ratios of the number of electrons to oxygen molecules upon reduction (discharge) and oxidation (charge) [29]

Cycle number	Discharge e^-/O_2	Charge e^-/O_2
1	2.01	1.98
5	1.99	2.04
10	2.02	1.98
100	2.03	2.01

From Science 337(6094):563-566 (2012). Reprinted with permission from AAAS

proportion increasing with cycle number. From the spectra at the end of each recharge, it is clear that the discharge product has been completely removed.

To further corroborate that the reaction is mainly Li_2O_2 formation/decomposition, it is important to establish the charge/mass balance and that no other gases than O_2 evolve on charging. The DEMS data listed in Table 2.2 show a value very close to the ideal $2e^-/O_2$ for every discharge and charge. No other gases were evolved. Taken together, the electrochemical and spectroscopic data in Figs. 2.21 and 2.22 and Table 2.1, obtained from an NPG electrode in contact with a $LiClO_4$/DMSO electrolyte, suggest that reversible formation/decomposition of Li_2O_2 is possible and can be sustained on cycling. Of course examining the cathode alone for solid products does not rule out the possibility of soluble or volatile side reaction products. We have probed the electrolyte by solution NMR and reported finding a small proportion of methylsulfone [92]. The small proportion is consistent with the close to $2e^-/O_2$ ratio observed in the DEMS. However, the extent of electrolyte degradation will depend on the surface area in contact with the electrode, and detection of side products in the electrolyte will be easier if the ratio of electrolyte volume to electrode surface is low, as pointed out recently this may be in part why NPG gives good results [93]. Of course there is no such thing as an electrolyte or electrode that is 100 % stable; indeed dimethyl sulfone is expected to form in DMSO electrolytes. The issues are the extent of side reactions and how detrimental the particular side products are to the reversible electrochemistry. It may be that in the case of DMSO the side reaction products do not passivate the electrode as severely as in some other electrolytes. So the issue of stability is more subtle and more complex than might at first appear. This key question is discussed on the section at the end of this chapter. Note that DMSO-based cells can only be charged to 4 V, above which it decomposes, perhaps explaining the capacity fading noted by some researchers [94]. However, a successful Li–O_2 cell would have to exhibit charging below 4 V to be viable [92]. DMSO electrolytes may not be stable in contact with Li metal. However, the use of a protected Li anode, which may be necessary in any case to avoid O_2 in the electrolyte oxidising the anode, would mitigate this problem.

2.3.6 Stability of Salts in Li–O_2 Batteries

Apart from providing the required conductivity, the Li salt used for nonaqueous battery electrolytes plays a vital role in interface phenomena, such as SEI stability and conductivity, and current collector passivation. As with the carbonate solvents, early studies used $LiPF_6$, following the precedent from lithium-ion batteries and is still the most widely used salt in Li–O_2 cells. Other candidate salts have been studied in Li–O_2 cells including $LiClO_4$, $LiBF_4$, $LiB(CN)_4$ and $LiSO_3CF_3$; Li imides like $LiN(SO_2CF_3)_2$ (TFSI); and orthoborate chelates like LiBOB and $LiNO_3$. Other anions have been described in the context of ionic liquids (Scheme 2.2). Analogous to the solvents, an initial requirement for the salts is stability against O_2^-. LiBOB and $LiBF_4$ have been shown to be unstable on discharge, forming LiB_3O_5, B_2O_3, Li oxalate and LiF, respectively, in substantial amounts rendering these salts unsuitable [51, 95, 96]. Cells comprising carbon cathodes, a Li metal anode and tetraglyme-based electrolytes with various salts have been discharged and the products analysed by XPS (Fig. 2.23) [14, 18, 51, 52, 66, 96]. The salt strongly influences the discharge capacity, which has been ascribed to different water contents (Fig. 2.23a) [52, 59]. Peaks in the C 1*s* and O 1*s* spectra can be assigned to different ratios of the side reaction products, i.e. Li_2CO_3, Li carboxylates and ethers (see Sect. 2.3.1). Although Li_2O_2 has been observed in these samples by Raman spectroscopy, O 1*s* spectra indicate that Li_2O_2 is not present in a significant proportion in the surface deposits. The halide spectra identify LiF, LiCl and other decomposition products (not specifically assigned). Based on XPS analysis, ClO_4^- has been identified as the most stable anion with respect to Li–O_2 chemistry; it is somewhat more stable than the fluorinated anions PF_6^- and $TFSI^-$. XPS is a valuable tool but it must be recalled

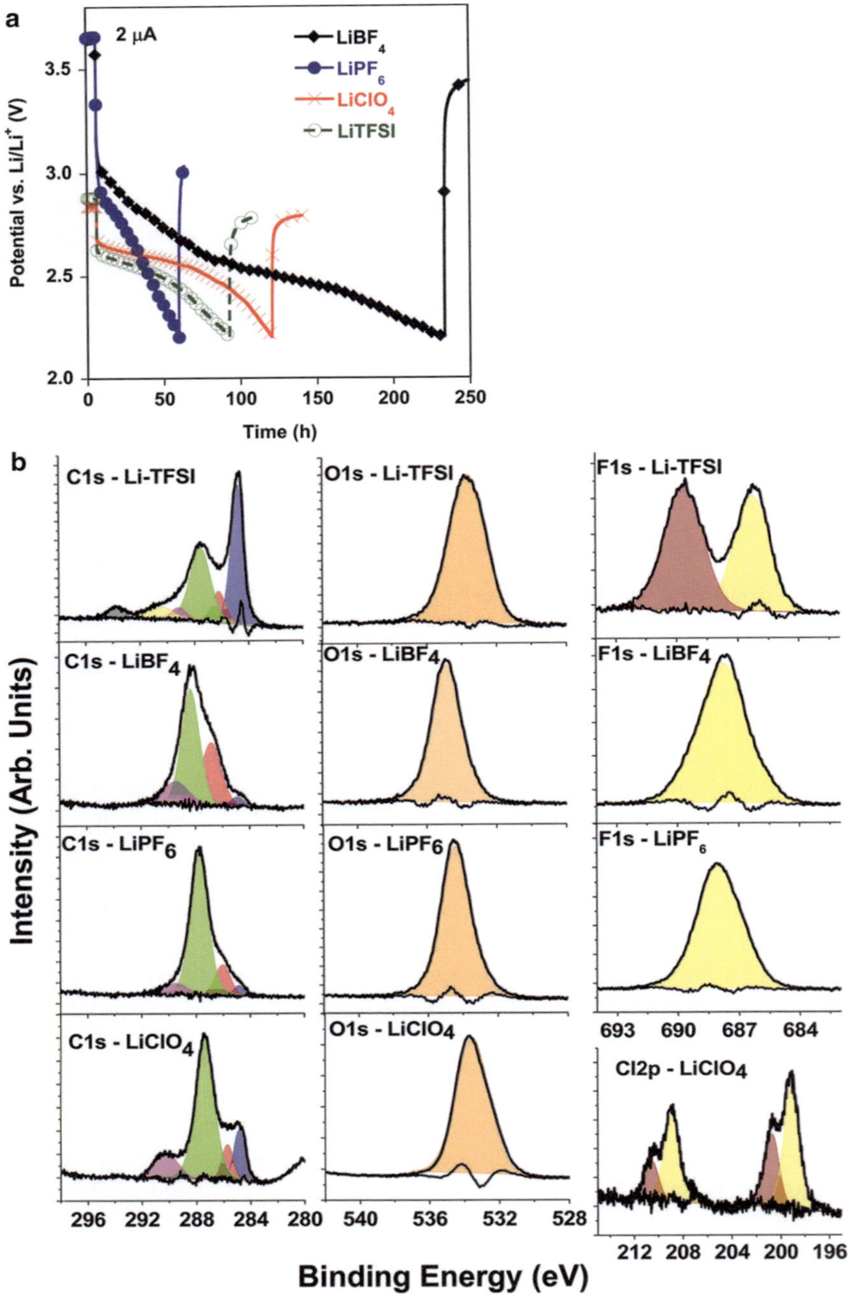

Fig. 2.23 (**a**) Discharge curves of graphitised carbon foam without any binder in tetraglyme containing 1 M of the specified salt. (**b**) C 1*s*, O 1*s*, F 1*s* and Cl 2p XPS spectra collected on the discharge products. The *dashed line* in the O 1*s* spectra marks the position where Li_2O_2 would occur [52]. Reprinted (adapted) with permission from [52]. Copyright (2012) American Chemical Society

that it is primarily a surface-sensitive method. Many species can form on the surface in complex electrochemical systems but not all will necessarily have a deleterious effect on the electrochemistry. Therefore, quantifying the bulk content of salt residuals is needed in future studies. A special case is the use of $LiNO_3$, which was recently reported to form a stable SEI on Li metal in DMA solution, which would otherwise continuously react [27]. No adverse effects on the cathode chemistry could be seen.

2.4 How Stable Is a Stable Electrolyte?

Reactions do not proceed without side reactions; therefore, the question is not whether an electrolyte exhibiting 100 % stability can be discovered, but rather how stable does it have to be to sustain the cycle and calendar life required for a particular cell design and application. The electrode, e.g. the porous carbon, will also contribute to parasitic reactions upon discharge or charge [31]. In this respect two effects of side reactions need to be considered for the Li–O_2 cathode: first, side reactions that decompose solvent, salt, electrode or binder to form solid, gaseous or dissolved side products, which deplete the particular component and will lead to failure once a threshold of consumption is exceeded and second, the question of the nature of the side reaction products and their effect on the reversible electrochemistry. Although oxidation of side reaction products like HCO_2Li, CH_3CO_2Li and Li_2CO_3 is possible, at least the latter is often incompletely decomposed and accumulates on cycling, which increasingly passivates the electrode leading to capacity fading and premature cell death. Here we confine the discussion to electrolyte stability alone. In Table 2.3 we estimate the effect of a small degree of electrolyte decomposition, 0.3 %, on every cycle based on the mass of Li_2CO_3 formed relative to Li_2O_2. This degree of purity can be achieved with a sufficiently stable combination of electrode material and electrolyte [29]. We assume the limiting case where none of the Li_2CO_3 formed in every cycle is decomposed on subsequent charging, i.e. all accumulates. The capacity of the cathode is taken as 5 mAh cm^{-2} and we assume the electrolyte comprises a film of 0.03 mL cm^{-2} initially. For such a system the fraction of the initial

Table 2.3 Estimation of the effect of a small degree of side reaction during every discharge over 100 cycles

Capacity per cycle	5	mAh cm^{-2}
Cycles	100	–
Fraction side product	0.3	%
Quantity of Li_2CO_3 after 100 cycles	1.28×10^{-5}	g cm^{-2}
Fraction of initial electrolyte decomposed[a]	2	%
Li_2CO_3 layer[b]	10	nm

Amount of Li_2O_2 and Li_2CO_3 is accumulative for 100 discharges
[a]Assuming 0.03 mL electrolyte per cm^2 cell area
[b]Assuming uniform distribution on the active carbon surface area

electrolyte decomposed after 100 cycles is ~2 %. Such an amount of electrolyte loss is unlikely to be detrimental to performance. However, accumulation of insulating Li_2CO_3 on the electrode could lead to passivation and capacity fading. Of course complete irreversibility of Li_2CO_3 is the worst-case scenario; generally some oxidation will occur. Li_2CO_3 can be oxidised below 3.5 V in the Li–O_2 cell [31].

2.5 Conclusions

The stability of nonaqueous electrolytes in rechargeable Li–O_2 batteries became a major focus of research following the realisation that organic carbonate electrolytes, used initially in such cells, decompose significantly due to nucleophilic attack by reduced oxygen species (O_2^-, LiO_2) at the cathode. The widely observed rechargeability of Li–O_2 cells using these electrolytes is, in fact, due to back-to-back irreversible reactions involving decomposition of the electrolyte on discharge and decomposition of the products of discharge on charging, rather than the reversible formation of Li_2O_2 at the cathode that is required to sustain true rechargeability. This realisation led to the establishment of a suite of experimental and computational methods capable of screening the stability of electrolytes. A range of electrolytes based on ethers, amides, sulfones, ionic liquids and DMSO have been investigated. All are more stable than the organic carbonates, but not all are equally stable. It is difficult to generalise, since each is a broad class of compounds in their own right. Even though it was soon realised, by a number of groups, that ethers exhibit side reactions on discharge and charge, they still remain the choice in many studies. As a consequence of the investigation of electrolyte stability, the importance of the stability of the electrode has been highlighted. Even if a stable electrolyte is identified, it must be combined with a stable electrode material. In fact there can be a synergy between them, such that carbon-based electrodes promote electrolyte decomposition, as well as decomposing in their own right. Chapter 3 deals in more detail with electrode issues. Although great progress has been made in only a few years, the work on aprotic electrolytes for Li–O_2 batteries is certainly not over. Doubtless we shall see other electrolytes explored, in the quest for yet lower cost, higher safety, stability and low volatility.

References

1. Mizuno F, Nakanishi S, Kotani Y et al (2010) Rechargeable Li-air batteries with carbonate-based liquid electrolytes. Electrochemistry 78(5):403–405
2. Gibian MJ, Sawyer DT, Ungermann T et al (1979) Reactivity of superoxide ion with carbonyl compounds in aprotic solvents. J Am Chem Soc 101(3):640–644
3. Aurbach D, Daroux M, Faguy P et al (1991) The electrochemistry of noble metal electrodes in aprotic organic solvents containing lithium salts. J Electroanal Chem 297(1):225–244

4. Freunberger SA, Chen Y, Peng Z et al (2011) Reactions in the rechargeable lithium–O_2 battery with alkyl carbonate electrolytes. J Am Chem Soc 133(20):8040–8047
5. Xu W, Xu K, Viswanathan VV et al (2011) Reaction mechanisms for the limited reversibility of Li–O_2 chemistry in organic carbonate electrolytes. J Power Sources 196(22):9631–9639
6. Xu W, Viswanathan VV, Wang D et al (2011) Investigation on the charging process of Li_2O_2-based air electrodes in Li-O_2 batteries with organic carbonate electrolytes. J Power Sources 196(8):3894–3899
7. Veith GM, Dudney NJ, Howe J et al (2011) Spectroscopic characterization of solid discharge products in Li-air cells with aprotic carbonate electrolytes. J Phys Chem C 115(29):14325–14333
8. Bryantsev VS, Blanco M (2011) Computational study of the mechanisms of superoxide-induced decomposition of organic carbonate-based electrolytes. J Phys Chem Lett 2(5):379–383
9. McCloskey BD, Bethune DS, Shelby RM et al (2011) Solvents' critical role in nonaqueous lithium–oxygen battery electrochemistry. J Phys Chem Lett 2(10):1161–1166
10. Tsiouvaras N, Meini S, Buchberger I et al (2013) A novel on-line mass spectrometer design for the study of multiple charging cycles of a Li–O_2 battery. J Electrochem Soc 160(3):A471–A477
11. Aurbach D, Gofer Y, Langzam J (1989) The correlation between surface chemistry, surface morphology, and cycling efficiency of lithium electrodes in a few polar aprotic systems. J Electrochem Soc 136(11):3198–3205
12. Curran HJ, Gaffuri P, Pitz WJ et al (1998) A comprehensive modeling study of n-heptane oxidation. Combust Flame 114(1–2):149–177
13. Roberts JL, Calderwood TS, Sawyer DT (1984) Nucleophilic oxygenation of carbon dioxide by superoxide ion in aprotic media to form the peroxydicarbonate(2-) ion species. J Am Chem Soc 106(17):4667–4670
14. Younesi R, Urbonaite S, Edström K et al (2012) The cathode surface composition of a cycled Li–o2 battery: a photoelectron spectroscopy study. J Phys Chem C 116(39):20673–20680
15. Peng Z, Freunberger SA, Hardwick LJ et al (2011) Oxygen reactions in a non-aqueous Li^+ electrolyte. Angew Chem Int Ed 50(28):6351–6355
16. Sawyer DT, Roberts JL (1966) Electrochemistry of oxygen and superoxide ion in dimethylsulfoxide at platinum, gold and mercury electrodes. J Electroanal Chem 12(2):90–101
17. Laoire CO, Mukerjee S, Abraham KM et al (2009) Elucidating the mechanism of oxygen reduction for lithium-air battery applications. J Phys Chem C 113(46):20127–20134
18. Younesi R, Hahlin M, Björefors F et al (2012) Li–O_2 battery degradation by lithium peroxide (Li_2o_2): a model study. Chem Mater 25(1):77–84
19. Laoire CO, Mukerjee S, Plichta EJ et al (2011) Rechargeable lithium/tegdme-$LiPF_6$/O_2 battery. J Electrochem Soc 158(3):A302–A308
20. Hassoun J, Croce F, Armand M et al (2011) Investigation of the O_2 electrochemistry in a polymer electrolyte solid-state cell. Angew Chem Int Ed 50(13):2999–3002
21. Lu Y-C, Gasteiger HA, Shao-Horn Y (2011) Catalytic activity trends of oxygen reduction reaction for nonaqueous Li-air batteries. J Am Chem Soc 133(47):19048–19051
22. Oh SH, Nazar LF (2012) Oxide catalysts for rechargeable high-capacity Li–O_2 batteries. Adv Energy Mater 2(7):903–910
23. Jung H-G, Hassoun J, Park J-B et al (2012) An improved high-performance lithium–air battery. Nat Chem 4(7):579–585
24. Freunberger SA, Chen Y, Drewett NE et al (2011) The lithium–oxygen battery with ether-based electrolytes. Angew Chem Int Ed 50(37):8609–8613
25. Mizuno F, Takechi K, Higashi S et al (2013) Cathode reaction mechanism of non-aqueous Li–O_2 batteries with highly oxygen radical stable electrolyte solvent. J Power Sources 228:47–56
26. Bardé F, Chen Y, Schaltin S et al (2013) Sulfone-based electrolytes for non-aqueous Li-O_2 batteries (submitted)

27. Walker W, Giordani V, Uddin J et al (2013) A rechargeable Li–O_2 battery using a lithium nitrate/n, n-dimethylacetamide electrolyte. J Am Chem Soc 135(6):2076–2079
28. Trahan MJ, Mukerjee S, Plichta EJ et al (2013) Studies of Li-air cells utilizing dimethyl sulfoxide-based electrolyte. J Electrochem Soc 160(2):A259–A267
29. Peng Z, Freunberger SA, Chen Y et al (2012) A reversible and higher-rate Li-O_2 battery. Science 337(6094):563–566
30. Ottakam Thotiyl MM, Freunberger SA, Peng Z et al (2013) A stable cathode for the aprotic Li–O_2 battery. Nat Mater. doi:10.1038/nmat3737
31. Ottakam Thotiyl MM, Freunberger SA, Peng Z et al (2013) The carbon electrode in nonaqueous Li–O_2 cells. J Am Chem Soc 135(1):494–500
32. Mizuno F, Nakanishi S, Shirasawa A et al (2011) Design of non-aqueous liquid electrolytes for rechargeable Li-O_2 batteries. Electrochemistry 79(11):876–881
33. Bryantsev V (2012) Calculation of solvation free energies of Li^+ and O_2^- ions and neutral lithium–oxygen compounds in acetonitrile using mixed cluster/continuum models. Theor Chem Acc 131(7):1–11
34. Bryantsev VS, Giordani V, Walker W et al (2011) Predicting solvent stability in aprotic electrolyte Li–air batteries: nucleophilic substitution by the superoxide anion radical ($O_2^{\bullet -}$). J Phys Chem A 115(44):12399–12409
35. Assary RS, Curtiss LA, Redfern PC et al (2011) Computational studies of polysiloxanes: oxidation potentials and decomposition reactions. J Phys Chem C 115(24):12216–12223
36. Laino T, Curioni A (2011) Toward the understanding of chemical degradation of aprotic solvents for Li-air batteries. Abstr Pap Am Chem Soc 242:111
37. Bryantsev VS, Faglioni F (2012) Predicting autoxidation stability of ether- and amide-based electrolyte solvents for Li–air batteries. J Phys Chem A 116(26):7128–7138
38. Bryantsev VS, Uddin J, Giordani V et al (2013) The identification of stable solvents for nonaqueous rechargeable Li-air batteries. J Electrochem Soc 160(1):A160–A171
39. Allen CJ, Mukerjee S, Plichta EJ et al (2011) Oxygen electrode rechargeability in an ionic liquid for the Li–air battery. J Phys Chem Lett 2(19):2420–2424
40. Laoire CO, Mukerjee S, Abraham KM et al (2010) Influence of nonaqueous solvents on the electrochemistry of oxygen in the rechargeable lithium–air battery. J Phys Chem C 114(19):9178–9186
41. Monaco S, Arangio AM, Soavi F et al (2012) An electrochemical study of oxygen reduction in pyrrolidinium-based ionic liquids for lithium/oxygen batteries. Electrochim Acta 83:94–104
42. Takechi K, Higashi S, Mizuno F et al (2012) Stability of solvents against superoxide radical species for the electrolyte of lithium-air battery. ECS Electrochem Lett 1(1):A27–A29
43. Zhang J-G, Xu W, Xiao J et al (2012) Critical components of rechargeable Li-air batteries. ECS Meeting Abstracts MA2012-02(11):1171
44. Herranz J, Garsuch A, Gasteiger HA (2012) Using rotating ring-disc electrode voltammetry to quantify the superoxide radical stability of aprotic Li-air battery electrolytes. J Phys Chem C 116(36):19084–19094
45. Shao Y, Xu W, Ding F et al (2012) Electrochemical investigation on stable nonaqueous electrolytes for rechargeable Li-air batteries. ECS Meeting Abstracts 584
46. Dietzel PDC, Kremer RK, Jansen M (2004) Tetraorganylammonium superoxide compounds: close to unperturbed superoxide ions in the solid state. J Am Chem Soc 126(14):4689–4696
47. Chen Y, Freunberger SA, Peng Z et al (2012) Li–O_2 battery with a dimethylformamide electrolyte. J Am Chem Soc 134(18):7952–7957
48. Hartmann P, Bender CL, Vračar M et al (2012) A rechargeable room-temperature sodium superoxide (NaO_2) battery. Nat Mater 12:228–232
49. Black R, Oh SH, Lee J-H et al (2012) Screening for superoxide reactivity in Li–O_2 batteries: effect on Li_2O_2/LiOH crystallization. J Am Chem Soc 134(6):2902–2905
50. Sharon D, Etacheri V, Garsuch A et al (2012) On the challenge of electrolyte solutions for Li–air batteries: monitoring oxygen reduction and related reactions in polyether solutions by spectroscopy and EQCM. J Phys Chem Lett 4(1):127–131

51. Nasybulin E, Xu W, Engelhard MH et al (2013) Effects of electrolyte salts on the performance of $Li–O_2$ batteries. J Phys Chem C 117(6):2635–2645
52. Veith GM, Nanda J, Delmau LH et al (2012) Influence of lithium salts on the discharge chemistry of Li–air cells. J Phys Chem Lett 3(10):1242–1247
53. Leskes M, Drewett NE, Hardwick LJ et al (2012) Direct detection of discharge products in lithium–oxygen batteries by solid-state NMR spectroscopy. Angew Chem 124(34):8688–8691
54. Huff LA, Rapp JL, Zhu L et al (2013) Identifying lithium–air battery discharge products through ^{6}Li solid-state MAS and $^{1}H–^{13}C$ solution NMR spectroscopy. J Power Sources 235:87–94
55. Abraham KM, Jiang Z (1996) A polymer electrolyte-based rechargeable lithium/oxygen battery. J Electrochem Soc 143(1):1–5
56. Read J (2006) Ether-based electrolytes for the lithium/oxygen organic electrolyte battery. J Electrochem Soc 153(1):A96–A100
57. Lu Y-C, Kwabi DG, Yao KPC et al (2011) The discharge rate capability of rechargeable $Li–O_2$ batteries. Energy Environ Sci 4:2999–3007
58. Mitchell RR, Gallant BM, Thompson CV et al (2011) All-carbon-nanofiber electrodes for high-energy rechargeable $Li–O_2$ batteries. Energy Environ Sci 4(8):2952–2958
59. Meini S, Piana M, Tsiouvaras N et al (2012) The effect of water on the discharge capacity of a non-catalyzed carbon cathode for $Li–O_2$ batteries. Electrochem Solid State Lett 15(4):A45–A48
60. Zhang Z, Lu J, Assary RS et al (2011) Increased stability toward oxygen reduction products for lithium-air batteries with oligoether-functionalized silane electrolytes. J Phys Chem C 115(51):25535–25542
61. Curtiss LA, Lau KC, Redfern P et al (2011) Computational studies of electrolyte stability for Li-air batteries. Abstr Pap Am Chem Soc 242:110
62. McCloskey BD, Speidel A, Scheffler R et al (2012) Twin problems of interfacial carbonate formation in nonaqueous $Li–O_2$ batteries. J Phys Chem Lett 3:997–1001
63. Wang H, Xie K (2012) Investigation of oxygen reduction chemistry in ether and carbonate based electrolytes for $Li–O_2$ batteries. Electrochim Acta 64:29–34
64. McCloskey BD, Scheffler R, Speidel A et al (2011) On the efficacy of electrocatalysis in nonaqueous $Li–O_2$ batteries. J Am Chem Soc 133(45):18038–18041
65. Ryan KR, Trahey L, Ingram BJ et al (2012) Limited stability of ether-based solvents in lithium–oxygen batteries. J Phys Chem C 116(37):19724–19728
66. Younesi R, Hahlin M, Treskow M et al (2012) Ether based electrolyte, $LiB(CN)_4$ salt and binder degradation in the $Li–O_2$ battery studied by hard X-ray photoelectron spectroscopy (HAXPES). J Phys Chem C 116(35):18597–18604
67. Li F, Zhang T, Yamada Y et al (2012) Enhanced cycling performance of $Li–O_2$ batteries by the optimized electrolyte concentration of LiTFSA in glymes. Adv Energy Mater 3(4):532–538
68. Freunberger SA, Chen Y, Drewett NE et al (2011) Die Lithium-Sauerstoff-Batterie mit etherbasierten Elektrolyten. Angew Chem 123(37):8768–8772
69. Clover AM (1922) The autoxidation of ethyl ether. J Am Chem Soc 44(5):1107–1118
70. Rein H (1950) Über das Tetrahydrofuran-peroxyd. Angew Chem 62(5):120
71. Aurbach D, Zinigrad E, Cohen Y et al (2002) A short review of failure mechanisms of lithium metal and lithiated graphite anodes in liquid electrolyte solutions. Solid State Ion 148(3–4):405–416
72. Assary RS, Lu J, Du P et al (2012) The effect of oxygen crossover on the anode of a $Li–O_2$ battery using an ether-based solvent: insights from experimental and computational studies. ChemSusChem 6(1):51–55
73. Ernst S, Aldous L, Compton RG (2011) The electrochemical reduction of oxygen at boron-doped diamond and glassy carbon electrodes: a comparative study in a room-temperature ionic liquid. J Electroanal Chem 663(2):108–112
74. Martiz B, Keyrouz R, Gmouh S et al (2004) Superoxide-stable ionic liquids: new and efficient media for electrosynthesis of functional siloxanes. Chem Commun 6:674–675

75. Buzzeo MC, Klymenko OV, Wadhawan JD et al (2003) Voltammetry of oxygen in the room-temperature ionic liquids 1-ethyl-3-methylimidazolium bis((trifluoromethyl)sulfonyl) imide and hexyltriethylammonium bis((trifluoromethyl)sulfonyl)imide: One-electron reduction to form superoxide. Steady-state and transient behavior in the same cyclic voltammogram resulting from widely different diffusion coefficients of oxygen and superoxide. J Phys Chem A 107(42):8872–8878
76. Cecchetto L, Salomon M, Scrosati B et al (2012) Study of a Li–air battery having an electrolyte solution formed by a mixture of an ether-based aprotic solvent and an ionic liquid. J Power Sources 213:233–238
77. Zhang D, Li R, Huang T et al (2010) Novel composite polymer electrolyte for lithium air batteries. J Power Sources 195(4):1202–1206
78. Garsuch A, Badine DM, Leitner K et al (2011) Investigation of various ionic liquids and catalyst materials for lithium-oxygen batteries. Z Phys Chem 225:1–13
79. Cui ZH, Fan WG, Guo XX (2013) Lithium–oxygen cells with ionic-liquid-based electrolytes and vertically aligned carbon nanotube cathodes. J Power Sources 235:251–255
80. Zhang T, Zhou H (2012) From Li–O_2 to Li–air batteries: carbon nanotubes/ionic liquid gels with a tricontinuous passage of electrons, ions, and oxygen. Angew Chem Int Ed 51(44):11062–11067
81. Sawyer DT, Valentine JS (1981) How super is superoxide? Acc Chem Res 14(12):393–400
82. Maricle DL, Hodgson WG (1965) Reduction of oxygen to superoxide anion in aprotic solvents. Anal Chem 37(12):1562–1565
83. Peover ME, White BS (1966) Electrolytic reduction of oxygen in aprotic solvents: the superoxide ion. Electrochim Acta 11(8):1061–1067
84. Oh SH, Black R, Pomerantseva E et al (2012) Synthesis of a metallic mesoporous pyrochlore as a catalyst for lithium–O_2 batteries. Nat Chem 4(12):1004–1010
85. Zhang X, Wang L (2012) A stable sulfone based electrolyte for high performance rechargeable Li–O_2 batteries. Chem Commun 48:11674–11676
86. Xu D, Wang Z-L, Xu J-J et al (2012) Novel DMSO-based electrolyte for high performance rechargeable Li–O_2 batteries. Chem Commun 48(55):6948–6950
87. Xu K, Angell CA (1998) High anodic stability of a new electrolyte solvent: unsymmetric noncyclic aliphatic sulfone. J Electrochem Soc 145(4):L70–L72
88. Sun X-G, Angell CA (2005) New sulfone electrolytes for rechargeable lithium batteries: part I. Oligoether-containing sulfones. Electrochem Commun 7(3):261–266
89. Abouimrane A, Belharouak I, Amine K (2009) Sulfone-based electrolytes for high-voltage Li-ion batteries. Electrochem Commun 11(5):1073–1076
90. Merritt MV, Sawyer DT (1970) Electrochemical studies of the reactivity of superoxide ion with several alkyl halides in dimethyl sulfoxide. J Org Chem 35(7):2157–2159
91. Andrieux CP, Hapiot P, Savéant JM (1985) Electron transfer coupling of diffusional pathways. Homogeneous redox catalysis of dioxygen reduction by the methylviologen cation radical in acidic dimethylsulfoxide. J Electroanal Chem Interfacial Electrochem 189(1):121–133
92. Chen Y, Freunberger SA, Peng Z et al (2013) Charging a Li–O_2 battery using a redox mediator. Nat Chem 5(6):489–494
93. Sharon D, Afri M, Noked M et al (2013) Oxidation of dimethyl sulfoxide solutions by electrochemical reduction of oxygen. J Phys Chem Lett 4:3115–3119
94. McCloskey BD, Valery A, Luntz AC et al (2013) Combining accurate O_2 and Li_2O_2 assays to separate discharge and charge stability limitations in nonaqueous Li–O_2 batteries. J Phys Chem Lett 4:2989–2993
95. Oh SH, Yim T, Ekaterina P et al (2011) Decomposition reaction of lithium bis(oxalato)borate in the rechargeable lithium-oxygen cell. Electrochem Solid State Lett 14(12):A185–A188
96. Younesi R, Hahlin M, Edström K (2013) Surface characterization of the carbon cathode and the lithium anode of Li–O_2 batteries using $LiClO_4$ or LiBOB salts. ACS Appl Mater Interfaces 5(4):1333–1341

Chapter 3
Cathode Electrochemistry in Nonaqueous Lithium Air Batteries

A.C. Luntz, B.D. McCloskey, S. Gowda, H. Horn, and V. Viswanathan

3.1 Introduction

There is now a well-recognized need to develop rechargeable batteries that are alternatives to lithium-ion batteries, e.g., the emergence of an annual conference called Beyond Li-ion (since 2009) and the extensive new US Department of Energy support for this topic (ARPA-E and JCESR). As this book illustrates, much of the interest in this arena focuses on trying to develop lithium air batteries, both aqueous and nonaqueous versions. The hope is that successful lithium air battery development would give a safe and cost-effective secondary battery with a much higher specific energy than current and even future lithium-ion batteries. This could enable electrification of road transport with much diminished "range anxiety." Lithium air battery's specific energy advantage over lithium-ion batteries could arise from two potential sources: (1) one of the reactants, O_2, is not stored in the battery but comes from breathing air as in a fuel cell and (2) the use of light materials for the electrodes, i.e., Li metal, as the anode rather than intercalated graphite (LiC_6) and possibly porous C for the cathode. The dramatic achievement that is theoretically possible is given in the much-abused Fig. 3.1. This shows that of the currently discussed alternate battery chemistries to lithium-ion, only lithium air is even

A.C. Luntz (✉)
IBM Research, Almaden Research Center, San Jose, CA 95120, USA

SUNCAT, SLAC National Accelerator Laboratory, Menlo Park, CA 94025, USA
e-mail: acluntz@slac.stanford.edu

B.D. McCloskey • S. Gowda • H. Horn
IBM Research, Almaden Research Center, San Jose, CA 95120, USA

V. Viswanathan
SUNCAT, SLAC National Accelerator Laboratory, Menlo Park, CA 94025, USA

Department of Mechanical Engineering, Stanford University, Stanford, CA 94305, USA

N. Imanishi et al. (eds.), *The Lithium Air Battery: Fundamentals*,
DOI 10.1007/978-1-4899-8062-5_3,

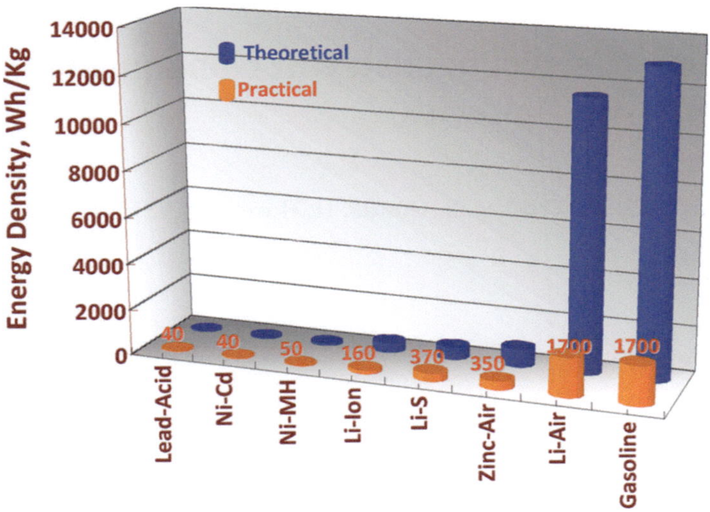

Fig. 3.1 The gravimetric energy densities (Wh/kg) for various kinds of rechargeable batteries compared to gasoline. The theoretical density, based strictly on thermodynamics of the active stored reactants, is shown as the *blue bars* while the practical achievable density is indicated by the *orange bars* and numerical values. For lithium air, the practical is an optimistic estimate. For gasoline, the practical value includes the average tank-to-wheel efficiency of automobiles (in the USA). From [15] with permission

remotely capable of competing with gasoline in terms of theoretical specific energy. However, despite their great promise, there are significant challenges to developing practical lithium air batteries [7, 15]. A practical limit to the specific energy of lithium air is still completely unknown so the one given in Fig. 3.1 is simply an optimistic estimate.

This chapter is devoted to summarizing what the authors believe represent our current understanding at the time of writing of the cathode chemistry in nonaqueous lithium air batteries. For the ideal nonaqueous lithium air battery, the net electrochemical reaction is $2Li + O_2 \leftrightarrow Li_2O_2$, with battery discharge described by the forward direction and charge described by the reverse direction. However, the actual battery chemistry can be considerably more complicated. All readers of this book should be aware that this is a dynamic young field and that many aspects of lithium air research are far from settled. This chapter summarizes the authors' opinionated perspective, largely based on experiments at IBM and theory at SLAC/Stanford, with comments relating this work to other contributions in the field. However, there is at present considerable differences in interpretations of the varied behavior of lithium air batteries amongst participants in the field (including authors in this book!), so that the reader should exercise their own judgment as to what is the truth based on the data and arguments presented.

Figure 3.2a shows a typical laboratory Swagelok™-type Li–O_2 battery used in our laboratory. This consists simply in clamping together a Li metal foil anode,

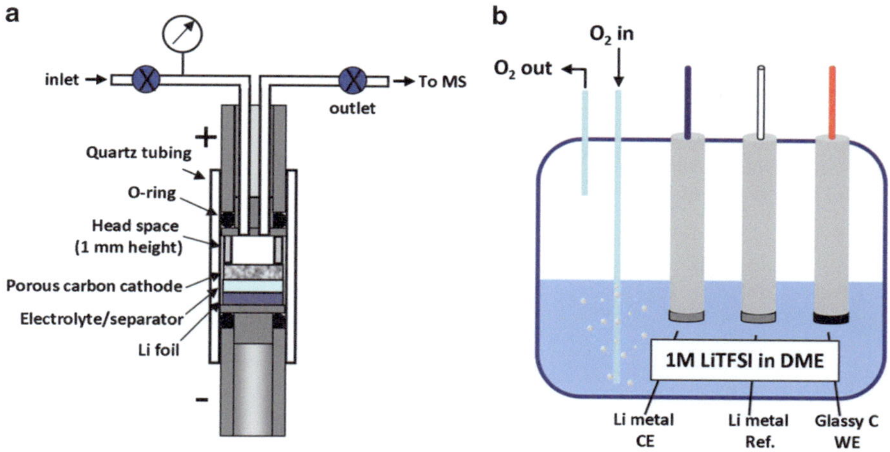

Fig. 3.2 (**a**) Swagelok™ DEMS cell (MS refers to differentially pumped mass spectrometer). (**b**) Bulk electrolysis cell. From [37] with permission

an electrolyte-soaked separator and a porous C cathode of ~0.1 m^2 total surface area (in ~1 cm^2 projected area), and a stainless steel mesh current collector. Pure O_2 gas is used instead of air to avoid deleterious contamination issues (see Sect. 3.5). Extreme care is taken to both clean and dry all cell components and for high hermetic integrity throughout the experiments. The only aspect of this cell that is unique compared to many other test cells in the field is that it is connected to a differentially pumped mass spectrometer (MS) for quantitative analysis of the O_2 consumed in discharge (ORR) and the O_2 evolved during charge (OER) via the technique called differential electrochemical mass spectrometry (DEMS). Thus, in addition to Coulometry measurements, the chemistry can be simultaneously measured quantitatively by the DEMS. More details of this cell construction are presented in Sect. 3.2.1.1. There can be many contributions to the time t-dependent Coulometry [cell potential $U(t)$ vs. current $i(t)$] observed in lithium air batteries: fundamental kinetic overpotentials, concentration polarizations due to reactant transport limitations (Li–ion [Li^+], O_2 gas or charge transport), iR drops from the cell impedance, cell component stability, etc. In order to isolate the fundamental Li–O_2 electrochemistry from all cell-dependent parameters, Li–O_2 electrochemistry has also been probed in the bulk electrolysis cell given in Fig. 3.2b. The hermetically sealed electrolysis cell uses flat nonporous polished glassy carbon (GC) as the working electrode and high-purity Li metal foil as the counter and reference electrodes. All electrodes have ~1 cm^2 surface area. A large volume (~15 mL) of electrolyte is vigorously stirred. Two gas ports are built into the cell lid and O_2 is bubbled through the electrolyte to insure saturation. The low current into the reference electrode minimizes the iR drops (with additional iR compensation to remove any residual secondary effects), and the rapid stirring minimizes Li^+ and O_2

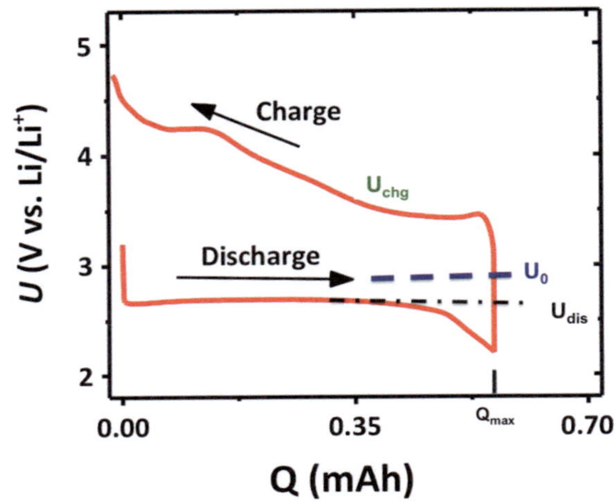

Fig. 3.3 Galvanostatic $Li–O_2$ discharge–charge cycle at 200 mA/g_C in a Swagelok™ cell with 1 M LiTFSI in DME as the electrolyte and a Vulcan XC72 carbon black cathode. From [35] with permission

transport limitations (concentration polarizations). The large volume of electrolyte minimizes some contributions from electrolyte decomposition during the discharge–charge (but see Sect. 3.3). Thus, the major aspects contributing to the electrochemistry in the electrolysis cell are the kinetic overpotentials and charge transport through the Li_2O_2 (see Sects. 3.6 and 3.7). However, when comparing results from the bulk electrolysis cell to those from the Swagelok™ cell, it is important to keep in mind the large difference in active surface areas of the two cathodes, i.e., 1 vs. ~10^3 cm^2.

Figure 3.3 shows a typical single galvanostatic discharge–charge cycle in the Swagelok™ cell at a current of 200 mA/g_{Carbon}, i.e., potential U vs. capacity Q, with a typical "good" electrolyte (1 M $LiN(CF_3SO_2)_2$[LiTFSI] in dimethoxyethane [DME]). There are many features to this cycle that we will attempt to explain throughout this chapter: the equilibrium potential U_0, a potential drop relative to that during discharge defined as U_{dis}, a sudden death to the cell characterized by the sudden drop in U_{dis} at some Q_{max}, and a complicated potential dependence during charging U_{chg}, with an initial small overpotential followed by a continuously increasing charging potential. U_{dis}, Q_{max}, and U_{chg} have significant dependences on current i. Both Q_{max} and U_{chg} depend on the type and amount of C in the cathode and on the extent of discharge Q_{dis}, i.e., if the discharge–charge cycle occurs for $Q_{dis} < Q_{max}$.

Before discussing our interpretation of the features of the galvanostatic discharge–charge cycle of Fig. 3.3, we first present an equivalent galvanostatic discharge–charge cycle in the bulk electrolysis cell in Fig. 3.4. This allows a separation of cell-dependent properties from those that are inherent in the electrochemistry. While the qualitative features in Fig. 3.4 are similar to those of the Swagelok™ cell, there are some important quantitative differences in the shape of $U_{dis}(Q)$ as well as in Q_{max}, and U_{chg}. For example, because of the much smaller surface area of the GC relative to the active cathode area in the Swagelok™ cells,

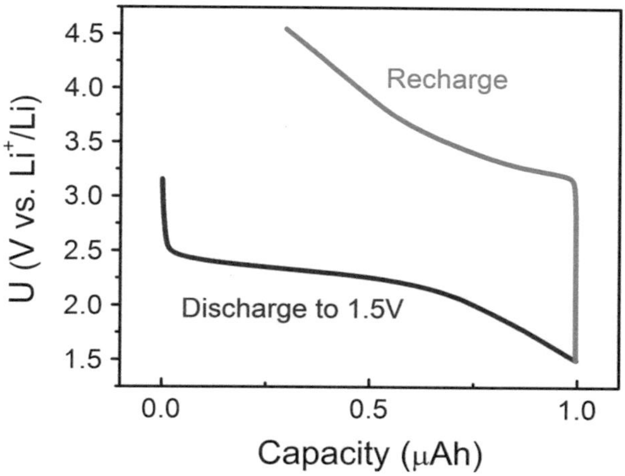

Fig. 3.4 Galvanostatic Li–O_2 discharge–charge at 100 μA current on GC in the electrolysis cell (same electrolyte as in Fig. 3.3)

Q_{max} is ~10^3 smaller and the cathode current density *j* (defined as the current normalized to the total electrochemically active surface area in the cathode) is ~10^3 higher than in Fig. 3.3. One other aspect that may drive some of the differences in the two types of cells is the morphology of Li_2O_2 growth. In the Swagelok™ cells, some authors report that Li_2O_2 grows as large (~500 nm) nanoparticles at low current densities but tends towards more uniform film growth at higher current densities (see Sect. 3.2.1.2). However, we do not observe these large nanoparticles in pure electrolytes at any current. On the other hand, in the bulk electrolysis cell the Li_2O_2 always grows as a continuous thickening film on the GC at all current densities (see Sect. 3.6.1). The discharge–charge properties in the electrolysis cell will be discussed in detail in Sect. 3.6 as they more properly relate to the fundamental Li–O_2 electrochemistry.

One aspect that we will stress throughout this chapter is that these galvanostatic discharge–charge cycles *do not measure the chemistry*; they simply measure the sum of all electrochemical processes. There is no guarantee at all that this cycle is dominated by the electrochemical reactions $2Li + O_2 \leftrightarrow Li_2O_2$. Many of the early authors in this field [9, 48] and even more recent ones [20, 64] measure repeated galvanostatic discharge–charge cycles and from this assume rechargeability of Li–O_2. We believe this conclusion is unwarranted without the Li–O_2 chemistry being simultaneously measured quantitatively. In fact, all early studies using carbonate solvents confused electrolyte decomposition with the desired Li–O_2 electrochemistry so that the cyclability claimed was related in a vague way to the amount of excess electrolyte compared to that consumed during the discharge–charge cycle. Therefore, we urge readers to be extremely cautious about claims of rechargeability simply from Coulometry cycles. Coulometry results must be compared to *quantitative* measures of the chemistry occurring, e.g., DEMS.

After this introduction, the plan for this chapter is as follows: Sect. 3.2 discusses in more detail some of the key parameters describing galvanostatic discharge–charge in Li–O_2 batteries; Sect. 3.3 discusses the important issues of electrolyte stability; Sect. 3.4 discusses the equivalent problem of cathode stability; Sect. 3.5 discusses some of the problems and prospects in converting Li–O_2 cells to lithium air cells; Sect. 3.6 discusses the fundamental cathode electrochemistry, i.e., the mechanism, kinetic overpotentials (Tafel plots), viability of electrocatalysis, etc.; Sect. 3.7 discusses the limitations imposed by charge transport and finally Sect. 3.8 presents a brief summary and outlook.

3.2 Characterization of the Li–O_2 Electrochemistry with C Cathodes

3.2.1 Galvanostatic Discharge–Charge Coulometry

There have been innumerable studies of galvanostatic discharge–charge Coulometry in cells similar to Fig. 3.2a with C cathodes. In this section, we only present some results using ether solvents (e.g., DME), as ethers currently appear to be the most stable Li–O_2 electrolytes that also exhibit stability to Li metal. Results for other solvents are presented in Sect. 3.3, while C cathodes with added "catalysts" are briefly discussed in Sect. 3.6.4.

3.2.1.1 Comments on Swagelok™ Cell Construction

Our experience, especially by combining DEMS with the Coulometry, is that cell cleanliness, dryness, and electrolyte purity are crucial in obtaining reliable and reproducible results when studying Li–O_2 batteries. We therefore summarize here a few of the procedures that we found necessary to ensure high cell cleanliness. A more extensive discussion of the details is available in some of our publications and in their Supplementary Information [34, 35, 37, 38]. Other researchers in the field will undoubtedly have found their own set of successful procedures as well.

The first experiments in our laboratory used typical Swagelok-type cells [2], with plastic ferrules to impart a seal around the battery stack. However, DEMS showed that the cells were difficult to hermetically seal and allowed atmospheric contaminants to compromise the feed gas purity. Our hermetic DEMS cell design was inspired to eliminate such contamination. Others also use modified cells with specially designed hermetic seals to eliminate air contamination [40]. Feed gases must obviously be of highest purity (we use Research Purity from Matheson Tri-gas). We also use an additional moisture trap (Pur-gas from Matheson) after the gas regulator as an additional precaution to ensure no H_2O contamination is present in our feed streams.

For most experiments, we have used two types of C cathodes. One is simply AvCarb P50 carbon paper and the other was constructed by air-spraying a slurry of carbon black particles (e.g., Vulcan XC72) and PTFE as a binder onto a stainless steel mesh. Slurries were usually prepared by mixing carbon black with a PTFE dispersion (60 wt% in water) in a 5:1 wt/wt ratio in a 20:80 isopropanol/water (IPA/H_2O) mixture. The wt% of carbon black in the IPA/H_2O can be modified to achieve a necessary slurry consistency for the air-spraying process. A Badger model 350 air-sprayer (1.4 atm. inlet pressure) was used to uniformly coat the SS mesh. Our DEMS cells employ 12 mm diameter cathodes and typically have a carbon loading of 6 mg for P50 cathodes and 0.5–2 mg for carbon black cathodes. After preparation of the cathode, they are rinsed in acetone and isopropanol multiple times and dried at high temperatures (~150 °C) under vacuum for at least one day prior to use. Cathodes are then transferred to an Ar-filled glove box (<0.1 ppm H_2O and O_2) and rinsed in an ultrapure solvent such as DME. This rinsing procedure is then followed by drying in the glove box at 200 °C for at least 2 h prior to cell assembly. We also find it essential to rinse and dry separators prior to using in the cell. Microporous polymer separators, such as Celgard 2500, are typically cut using a punch and then transferred to the Ar glove box where they are rinsed in an ultrapure solvent, dried under vacuum, and then further dried at 75 °C overnight in the Ar glove box. As a result of large amounts of material leaching, glass fiber separators, such as Whatman filters, should be avoided if at all possible in favor of microporous polymer separators that don't exhibit such leaching. However, some solvents, such as propylene carbonate and dimethyl sulfoxide, do not wet microporous polymer separators. Therefore, glass fiber separators that have been rinsed multiple times in acetone and isopropanol and vacuum dried at >150 °C are used for these solvents.

Electrolyte purity is absolutely essential. We always use battery-grade materials as received when available or extensively purify solvents by molecular sieve adsorption followed by a distillation procedure. Solvent water content is always tested using a Karl Fischer titration and we always strive to have an H_2O content of <20 ppm in the electrolyte. We also characterize solvent purity using NMR and GC-MS and use only those solvents in which no observable contaminants are detected. If molecular sieves are necessary to purify an electrolyte solvent, the solvent should always be distilled after the molecular sieve treatment, as ionic impurities leaching from the sieves will impart significant electrolyte contamination.

All cell components that come into direct contact with the battery stack are made from metals, plastics, or glassware that can be easily cleaned. After cleaning using sonication in soap water and acetone, cell components are dried at 130 °C prior to being transferred to the glove box while still hot. Markez o-rings (Marco Rubber) are used for the DEMS cell seals because they are not susceptible to solvent attack and can be baked at high temperatures.

Most importantly, the efficacy of a protocol designed to impart the highest possible cell cleanliness can be tested by performing an oxidative potential scan on an assembled cell from the cell's open circuit voltage (OCV) (~3 V vs. Li/Li+) up to the inherent oxidation peak of the solvent (typically ~4–5 V). Contaminants are present in the cell if small oxidative peaks arise at potentials lower than the inherent oxidation potential of the solvent.

The DEMS cell is a conventional Swagelok™-type cell as described above but is built to provide high hermetic integrity. Evolved gases are sent to a calibrated differentially pumped residual gas analyzer (Stanford Research Systems) for quantitative identification using a purge of the cell headspace (~1.5 mL) into a much larger evacuated volume using a high-pressure pulsed gas valve (HPGC), with O_2 gas as the purge for discharge and Ar for charge. During charge, the quantitative composition of the headspace gas swept out of the cell is analyzed using a calibrated residual gas analyzer, with the absolute quantities of the different gas components determined by comparing the calibrated mass spectrometer intensity for the various masses to the ^{36}Ar peak of known Ar headspace pressure and volume. Alternatively, the carefully calibrated ~1.5 mL cell headspace volume can be isolated, and pressure decay/rise (ΔP) during cell discharge/charge can be monitored using an in-line pressure transducer. These ΔP measurements allow accurate quantification of the number of electrons per O_2 consumed in discharge or produced during charging (e^-/O_2) when corrections are made for the small pressure rises due to gas evolution from parasitic reactions. During charging, the e^-/O_2 ratio obtained from the calibrated mass spectrometer measurement agrees within 3 % of that from the pressure rise.

3.2.1.2 Features of the Galvanostatic Discharge–Charge Cycle

In this section we discuss some properties of the key descriptors of the galvanostatic discharge–charge cycle in Fig. 3.3, i.e., U_0, Q_{max}, U_{dis}, and U_{chg}.

3.2.1.2.1 U_0

The OCV or equilibrium potential U_0 in both the Swagelok™ and electrolysis cells is ~3.1 V (vs. Li/Li^+) prior to discharge but is ~2.85 V after discharge to sudden death. The thermodynamic standard potential for formation of bulk Li_2O_2 is 2.96 V. However, U_0 is in reality the equilibrium potential for the surface reaction of $2Li + O_2 = Li_2O_2$ on whatever surface is present, and this need not be the same as the standard potential for bulk Li_2O_2 formation. For example, the equilibrium potential for the reaction at a C surface (initial OCV) may be different than that on a Li_2O_2 surface (after sudden death). Even the reaction on the Li_2O_2 surface may depend on the crystal facet, termination of the facets, extent of steps, kinks, etc. since the free energy of the reaction $2Li + O_2 \rightarrow Li_2O_2$ depends on the detailed state of the surface.

3.2.1.2.2 Q_{max}

A high Q_{max} given the active component weights is of course the main reason for interest in lithium air batteries, so naturally there have been many studies of the dependence of Q_{max} (at low *i*) on the type or properties of the C cathode, especially where Q_{max} is large. Many authors have attributed this high capacity to various microstructure of the C cathodes, e.g., porosity, mesoporosity. However, the most convincing argument is simply that Q_{max} depends on the total surface area of the C cathode [39] (and hence approximately on the carbon weight $g^{2/3}$).

The morphology of the Li_2O_2 growth on the cathode with discharge capacity *Q* is quite complicated and depends strongly upon on current and apparently on electrolyte purity. At all currents, we observe that the Li_2O_2 deposit is a nearly uniform thin deposit over the entire cathode surface area at Q_{max}. In all cases the crystallinity of the Li_2O_2 deposit is relatively high (>15 nm) as judged by XRD line-widths of the major diffraction peaks [34]. On the other hand, SEM and TEM studies by others observe that discharge on vertically aligned carbon nanotube (VACNT) cathodes at low *i* produces toroid- or abacus-shaped beads of ~200–500 nm diameter of Li_2O_2 [10, 13]. During discharge, both the number of beads and the size of the beads increase and this accounts for the discharge capacity. The exact dependence of the number and size with *Q* is not yet determined, but it appears that new beads nucleate after initial ones grow in size. Ultimately, a moderately dense set of large beads is observed at Q_{max}. In addition to the beads of Li_2O_2, a rust-like material coats the C nanotubes as well and this has been identified as Li_2CO_3 by small-angle electron diffraction. The rust-like coating of the nanotubes increases with cycling at the expense of the number of the toroid-shaped beads that are formed. This suggests that the decrease in Q_{max} with cycling is caused by the buildup of Li_2CO_3 on the surface that isn't totally removed during charging, even to high potentials of >4.2 V (see Sects. 3.3 and 3.4).

Figure 3.5 shows a set of galvanostatic discharges at different currents. Both Q_{max} and U_{dis} are strong functions of *i*. We discuss below the dependence of U_{dis} on *i* and in Sect. 3.7 that the dependence of Q_{max} with *i* is likely due to charge transport limitations. Figure 3.5 points out a serious issue for current lithium air batteries: a very unfavorable trade-off between *i* and Q_{max}. This trade-off may make it difficult to take advantage of the potentially high specific energy of lithium air batteries at reasonable power densities.

3.2.1.2.3 U_{dis}

Figure 3.5 already demonstrates that U_{dis} is a strong function of *i*. Figure 3.6 elaborates by plotting cell U_{dis} vs. *i* for cells with two different types of C cathodes. The linear dependence of U_{dis} with *i* suggests that U_{dis} is dominated by *iR* drops in the cell, with ~40 Ω for XC72 and ~80 Ω for P50 cathodes. Measurements of the cell resistance using electrochemical impedance spectroscopy (EIS) give similar values of *R* to that from the slopes in Fig. 3.6. *R* appears to be dominated by the

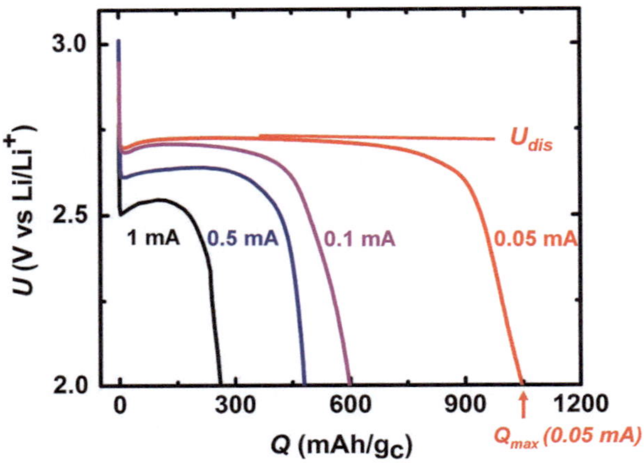

Fig. 3.5 Galvanostatic discharges in Swagelok™ cell as a function of current, with XC72 cathode and 1 M LiTFSI/DME electrolyte. U_{dis} is the average plateau discharge potential at a given current and Q_{max} is the capacity to a discharge voltage of 2 V. Q is normalized to the weight of C in the cathode since this is typical in the field. From [54] with permission

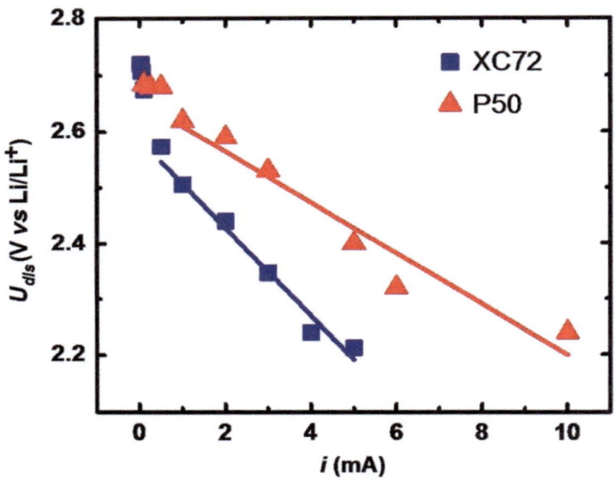

Fig. 3.6 Measurement of Li–O_2 discharge potential in the plateau region (U_{dis}) as a function of galvanostatic current i imposed during discharge. Swagelok™ cells are used with Li metal anode, 1 M LiTFSI/DME electrolyte-soaked Celgard separators (~50 μm thick), and either P50 carbon paper cathode (*red triangles*) or XC72 C with PTFE binder on SS mesh as the cathode (*blue squares*). From [54] with permission

solid–electrolyte interface (SEI) on the Li anode, although there are undoubtedly other contributions as well. This also suggests that any kinetic overpotential in discharge due to the surface electrochemical reaction is significantly smaller than that observed in Fig. 3.6. We discuss the kinetic overpotentials in Sect. 3.6.2.

3.2.1.2.4 U_{chg}

The complicated dependence of U_{chg} on the charging capacity Q_{chg} following galvanostatic discharge is one of the most important (and controversial) aspects of Li–O_2 batteries. The high potential rise occurring during Q_{chg} underlies many important limitations in current Li–O_2 batteries: a poor discharge–charge cycle

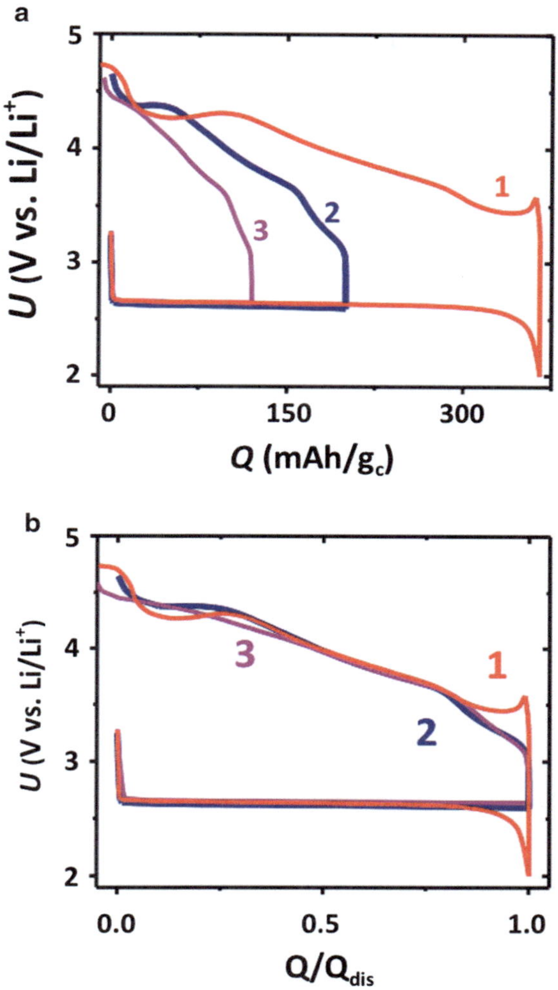

Fig. 3.7 (**a**) Galvanostatic discharge–charge at 400 mA/g C of $LiBF_4$ in DME for different capacities of discharge Q_{dis}. (**b**) Same results but plotted relative to the total discharge capacity Q/Q_{dis}

electrical efficiency, limitations in electrolyte electrochemical stability, and poor cyclability (see Sect. 3.3).

Early experiments using carbonate solvents in the electrolytes observed that U_{chg} rose almost immediately to >4 V, and this was interpreted as a kinetic overpotential and has led to a focus on electrocatalysis. However, as will be pointed out in Sect. 3.2.2, this rapid rise is related to electrolyte decomposition and more stable ether electrolytes typically produce an initial low U_{chg} that rises with Q_{chg} in some manner as in Fig. 3.3. The exact manner in which U_{chg} rises depends upon many factors: the type of C [35] and loading of C in the cathode, the extent of discharge, etc.

Figure 3.7a shows how U_{chg} varies with the extent of discharge Q_{dis}. It appears that the rate of rise in U_{chg} is also strongly dependent on Q_{dis}, but in Fig. 3.7b the

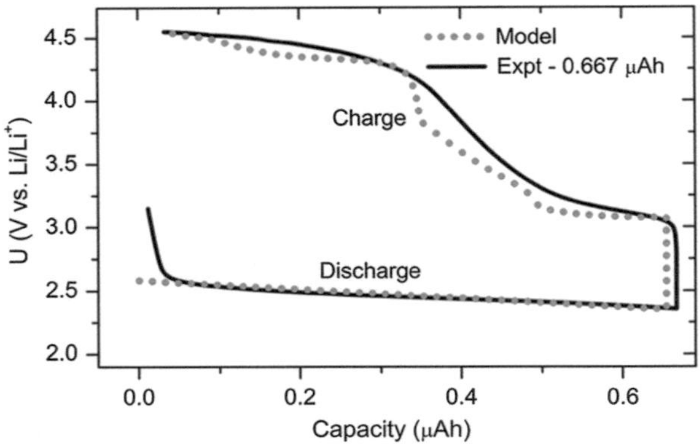

Fig. 3.8 Galvanostatic discharge on GC in the bulk electrolysis cell at 10 μA followed by charge at 5 μA. The points are from a theoretical kinetic model. From [38] with permission

same results are plotted as a function of Q/Q_{dis}, i.e., normalized to the extent of discharge. Except for very small differences at the outset (which are likely related to morphology differences), the rises in U_{chg} superimpose. This suggests to us that the process responsible for the rise in U_{chg} depends only on Q_{chg} and is independent of the extent (or details) of the discharge. In addition, the O_2 evolution measured by DEMS during Q_{chg} for different Q_{dis} is also only a function of Q/Q_{dis} as well. We also find that the rate of increase in U_{chg} is independent of i_{chg} (and hence t_{chg}), again depending only upon Q_{chg}. This argues that the rise in U_{chg} is a purely electrochemical phenomenon occurring during charge. There is no contribution from any process which is only time related, e.g., thermal chemistry.

Similar experiments on the GC electrodes in the bulk electrolysis cell show qualitatively the same rise in U_{chg} as shown in Fig. 3.8. In addition, the scaling with Q/Q_{dis} and independence with t_{chg} are also identical phenomena in the bulk electrolysis cell to that observed in the Swagelok™ cell. Since the electrolysis cell only depends on the fundamental electrochemistry and possible parasitic reactions at the Li_2O_2–electrolyte interface, this similarity shows that the origin of the rise in U_{chg} is not some cell-dependent phenomenon.

We have argued previously that the potential rise in U_{chg} arises from the deposition of carbonate/carboxylates at the Li_2O_2–electrolyte interface due to electrochemical decomposition of the electrolyte during the charging process [38]. During galvanostatic charging, as the Li_2O_2–electrolyte interface becomes covered by solid electrolyte decomposition products, the only way that the charging rate (current) remains constant is by increasing the overpotential. As U_{chg} increases, this causes an increase in electrolyte electrochemical decomposition and even more deposition at the interface, etc. This process is schematically shown in Fig. 3.9. The points in Fig. 3.8 are a simple kinetic model representing simultaneously the Li_2O_2 OER, the electrochemical decomposition of the electrolyte, and the oxidation of the decomposition products as indicated schematically in Fig. 3.9. The unfortunate implication from this is that deposition of a few monolayer of solid electrolyte decomposition products at the interface causes the rise in U_{chg}, and this is very difficult to avoid given the stability issues of electrolytes with Li_2O_2 (Sect. 3.3).

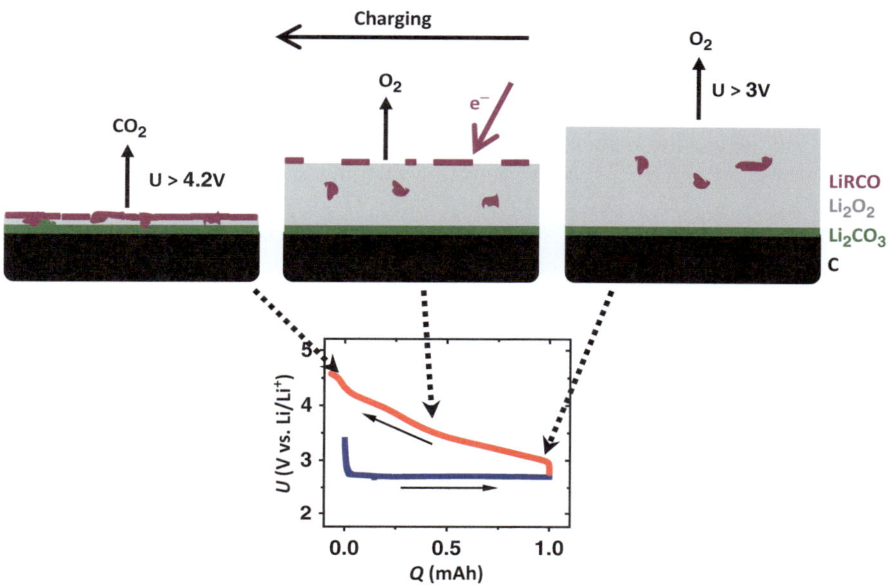

Fig. 3.9 *Upper panels* indicate schematically the cathode (*black*) and a picture of what happens in the deposit during charging to cause the rising charging potential. ~1 monolayer of Li_2CO_3 forms at the C interface; some dispersed carbonate possibly forms as a result of electrolyte decomposition in the Li_2O_2 deposit and ~1 monolayer at the electrolyte interface that forms during charging. The e^- represents some unspecified electrochemical reaction producing carboxylate or other solid deposit at the interface during charging. The proportions are not to scale. The *inset* shows the potential observed for galvanostatic discharge–charge. The *dashed arrows* indicate qualitatively the charging potential appropriate for the three panels. From [38] with permission

Given this picture, the dependence of U_{chg} on Q/Q_{dis} is simply due to the fact that the ratio R_{OER}/R_d is independent of the amount of Li_2O_2 formed, where R_{OER} is the rate of Li_2O_2 OER and R_d is the electrochemical rate of electrolyte decomposition. We note that when CO_2 is present in the O_2 headspace and forms Li_2CO_3, U_{chg} increases much faster than when it is not present (see Sect. 3.5). This provides qualitative support to the picture that it is concentration of carbonates or other insoluble decomposition products at the Li_2O_2–electrolyte interface that is responsible for the rise in U_{chg}. Note that if the electrolyte (and cathode) decomposition results in soluble decomposition products, then a rapid rise in U_{chg} may not be observed. This does not mean, however, that there is no decomposition occurring in the electrolyte. It is our belief that this may in part explain the spectacular suggestions of extensive rechargeability with a nanoporous Au cathode and DMSO solvent [45].

Lu and Shao-Horn [29] have speculated that the rising potential in U_{chg} is due to two mechanisms for charging: a low overpotential one due to Li de-intercalation at the surface $Li_2O_2 \rightarrow 2LiO_2 + 2(Li^+ + e^-)$ followed by chemical disproportionation $2LiO_2 \rightarrow Li_2O_2 + O_2$ and a higher overpotential mechanism for an unspecified bulk process $Li_2O_2 \rightarrow O_2 + 2(Li^+ + e^-)$. However, since all electrochemistry must occur at the Li_2O_2–electrolyte interface, we do not understand why the low overpotential surface OER would not dominate the entire OER process nor what constitutes a separate bulk OER process. However, the most compelling argument against this suggestion is simply that it does not agree with DEMS results for charge which shows an e^-/O_2 ~2.0 at the outset of charging, consistent only with the full process $Li_2O_2 \rightarrow 2Li + O_2$ (see, e.g., Fig. 3.25). Their suggestion would imply that the DEMS would see a region with current but no O_2 evolution. We note that U_{chg} also rises before the end of the charge cycle for a Na-O_2 battery as well [17]. In this case the discharge–charge cycle involves a single-electron transfer forming/from NaO_2, so that there is only a single mechanism possible. The importance of the rise in U_{chg} to Li–O_2 batteries begs for more detailed experiment and theory to more clearly delineate its origin and cure.

3.2.2 Discharge–Charge Chemistry

As pointed out in the introduction, it is essential that chemical measurements be added to Coulometry to correctly identify the electrochemical processes responsible for the currents. A wide variety of spectroscopic techniques have now been employed to identify the products formed during discharge, e.g., XRD, Raman, IR, DEMS, NMR, and XPS. The necessity of adding spectroscopy to Coulometry is emphasized by the now well-known misconceptions derived from using carbonate solvents in the electrolyte. Most early studies used these carbonate solvents as they are pervasive in lithium-ion batteries and form good SEI with Li metal. These early studies simply used Coulometry to characterize the electrochemistry, focusing on cyclability and the importance of catalysts to lower a high overpotential for charging [9, 28, 30]. However, when spectroscopic investigations were also added, it became apparent that Li_2O_2 that may be initially formed in the electrochemistry is chemically unstable in the carbonate solvents, rapidly forming principally Li_2CO_3 and carboxylates $LiRCO_3$ instead [12, 34, 57] (see Sect. 3.3). Figures 3.10 and 3.11 show XRD and Raman spectra of discharged cathodes with ethers, carbonates, and mixed carbonate/ethers as electrolyte solvents. Only the pure ether shows any evidence for Li_2O_2 in the cathode.

Figure 3.12 also shows DEMS results for cells using the same solvents during a galvanostatic charge after the discharge. Only the pure DME shows any significant O_2 evolution, while it is CO_2 from Li_2CO_3 and $LiRCO_3$ oxidation that dominates charging in the mixed ether-carbonate and carbonate solvents. We believe it is the electrochemistry of Li_2CO_3 and $LiRCO_3$ that causes the high overpotentials for charging and apparent catalysis effects originally observed by others [36]. This is discussed more fully in Sect. 3.6.4. However, IR measurements have qualitatively shown that even with the ether-based electrolytes, several contaminants are also produced during discharge [11, 16, 49, 52, 59], e.g., carbonates, carboxylates,

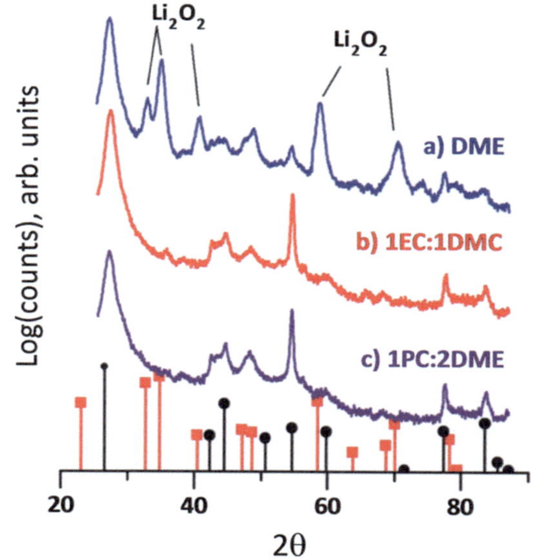

Fig. 3.10 XRD patterns of discharged carbon cathodes using the different electrolyte solvents as indicated. The *squares* mark Li_2O_2 peaks and the *circles* mark graphite peaks in the XRD. From [34] with permission

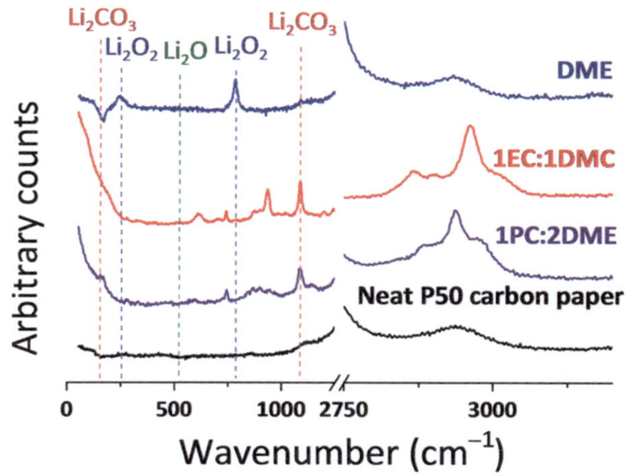

Fig. 3.11 Raman spectra of discharged carbon cathodes using various solvents for the electrolyte: *DME*, mixed carbonates, and a mixed carbonate/DME. Discharge conditions of ~−0.1 mA/cm^2 under ~1 bar O_2 to 2V. From [34] with permission

lithium formate, and perhaps salt decomposition products such as LiF. The general issue of electrolyte stability is discussed more fully in Sect. 3.3.

It should be emphasized that it is only with great difficulty that the spectroscopic techniques can become fully quantitative, i.e., that they can determine absolutely the amount of Li_2O_2 formed during discharge or consumed during charge relative to that expected from the currents. The one technique that has been well quantified is DEMS [35]. In this case the absolute moles of O_2 consumed during discharge (ORR) or formed during charging (OER) is obtained to ~3 % accuracy. By comparing ORR and OER to the Q used in the electrochemistry, a number of important

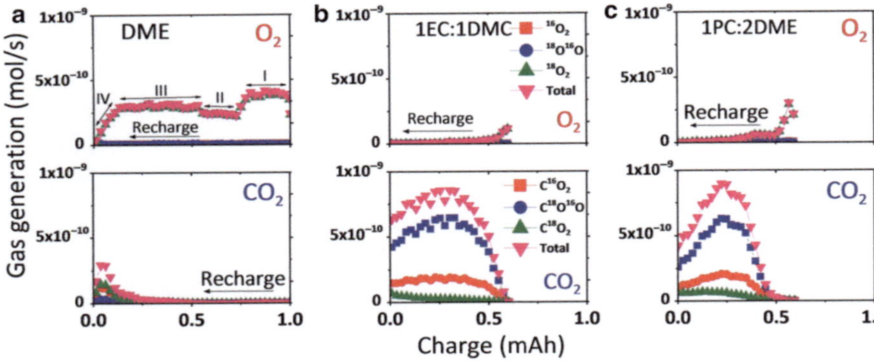

Fig. 3.12 Isotopically labeled O_2 and CO_2 gas evolution during charging following discharge in electrolytes using different solvents. From [34] with permission

parameters defining the rechargeability of the cell are obtained [35], i.e., $(e^{-}/O_2)_{dis}$, $(e^{-}/O_2)_{chg}$, and OER/ORR. Unfortunately, the true measure of rechargeability, OER/ORR, only quantifies the combined discharge–charge process, so it cannot easily distinguish chemical problems occurring during discharge from electrochemical ones during charge. The DEMS results are discussed in more detail in Sect. 3.3 and imply that no electrolyte is yet capable of giving a rechargeability on a single cycle of >90 %! There have also been attempts to quantify both IR [11] and XRD [26] measurements of the Li_2O_2 formed during discharge by comparing spectroscopic intensities to that of Li_2O_2-packed cathodes. If these calibrations ultimately prove to be accurate, they provide important additional quantitative chemical information, i.e., the amount of Li_2O_2 produced relative to Q_{dis}. Results from these studies suggested that even using the more stable ether solvent-based electrolytes that Li_2O_2 comprised only ~60 % of the discharge product. However, the absolute accuracy of the quantification is somewhat unknown in our view and as discussed previously the amount of Li_2O_2 observed may be very dependent upon the details of cell integrity, e.g., H_2O content, etc. Our DEMS results indicate that Li_2O_2 comprises at least ~80 % of the discharge product, as we obtain an OER/ORR of ~0.8 in DME-based electrolytes and find, using isotopic labeling, that O_2 is only evolved from Li_2O_2 (and not decomposition products) during charge. Therefore, measurements of a smaller fraction of Li_2O_2 in the discharge product imply some kind of extra decomposition path not present in our cells, likely through more H_2O or other contaminants.

Of note when comparing Figs. 3.10 and 3.11 is that qualitative XRD by itself is not a sufficient spectroscopic technique to identify all discharge products, as some solvent decomposition products (e.g., Li alkyl carbonates) form amorphous deposits in the cathode. Many current studies imply that stable cycling in solvents is possible by showing XRD patterns of Li_2O_2 formation after either the first discharge or a few cycles. Nevertheless, in most studies the XRD is neither quantitative (although efforts to make it quantitative have been explored [26]) nor sensitive to decomposition products that form only amorphous phases. As a result, Li_2O_2 could be a minor fraction of the actual discharge product, yet appear to be the only discharge product in an XRD spectra. Therefore, XRD always has to be coupled with other techniques that are not sensitive to crystallinity and are quantitative to assess cyclability.

3.3 Electrolyte Stability

This section will review efforts to understand the cathode electrochemistry in various nonaqueous electrolytes. Particular emphasis will be placed on electrolyte stability during a single cell cycle. We will show that no electrolyte presently gives a cycle rechargeability >90 %. Therefore, the search for stable nonaqueous Li–O_2 electrolytes has become a major focus in the field. We will discuss that there are two difficult and generally competing stability issues for the electrolyte: chemical stability to nucleophilic attack (and hydrogen abstraction) by Li_2O_2 and electrochemical stability during charge. We have previously discussed this electrolyte dilemma as being caught between a rock and a hard place [35]. The well-known difficulties with the carbonates as solvents stem principally from the former issue.

3.3.1 DEMS Experiments of Electrolyte Stability

From a system perspective, the chemical requirement for a truly rechargeable Li–O_2 battery is simply that any oxygen reduced during discharge (ORR) is completely released during charge (OER) and that this OER/ORR ratio remains unity for many cycles. *To date, we are aware of no electrolyte system for which this statement is true*. One important limitation in this DEMS definition of rechargeability is if OER/ORR < 1, there is no definitive means to distinguish whether the problem is due to chemical stability of the electrolyte, i.e., non-electrochemical reaction of Li_2O_2 with the solvent or electrochemical stability during charging. Since the ideal nonaqueous Li–O_2 reaction is $2Li + O_2 \leftrightarrow Li_2O_2$, this implies ideally that oxygen consumption during discharge $(e^-/O_2)_{dis}$ and oxygen evolution during charge $(e^-/O_2)_{chg}$ are both 2.00 and that no other gases, e.g., CO_2 and H_2, are evolved during discharge or charge. The ideal electrochemical reaction also implies that the absolute number of moles of Li_2O_2 produced during discharge, $n(Li_2O_2)$, should be given by the Coulometry, i.e., $n(Li_2O_2) = Q_{dis}/2F$, with Q_{dis} in Coulombs and F is Faraday's constant. The combination of measuring both $n(Li_2O_2)$ and OER/ORR would give a complete picture of electrolyte chemical *and* electrochemical stability.

In our laboratory, we extensively use quantitative in situ DEMS to probe the stability of electrolytes during the Li–O_2 electrochemistry, often coupled with other (mostly qualitative) spectroscopic techniques, e.g., NMR, FTIR, Raman, and XRD.

The quantitative DEMS system is described in detail in previous publications [34, 35, 37] and is described briefly in Sect. 3.2.1.1. The main feature of this DEMS is a cell that is designed to ensure high hermetic integrity and to incorporate a low-headspace volume to give high sensitivity both for the mass spectrometry and for molar consumption via pressure decay measurements. We also use a gas chromatography valve sampling procedure to precisely, and accurately, quantify gases evolved during some adjustable time period.

Typical quantitative DEMS results are shown in Fig. 3.13 which compares galvanostatic discharge–charge curves for Li–O_2 cells employing two electrolytes: 1 M LiTFSI dissolved in a 1:1 mixture of PC/DME and in pure DME. These are two of the electrolytes discussed qualitatively in Sect. 3.2. The $(e^-/O_2)_{dis}$ values are simply obtained by dividing the charge (Q) passed by the moles of O_2 consumed during a given time period in the discharge. Figure 3.13a shows that the Coulometry profile is qualitatively different for the two electrolytes, with the PC/DME rising almost immediately to >4.5 V during charging. It was this early observation that forced many in the field to focus on electrocatalysis as a means to lower this overpotential. However, it now well accepted that this high overpotential is in fact simply the overpotential for releasing CO_2 from the carbonate deposits formed in this electrolyte. In Fig. 3.13b, $(e^-/O_2)_{dis}$ >2.0 is observed for most of the discharge with carbonate-containing electrolytes. This implies that some additional electrochemical processes other than pure Li_2O_2 formation are also occurring during discharge. However, more importantly, XRD, FTIR, Raman, NMR, and theoretical calculations have all shown that the carbonate solvents are not stable to nucleophilic attack by Li_2O_2 [3, 4, 12, 23, 34, 60]. Therefore, during cell charge, predominantly CO_2 is evolved from the solid carbonate, carboxylate, and formate cathode deposits. Little O_2 is evolved and this indicates negligible electrochemical rechargeability. On the other hand, discharge in a pure DME-based electrolyte (Fig. 3.13c) exhibits $(e^-/O_2)_{dis} \approx 2.00$, indicating minimal parasitic electrochemical process occurring during discharge. Li_2O_2 formation is shown to be the primary, although not exclusive, discharge product [11, 34, 49]. In addition, since $(e^-/O_2)_{dis} \approx 2.00$, any parasitic thermal chemistry must occur from reaction with Li_2O_2 and not any intermediates formed after a one-electron process (e.g., LiO_2). O_2 evolution from Li_2O_2 is the primary electrochemistry during cell charge; however, parasitic reactions limit the OER/ORR ratio to only ~0.8.

Numerous solvent/salt combinations have been explored by our laboratory using DEMS in our (to now) fruitless attempts to find a truly stable electrolyte [35]. DEMS results for three of these electrolytes, DMSO, MPP-TFSI, and CH_3CN, are shown in Fig. 3.14. The different parameters obtained from DEMS for a wider range of electrolytes are given in Table 3.1. The most important conclusion from Table 3.1 is that although some electrolytes have an $(e^-/O_2)_{dis} \approx 2.00$ (e.g., ethers and CH_3CN), all electrolytes have $(e^-/O_2)_{chg}$ >2.0 and OER/ORR < 0.9. This naturally severely limits cyclability.

Several possibilities exist for why OER/ORR < 1: parasitic electrochemistry occurring during discharge, thermal chemistry between Li_2O_2 and the solvent, and parasitic electrochemistry during charging. Although we cannot quantify the

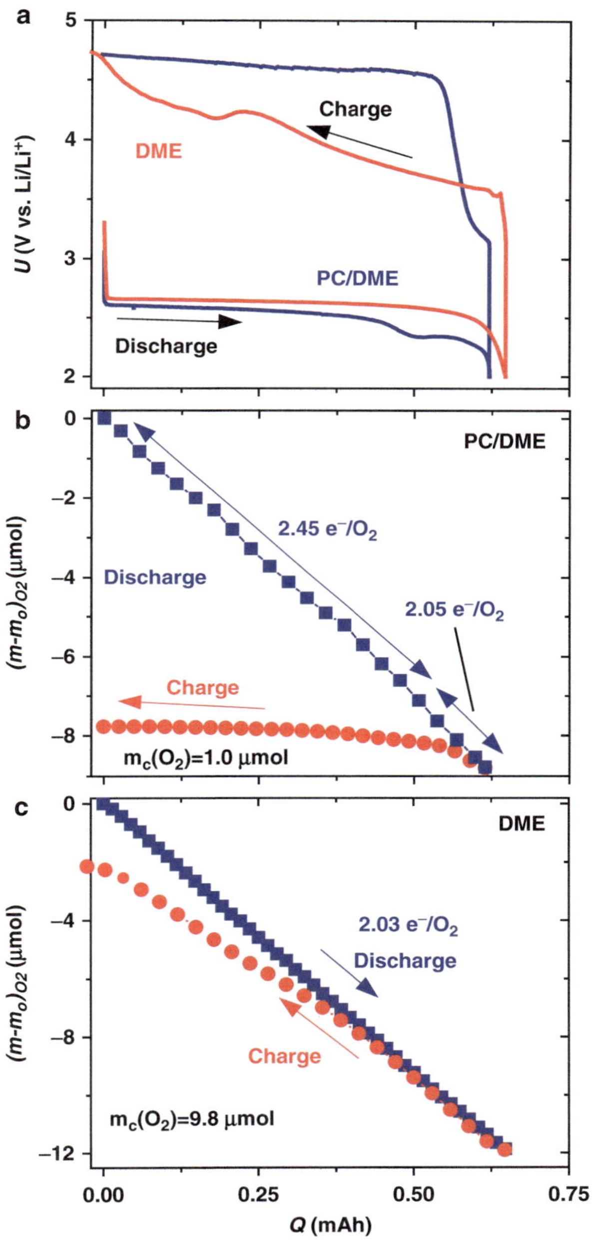

Fig. 3.13 (**a**) Galvanostatic discharge–charge curves at 250 mA/g_C (~1.5 mg XC72 typical loading rate) for cell employing 1 M LiTFSI in DME and 1:1 (vol) PC/DME as the electrolyte, (**b**) O_2 consumption and evolution during discharge–charge of the PC/DME-based cell; 10 μmol CO_2 was evolved during charge, (**c**) O_2 consumption and evolution during discharge–charge of the DME-based cell; 0.4 μmol CO_2 was evolved during charge. *PC* propylene carbonate, *DME* 1,2-dimethoxyethane

various contributions to this rechargeability loss purely from DEMS (current efforts in our laboratory are focused on trying to find accurate methods for additionally measuring $n(Li_2O_2)$ that would allow such characterization), we can suggest some qualitative arguments as to the dominant problems for the various electrolytes.

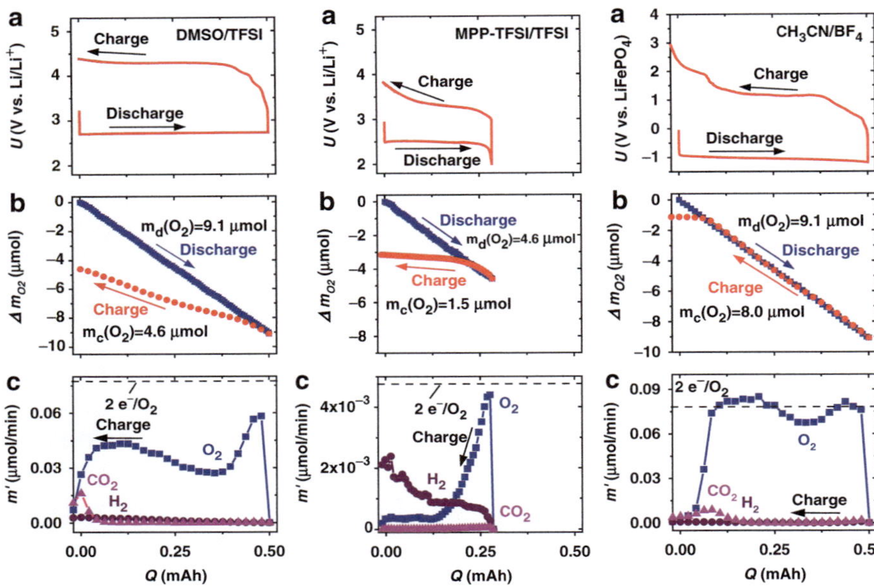

Fig. 3.14 (**a**) Galvanostatic discharge–charge curves for cells employing 1 M LiTFSI in DMSO, 0.2 M LiTFSI in MPP-TFSI, and 1 M $LiBF_4$ in CH_3CN. P50 was used as the cathode for all cells, Li metal was the anode in the DMSO and MPP-TFSI cells, and $LiFePO_4$ was the anode in the CH_3CN cell. Different rates were used for each cell: DMSO (200 μA discharge–charge), MPP-TFSI (15 μA discharge–charge), and CH_3CN (500 μA discharge, 250 μA charge). Cells were discharged under O_2 and charged under Ar. (**b**) Oxygen consumption (measured using pressure decay) during discharge and evolution (measured using DEMS) during charge. (**c**) Gas evolution rates for H_2, CO_2, and O_2 (these were the only gases found to evolve during charge) during cell charge. From [35] with permission

With the exception of BOB Li^+ salt, all OER/ORR results with DME as the solvent are nearly independent of the exact identity of the Li^+ salt used. We therefore suggest that any problems with the other salts are minimal or at least constant and focused on the dependence of rechargeability on solvent using LiTFSI or $LiBF_4$ as the salt. The ethers, which show the best rechargeability for electrolytes stable to Li metal, show an initial $(e^-/O_2)_{chg} \sim 2.0$ and OER/ORR ~ 1 at low charging potentials, both of which deviate significantly from the ideal Li–O_2 values at higher charging potentials [35]. This suggests that the ethers are reasonably chemically stable to Li_2O_2 but that electrochemical stability issues leading to an OER/ORR < 1 occur at high potentials during charging. Our findings on the stability of ethers are consistent with Xu et al., who found that Li_2O_2 was predominantly formed during discharge in an ether [59]. Some small amount of ether decomposition does occur on discharge, which can be ascribed to a chemical reaction between Li_2O_2 and the ether (most likely αβ-hydrogen abstraction followed by chain scission) [11, 34, 49]. However, as stated earlier, there is still no reliable simultaneous measure of $n(Li_2O_2)$ and ORR/OER for the ethers that would separate the two principal parasitic processes.

Table 3.1 Summary of DEMS results for various salt and solvent combinations

Cathode	Solvent[a]	Li^+ salt[b]	OER\ORR[c]	$(e^-/O_2)_{dis}$	$(e^-/o_2)_{chg}$	CO_2/ORR	H_2/ORR
XC72	DME	TFSI	0.78	2.01	2.59	0.07	0.03
		Trif	0.74	2.00	2.71	0.05	0.08
		ClO_4	0.77	2.00	2.59	0.05	0.08
		BF_4	0.78	2.06	2.65	0.04	0.08
		BOB	0.36	2.33	6.41	1.26	0.01
P50	MPP-TFSI	TFSI	0.33	2.30	7.04	0.01	0.28
P50	DMSO		0.51	2.05	4.05	0.03	0.02
XC72	1NM3		0.48	2.14	4.44	0.11	0.04
XC72	NMP		0.58	1.96	3.35	0.03	0.02
XC72	THF		0.72	2.01	2.80	0.03	0.09
XC72	DME		0.78	2.01	2.59	0.06	0.01
P50	CH_3CN[d]	BF_4	0.88	2.05	2.33	0.04	0.01
XC72	Triglyme		0.75	2.04	2.71	0.03	0.08

From [35] with permission
[a]*DME* dimethoxyethane, *THF* tetrahydrofuran, *TGE* triglyme, *CH_3CN* acetonitrile, *DMSO* dimethyl sulfoxide, *NMP* *N*-methyl pyrrolidone, *$1NM_3$* tri(ethylene glycol)-substituted trimethylsilane, *MPP-TFSI* *N*-methyl-*N*-propylpiperidinium bis-(trifluoromethylsulfonyl) imide
[b]*LiTFSI* lithium bis-(trifluoromethane sulfonyl) imide, *$LiBF_4$* lithium tetrafluoroborate, *LiTrif* lithium triflate, *LiBOB* lithium bis(oxalato) borate, *$LiClO_4$* lithium perchlorate
[c]Ideal values: OER/ORR = 1, $(e^-/O_2)_{dis} = 2$, $(e^-/O_2)_{chg} = 2$, $CO_2/ORR = 0$, $H_2/ORR = 0$
[d]Experiment performed using lithium iron phosphate ($LiFePO_4$) as the anode. Otherwise, Li metal was used as the anode for all experiments

In contrast, DMSO and NMP exhibit an $(e^-/O_2)_{dis} \sim 2$, but OER/ORR $\ll 1$ and $(e^-/O_2)_{chg} \gg 2.0$ from almost the outset of charging [35]. We believe that some thermal chemistry with the Li_2O_2 during its formation occurs for both. In the case of DMSO, dimethylsulfone is readily observed by NMR in the electrolyte after discharge. There is also likely parasitic electrochemical reaction of DMSO and NMP (or their chemical decomposition products) with Li_2O_2 during charging. These results are consistent with Xu et al. (DMSO) and Chen et al. (NMP), who also saw significant decomposition of DMSO and NMP during Li–O_2 discharge [6, 59]. Peng et al. reported that DMSO is highly stable when using a pure nanoporous gold electrode instead of carbon [45]. However, ongoing attempts to reproduce these results in our laboratory have not proven successful (see Sect. 3.4).

For the ionic liquid MPP-TFSI electrolyte, parasitic electrochemistry occurs during discharge, as exhibited by its $(e^-/O_2)_{dis} > 2$ and discharge gas evolution. We suggest from the abundant H_2 evolution (which is consistent with results from Mizuno et al. [41]) that Li_2O_2 extracts protons from the MPP-TFSI structure. It should be noted that MPP-TFSI is chemically stable to Li metal because no gases are evolved and no discoloration of the solution occurs without Li–O_2 discharge. 1NM3, LiBOB salt, and previously studied carbonate solvents also show $(e^-/O_2)_{dis} > 2$, and because of this parasitic electrochemistry, during discharge these electrolytes exhibit very poor chemical rechargeability.

From our DEMS results, we conclude that stability of organic solvents in the presence of Li–O_2 electrochemistry is tightly constrained by two competing molecular properties, namely, susceptibility to Li_2O_2 nucleophilic attack (and hydrogen abstraction) during discharge and electrochemical oxidative stability in the presence of Li_2O_2 during charge. Because Li_2O_2 is a strong nucleophile, a truly stable solvent must resist nucleophilic attack. This requirement has been studied for a variety of typical solvents by Black et al. [3] and Takechi et al. [50] via chemical reactions between KO_2, Li_2O_2, and LiOH and by Bryantsev using DFT calculations [5]. However, we wish to stress again that because $(e^-/O_2)_{dis} \approx 2.0$ for most solvents, the nucleophile responsible for the thermal chemistry must be Li_2O_2 and not some nucleophile formed by a one-electron electrochemical reduction (O_2^- or LiO_2) as often discussed in the literature. Stability to KO_2 or O_2^- must simply be regarded as proxies for stability to nucleophilic attack by Li_2O_2. We believe that Li_2O_2 is in fact a stronger nucleophile than either because of the local O_2^{2-} character of the solid. In addition to resisting nucleophilic attack, the solvent must be stable to electrochemical oxidation in the presence of Li_2O_2. This requires a very high oxidation potential (U_{ox}). For reasons that are not particularly well understood, the oxidation potential of the solvent in the presence of Li_2O_2 is unfortunately ~0.5 V lower than on an inert electrode (C or Au).

These two requirements (stability to nucleophilic attack and electrochemical oxidative stability) are, in general, inversely correlated (see below). In addition, any successful solvent must fulfill many other requirements as well: high Li^+ mobility and O_2 solubility and diffusion, the formation of a stable solid-electrolyte interface on Li (if no Li-protecting membrane is used), low vapor pressure, and so on. However, given the richness of organic chemistry, finding suitable compromises to the various constraints should eventually be possible.

3.3.2 *Cyclability*

Many reports of battery cyclability using a given electrolyte are based exclusively on Coulometry without being coupled to measures of O_2 consumption and evolution. These follow the loss of discharge capacity with repeated discharge–charge cycles or limit the discharge capacity of the cell to enable extended cycle life. This approach simply assumes that Li_2O_2 formation and O_2 evolution from Li_2O_2 are the dominant electrochemical processes occurring, with secondary parasitic processes contributing to any losses in cyclability. Often reports of rechargeability also include qualitative XRD evidence of Li_2O_2 formation and dissolution after a few discharge/charge cycles without any quantitative characterization of the discharge/charge processes. Most of these reports are simply measuring the amount of excess electrolyte present in the cell that is being consumed. We cannot stress enough that in order to truly measure the rechargeability of Li–O_2 cells, Coulometry has to be coupled to *quantitative* measurements of chemical processes occurring during battery operation.

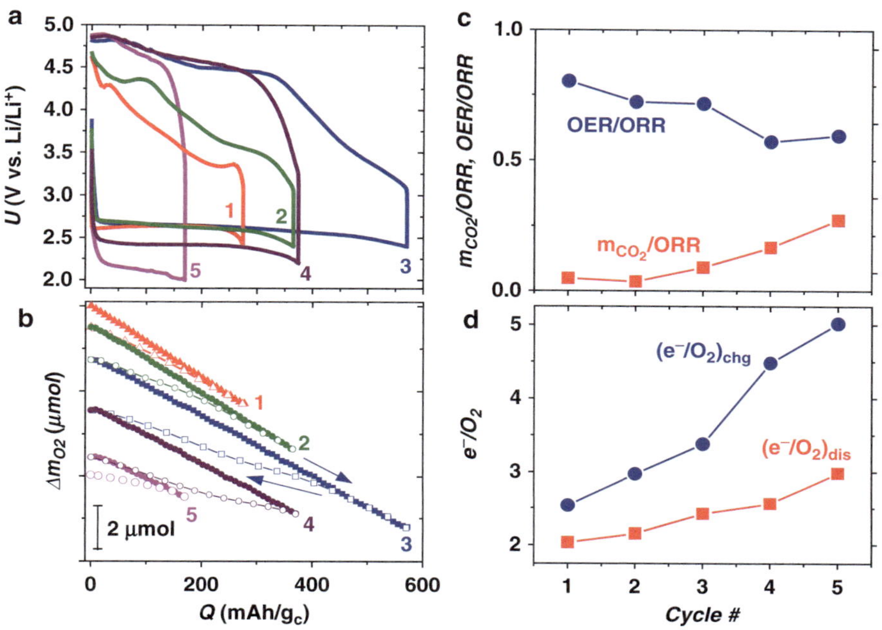

Fig. 3.15 (**a**) Five galvanostatic discharge–charge cycles for a DME-based cell (1 M LiTFSI, 400 mA/g XC72, discharge under O_2, charge under Ar). (**b**) Oxygen consumption during discharge (measured using pressure decay) and evolution (measured using DEMS) during charge for the five cycles. (**c**) OER/ORR and CO_2 evolved during charge (m_{CO_2}) divided by ORR. (**d**) e^-/O_2 for discharge and charge of each cycle. From [35] with permission

Since we do not find by DEMS that any electrolyte gives an OER/ORR > 0.9, we do not believe any system is truly cyclable on a C cathode. As an example, Fig. 3.15 shows cycling behavior coupled to DEMS measurements for a LiTFSI/DME electrolyte and C cathode. All quantitative measures of the electrochemistry deviate more from the ideal Li–O_2 ones with cycle number from the outset, particularly OER/ORR. This clearly indicates that significant decomposition occurs during a discharge–charge cycle for this cell. We show in Sect. 3.4 that the C used as a cathode is also not fully stable during the discharge–charge cycle, so that part of the stability issue is the C cathode and part is the electrolyte. Ottakam Thotiyl et al. also demonstrate some decomposition of the C cathode [44]. In fact they suggest that carbonate buildup on the cathode with cycling is the principal reason for the eventual loss in capacity. However, even with the clear DEMS evidence of decomposition processes occurring in the cell from the outset of cycling, the Coulometric capacity for discharge initially increases before decreasing. Clearly, pure Coulometry is a poor measure of rechargeability. A battery could appear rechargeable by Coulometry as long as the deposited products are oxidized off the cathode during the charge. However, any electrolyte decomposition must ultimately lead to depletion of the electrolyte and buildup of contamination products. If there is large excess of electrolyte, then the Li–O_2 cell may appear rechargeable even when it is not. Also, a large excess of electrolyte is totally unacceptable if Li–O_2 is to maintain its specific energy advantage over Li–ion batteries.

3.3.3 Theoretical Aspects of Electrolyte Stability

As discussed previously, electrolytes have many (possibly conflicting) requirements. Some of the most obvious requirements for a solvent are as follows:

1. Solvent must have sufficient electrochemical stability in the Li–O_2 potential window.
2. Solvent must be chemically stable to Li_2O_2.
3. Solvent must have sufficient Li^+ ionic solubility and give appreciable Li^+ mobility.
4. Ideally, solvent must form a stable SEI with Li metal.
5. Solvent must have sufficient O_2 solubility and diffusion.

As discussed previously, organic chemistry is a very rich field and there are almost infinitely many possible solvents, most of which are not commercially available or even well-known chemicals. Since synthesis of each possible solvent for testing in Li–O_2 batteries is a time- and labor-consuming process, it is essential to develop some facile theoretical predictors for their acceptability in terms of the criteria 1–5. In this section, we briefly sketch how quantum chemistry can contribute to our understanding of the first three items, with the ultimate aim of trying to develop theoretical predictors for solvent electrochemical and chemical stability, as well as Li^+ solubility. This effort, which is still in its infancy, grew out of the attempts at IBM to synthesize unknown potentially more stable solvents for lithium air.

Figure 3.16 gives an intuitive picture of possible solvent redox processes at an electrode. Oxidation occurs when the solvent loses an electron and reduction occurs when the solvent acquires an extra electron. Koopmans' theorem provides an intuitive physical interpretation, associating the oxidation (=anode) process with the ionization potential and the reduction (=cathode) process with the electron affinity. However, Koopmans' theorem already being a crude approximation in vacuum totally breaks down in the condensed phase (where the electrochemistry occurs).

Computational quantum chemistry lets you look at many of these aspects separately and thus allows a more thorough understanding of the underlying chemistry of the charge transfer reactions. In the following, we assume that the reader has some basic knowledge of quantum chemistry and we will simply discuss the basic features of the calculations. However, the details in the level of calculations and the basic acronyms for the techniques can be ascertained by talking to a quantum chemist friend. The properties of the liquid solvent employed in the lithium air batteries are modeled computationally using state-of-the-art density functional theory, DFT (e.g., DFT-B3LYP + D3, DFT-M11), using large basis

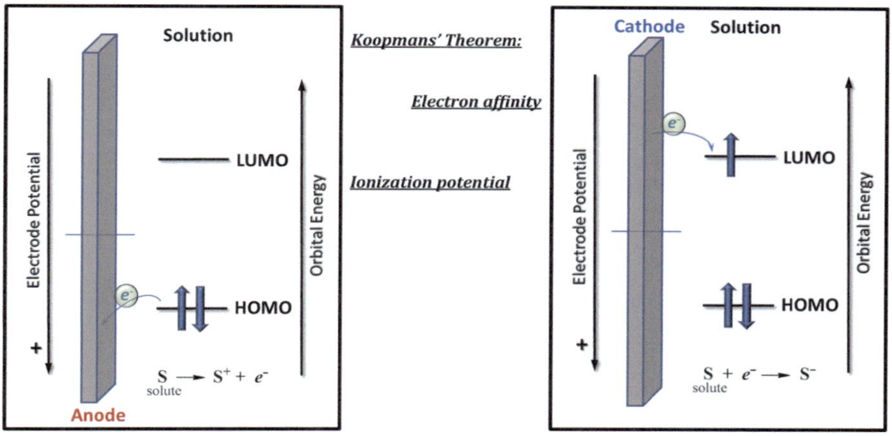

Fig. 3.16 An intuitive picture of the redox process

Fig. 3.17 Polarizable continuum model (PCM)

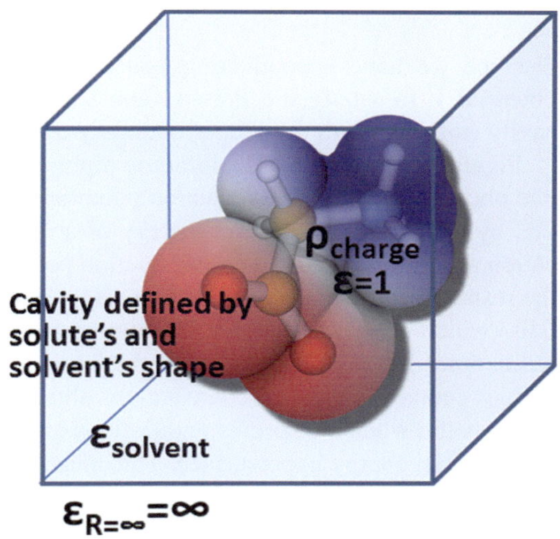

sets in a continuum representation of the solvent molecule's bulk liquid utilizing PCM ("polarizable continuum model," PCM-SMD) techniques. The DFT approach is a computationally economical means to approximately incorporate electron correlation into molecular orbital theory. The PCM approach is a popular way for modeling reactions in solution, i.e., it gives the corrections for the reaction in solution relative to that calculated by DFT for the gas phase. The underlying idea is depicted in Fig. 3.17 and outlined as follows: Embed a solute molecule inside a cavity defined by the solute molecule and the solvent molecule shapes. Inside the solute molecule (having a charge density ρ as determined by DFT), a dielectric of 1 is assumed, and outside the cavity the dielectric is that of the solvent; at infinite

$$\text{Cation} \xleftarrow[\Delta G^{ox}]{-\,e^-} \text{Neutral} \xrightarrow[\Delta G^{red}]{+\,e^-} \text{Anion}$$

$$\Delta G^{ox} = G^{cation} - G^{neutral} + G^{e} \qquad \Delta G^{red} = G^{anion} - G^{neutral} - G^{e}$$

Fig. 3.18 A stoichiometric representation of an oxidation and reduction process. Note that G^{e} is the (unknown) free energy of a solvated electron and is typically neglected; also note that G^{e} terms cancel for the operating window

Table 3.2 Calculated redox potentials of various solvents

Solvent	Dielectric[a]	Dipole moment [Debye]	Oxidation potential ϕ_{ox} [V]	Reduction potential ϕ_{red} [V]	Operating window $\phi_{ox} + \phi_{red}$ [V]
DME	7.2	2.24	7.03	−0.69	6.3
DMSO	47.2	5.79	6.40	−0.79	5.7
CH_3CN	36.6	5.13	9.21	−1.38	7.8
PC	66	11.01	8.31	−1.50	6.8

[a]Experimental values

distance we have a conducting boundary ($\varepsilon = \infty$). Solve for the electrostatic potential V to satisfy the Poisson and Laplace equations inside and outside the cavity (together with the proper boundary conditions).

Figure 3.18 shows a stoichiometric representation of the redox process and how one obtains oxidation and reduction potentials computationally using such models, i.e., by calculation of the free energy of a neutral molecule and of its respective cation and anion. The oxidation/reduction potentials are then simply obtained from the respective free energy differences. Note that, unlike experimental redox potentials (which are always measured relative to some standard electrode), computed redox potentials are absolute. Also, the free energies of the above neutral and ionic species contain vibrational contributions obtained from a vibrational analysis which is only valid when the species in question is at a stationary point (i.e., minimum) of the potential energy hypersurface. This means that the species' geometries must be relaxed. Any form of dissociation the cation and/or anion during the relaxation is an indication that the ion is unstable and spontaneously dissociates. In addition, low oxidation and/or reduction potentials may be indicative for the neutral species' instability towards chemical oxidation or reduction processes, although this correlation is less definitive.

In the following we present a few computational results for redox potentials obtained for four of the solvents discussed earlier; DME, DMSO, CH_3CN, and PC.

The results using the computational protocol described above are shown in Table 3.2.

Low oxidation potentials indicate that the corresponding solvent is prone to electrochemical oxidation and may be easily chemically oxidized as well (e.g., DMSO is readily oxidized chemically to dimethylsulfone). Large reduction potentials indicate that the corresponding solvent is easily reduced, both electrochemically and again possibly chemically. These calculations show that CH_3CN is the most stable solvent to oxidation, in good agreement with the DEMS (and other electrochemical oxidation) experiments. They also show that DME is more stable to oxidation than DMSO, again in agreement with experiment. This also suggests that PC is easily "reduced," although it is fairly stable to oxidation.

Although the redox potentials do determine the electrochemical windows of the solvent (on an inert electrode), they do not uniquely determine the chemical stability of a solvent to Li_2O_2. This is a difficult quantum chemical problem requiring the calculation of many barriers to possible chemical reactions of Li_2O_2 solid with different sites on the solvent, e.g., hydrogen abstraction or nucleophilic attack. One such calculation has been reported for PC [23]. Others look at the barriers for O_2^- reaction with the solvent as a measure of the chemical stability [4, 5]. However, as pointed earlier, O_2^- must only be thought of as a proxy for Li_2O_2 solid in order to study stability of the solvent to nucleophilic attack since there is no O_2^- formed in solution (see Sect. 3.6.1). To survey all possible chemical paths and extract barriers on a wide variety of solvents is too computationally demanding to use as a screening tool for possible solvents. Our approach instead is to try and find simple computational parameters that can give some insight into solvent stability, and the first of these are ϕ_{ox} and ϕ_{red} that define the electrochemical stability, and to use these parameters and others that can be easily derived from them, e.g., chemical potential $\mu = ½(\phi_{ox} - \phi_{red})$, chemical hardness $\eta = ½(\phi_{ox} + \phi_{red})$, or electrophilicity index $\omega = \mu^2/\eta$ to try to gain insight into chemical stability. While this effort is still in its infancy, the hope is that by combining insight from the calculations with DEMS measurement of OER/ORR, we can learn to design a new stable solvent.

Some lessons that have been learned so far are as follows:

- The presence of low-lying hyper-valent lone pairs (as in DMSO) causes low oxidation potentials.
- Fluorination raises the oxidation potential but lowers the reduction potential (i.e., makes it more negative); this can be beneficial (wider operating window) if applied in moderation.
- The presence of multiple bonds, which are associated with the presence of low-lying antibonding π-orbitals (as in olefins, amides, nitriles), makes solvent molecules more prone to reduction (e.g., CH_3CN, DMA) and causes lower (i.e., more negative) reduction potentials.

Li-salt solubility is another key requirement and not fulfilled for some of the "new" solvents originally synthesized. In order to provide some predictions from quantum chemistry, we have found a good correlation between the measured solubility and average solvent-Li^+ bond strength for Li^+ complexes with 1, 2, or 3 solvent molecules.

3.4 Cathode Stability

Carbon paper or carbon blacks bound to a metal current collector using PVDF, PTFE, or lithiated Nafion have typically been used as the porous cathode for most Li–O_2 battery studies to date because of carbon's low cost and high surface area. Many studies also incorporate small amounts of metal or metal oxide additives as "catalysts" for the ORR and OER reactions. Our view of the electrocatalysis issue is discussed in Sect. 3.6.4. Since there are stability issues with the electrolytes, it is necessary to also consider stability issues with all of the cathode components as well.

The stability of carbon to the highly oxidizing environment during charge is obviously questionable from well-known experiences in fuel cells. In addition, Li_2O_2 can chemically (or electrochemically) react with carbon via highly exothermic reactions [34]:

$$Li_2O_2 + C + \frac{1}{2}O_2 \rightarrow Li_2CO_3 \quad \Delta G = -542 \text{ kJ/mol}$$

$$2Li_2O_2 + C \rightarrow Li_2O + Li_2CO_3 \quad \Delta G = -533 \text{ kJ/mol}$$

where ΔG is the free energy change at 300 K. If the activation energy of these reactions are small, a thin layer (~no more than a few monolayers) of Li_2CO_3 and possibly Li_2O forms at the cathode/Li_2O_2 interface. The reaction between Li_2O_2 and C will self-limit, as already formed Li_2CO_3 blocks further access of solid Li_2O_2 from the C surface. The Li_2CO_3 formed from this parasitic reaction should be electrochemically decomposed to evolve CO_2 at high potentials (>4 V) during charge [34], although this process may not be fully complete for high-capacity discharge–charge cycles if the cathode is not kept for a long time at the high potential [13].

To understand the reaction between the carbon cathode and Li_2O_2, we used DEMS and carbon and oxygen isotope labeling. Figure 3.19a shows a galvanostatic discharge–charge cycle using a cathode of 99 % ^{13}C. Similar discharge–charge potential profiles were observed with other forms of C [36]. Figure 3.19b shows the gases evolved during cell charge. Both $^{13}CO_2$ and $^{12}CO_2$ are evolved in the 4–4.5 V peak, implying that solid carbonate deposits come both from reaction of Li_2O_2 with the C cathode ($^{13}CO_2$) and from decomposition of the electrolyte ($^{12}CO_2$). Roughly ~40 % of the CO_2 evolved in the high potential peak at the end of charging is $^{13}CO_2$. In addition, some $^{12}CO_2$ is evolved between 3 and 4 V. Figure 3.19c shows the isotopic distribution of CO_2 evolved from the ^{13}C cathode during an oxidative potential scan following a 1 mAh (200 μA) discharge under $^{18}O_2$. Both the $^{13}CO_2$ and $^{12}CO_2$ evolved during the peak at 4–4.5 V exhibit scrambled oxygen isotopic mixtures, in a similar manner to that observed previously during charging following a discharge in a carbonate electrolyte (where solid carbonates or carboxylates are known to dominate the deposits) [34]. This CO_2 evolution is therefore due to the electrochemical oxidation of Li_2CO_3 (or $LiRCO_3$) produced on the cathode during the discharge–charge. The CO_2 evolution that mirrors the electrochemical current

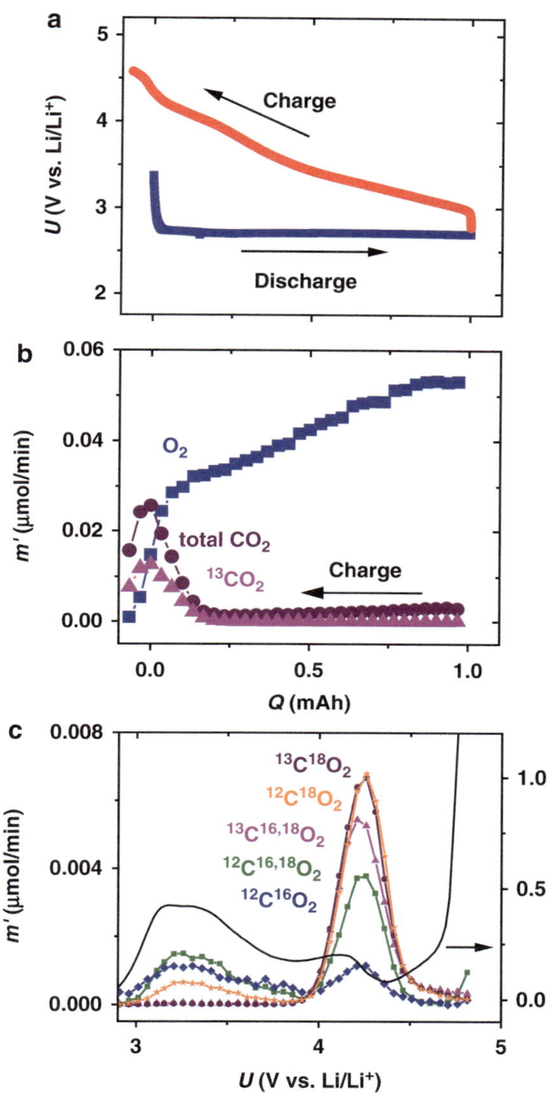

Fig. 3.19 (**a**) Potential (U) vs. Li/Li$^+$ vs. discharge capacity Q for Li–O_2 galvanostatic discharge and charge at 200 μA current on a ^{13}C cathode. (**b**) Molar evolution rates m' for O_2, $^{13}CO_2$, and $^{12}CO_2$ + $^{13}CO_2$ during charge as measured by the quantitative DEMS for the same discharge–charge in (**a**). (**c**) Quantitative molar evolution rates m' of various CO_2 isotopes as a function of charging potential U following 1 mAh discharge at 200 μA current with 1.2 bar $^{18}O_2$ and a ^{13}C cathode. The potential U was scanned up from the equilibrium potential $U_0 = 2.85$ V at a rate of 0.5 mV/s. The *solid line* is the current. From [38] with permission

(and O_2 evolution, not shown) between 3 and 4 V is <2 % of the O_2 OER and is composed almost exclusively of $^{12}CO_2$, indicating that it is evolved from electrolyte decomposition. Since there is no measurable electrochemical electrolyte oxidation between 3 and 4 V under O_2 and in the absence of Li_2O_2 [38], we believe that this CO_2 represents some small parasitic electrochemical reaction of Li_2O_2 with the electrolyte during charging.

We previously suggested that a thin layer of Li_2CO_3 forms on discharge due to the highly exothermic nature of the thermal chemical reaction between Li_2O_2 and carbon (and its observation in XPS on discharge) [38]. However, chemical titration studies by Ottakam Thotiyl et al. [44] suggest that the reaction between C and Li_2O_2 actually occurs at low charging potentials, $U > 3.5$ V. Since the DEMS results only measure the results following discharge–charge, this is certainly possible, and in our laboratory we have also recently confirmed that the reaction with C is mostly during charging. However, the net problem is the same regardless of the details of when the carbonate is formed; Li_2CO_3 or carboxylates are formed on the C during battery discharge–charge and this represents a loss in rechargeability.

Along with carbon, the stability of the electrode binder, particularly PVDF, has also been called into question. PVDF instability in the presence of Li_2O_2 was initially called into question by Xu et al. [60], who visually observed a chemical reaction when mixing Li_2O_2 powder into a NMP/PVDF solution. Black et al. [3] published a more detailed analysis of this reaction. We have also observed the decomposition of PVDF in the presence of Li_2O_2, although only when moisture is present. Reaction products of Li_2O_2 and PVDF after discharge in the presence of some moisture are clearly observed using Raman as two new unassigned peaks. However, these peaks do not appear if significant care is taken to eliminate H_2O contamination from all cell components. Furthermore, no significant difference in O_2 consumption or evolution during discharge and charge, respectively, is observed between cells that employ pure C cathodes compared to using PTFE or PVDF as a cathode binder. We have not observed decomposition of the PTFE binder (via XRD, Raman, NMR, or FTIR of discharged cathodes), although we have not made any systematic effort to set lower limits on possible decomposition for either binder. Because of the possible issue with binder stability, some authors [10, 13] resort to using binderless C cathodes, e.g., VACNT. However, our experience is that the stability of C itself to Li_2O_2 is a much more severe limitation than any possible small parasitic reaction with the binder when all cell components are sufficiently free of H_2O.

Because of the chemical and electrochemical reactions of C with Li_2O_2, porous C by itself will never form a stable cathode, and there are now many efforts to find other suitable cathode materials or coatings. Recently Peng et al. [45] have reported that a Li–O_2 electrochemical cell comprising a 100 nm thick nanoporous gold cathode and DMSO as the electrolyte solvent is able to sustain reversible cycling for 100 cycles with only ~5 % loss in capacity. Encouraged by this result, we have also tried DEMS experiments on the nanoporous Au/DMSO system. We also use different nanoporous Au thicknesses (150 nm–15 μm) to probe the efficacy at higher discharge capacities. The first ~10 discharge–charge cycles have low charging overpotentials similar to those of Peng et al., but later cycles require a significantly higher charging potential. In addition, all our experiments using high-capacity discharges (~0.2 mAh) have resulted in low OER/ORR ratios of <50 % as well as the limited cycling. We believe our limitations are due to the formation of a soluble electrolyte decomposition product, dimethylsulfone, which was readily detected by NMR in the extracted electrolyte. Therefore, we are not yet able to confirm the high rechargeability of the Au/DMSO combination. Our inability to reproduce the claims of Peng et al. should in no way be taken as evidence that their result is necessarily invalid. However, we simply have not yet been able to reproduce it and do not fully understand how the Au/DMSO system could circumvent the known electrolyte instability of DMSO to Li_2O_2. Certainly, the use of Au rather than C overcomes the issue of carbonate formation by reaction of Li_2O_2 with C. Since the electrolyte decomposition product dimethylsulfone is soluble rather than depositing at the Li_2O_2–electrolyte interface as carbonate or carboxylates do with other solvents, a low charging potential U_{chg} is expected on each cycle until significant electrolyte decomposition occurs to change the whole electrochemistry to one with higher overpotentials (see Sect. 3.2.1.2).

3.5 Role of Atmospheric Contaminants in Lithium Air

It is likely that O_2 will need to be harvested from air in a practical lithium air battery. Most experiments to date on rechargeable nonaqueous lithium air batteries have used pure O_2 rather than air in order to avoid unwanted parasitic reactions with air contaminants, e.g., H_2O, CO_2, etc., that may obscure or confuse the early studies on the basic Li–O_2 electrochemistry. Only a few studies exist on the effects of contaminants on the Li–O_2 electrochemistry, e.g., both H_2O and CO_2 were found to increase the capacity of Li–O_2 batteries, with H_2O contamination having a much larger effect on the cell capacity than CO_2 [40, 51]. We explored the effect of both H_2O and CO_2 contamination on Li–O_2 battery chemistry using both DEMS and FTIR [16].

3.5.1 H_2O

Results for H_2O contamination in a DME/LiTFSI electrolyte are presented in Fig. 3.20. Here, small amounts of water were added to the electrolyte before assembling a cell (the "ppm" values given in the figures are the water concentrations in the prepared electrolytes; 12 ppm was the "pure" electrolyte H_2O concentration as measured by Karl Fischer titration). As observed elsewhere [40], the inclusion of even a small amount of H_2O (500 ppm) leads to a large increase in cell capacity (Fig. 3.20a). However, regardless of the amount of water added to the electrolyte (up to 25,000 ppm H_2O), ~2.05 $(e^-/O_2)_{dis}$ is observed using pressure decay measurements, which is nearly identical to the values $(e^-/O_2)_{dis}$ in pure DME/LiTFSI given in Table 3.1. This implies that even in the presence of minor fractions of H_2O in the solvent, Li_2O_2 formation is still the dominant discharge electrochemistry. However, cathodes extracted after discharge and observed using FTIR show qualitatively that the amount of Li_2O_2 on the cathode surface decreases

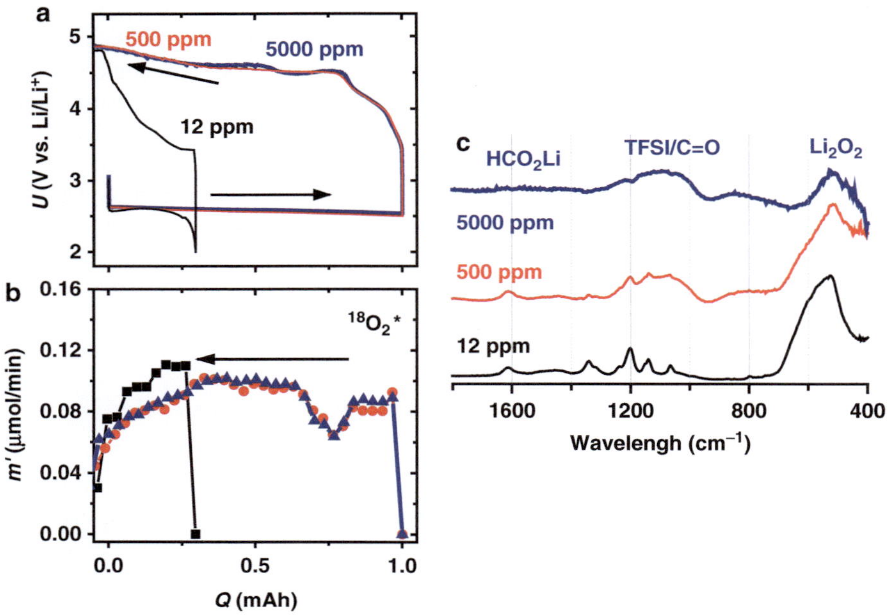

Fig. 3.20 (**a**) Galvanostatic discharge–charge curves for cells with 12, 500, and 5,000 ppm H_2O contamination in the electrolyte (XC72 carbon-based cathode, 1 M LiTFSI in DME electrolyte, $i = 0.47$ mA/cm^2, discharge under $^{18}O_2$, charge under Ar); (**b**) $^{18}O_2$ evolution for cells studies in (**a**) ($^{18}O_2$ was the only O_2 isotope evolved); (**c**) FTIR of extracted P50 cathodes discharged for 5 mAh in the H_2O-contaminated electrolytes

with increasing H_2O concentration (Fig. 3.20c). Therefore, Li_2O_2 reacts with H_2O to form solution soluble products (such as LiOH or H_2O_2). Since electrical passivation of the electrode surface by the Li_2O_2 insulator determines the discharge capacity (Sect. 3.7), its removal by chemical reaction leads to the observed increase in cell capacity. On charge, we monitor isotopic O_2 evolution after a discharge under $^{18}O_2$ (Fig. 3.20b) and using $H_2{}^{16}O$ as the contaminant. We observe only $^{18}O_2$ evolution, which indicates that O_2 evolution only occurs from Li_2O_2 formed during discharge and not from some other species formed during the reaction of H_2O with Li_2O_2 (such as LiOH). Furthermore, O_2 evolution predominantly occurs at very high potentials in electrolytes with significant amounts of H_2O, and this of course presents issues for electrolyte stability. The average O_2 evolution rate decreases at high H_2O concentrations. All of the above indicate that H_2O contamination significantly reduces the cell rechargeability. It should be noted that H_2 evolution occurred from all cells with H_2O concentrations >500 ppm because of the H_2O reaction with the Li anode. All pressure decay calculations to quantify e^-/O_2 values took into account this H_2 evolution.

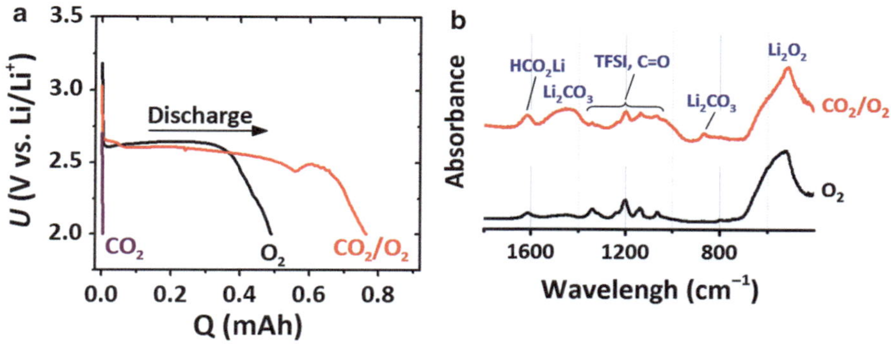

Fig. 3.21 (**a**) Galvanostatic discharge profiles (0.47 mA/cm^2) of Li cells discharged under three atmospheres: pure CO_2, pure O_2, and a 10:90 CO_2/O_2 mixture. XC72-based cathodes were used. (**b**) FTIR of cathodes extracted from cells discharged (0.9 mA/cm^2, 4.7 mAh/cm^2) under pure O_2 and a 10:90 CO_2/O_2 mixture. In the pure O_2 spectra, peaks at 1,064, 1,138, 1,202, and 1,340 cm^{-1} can all be attributed to the electrolyte salt (LiTFSI). P50 carbon paper was used as the cathode. From [16] with permission

3.5.2 CO_2

Carbon dioxide is known to react with lithium peroxide to form lithium carbonate and oxygen, i.e., $CO_2 + Li_2O_2 \rightarrow Li_2CO_3 + ½O_2$. Hence, any carbonate formation formed from CO_2 contamination in the feed gas could also impact battery performance in a similar fashion to the adverse effects of carbonates from electrolyte and C cathode instabilities.

The effect of CO_2 on Li–O_2 battery discharge chemistry was studied by discharging cells under three different atmospheres—pure CO_2, pure O_2, and a 10:90 CO_2/O_2 mixture (Fig. 3.21a). The cell discharged under a mixture of CO_2/O_2 has a similar discharge voltage plateau (~2.6 V) as the pure O_2 cell, although showing a larger discharge capacity. The 2.6 V voltage plateau during discharge has a ~2.0 $(e^-/O_2)_{dis}$, regardless of whether CO_2 contamination is present. This shows that the dominant electrochemistry during discharge is still Li_2O_2 formation even if some CO_2 is present. However, FTIR of discharged cathodes indicates that significant Li_2CO_3 deposits are formed during discharge under a CO_2/O_2 atm. Figure 3.21b compares FTIR spectra of cathodes discharged under a pure O_2 and a CO_2/O_2 atm. A large Li_2O_2 absorbance is observed in each case, with smaller amounts of HCO_2Li, Li_2CO_3, and LiTFSI also present. For the CO_2/O_2 discharge, Li_2CO_3 and HCO_2Li absorbance are significantly higher than in the pure O_2 case. These results confirm that some Li_2O_2 formed during discharge spontaneously reacts with CO_2 to form Li_2CO_3.

It is also interesting to note the higher capacity exhibited by the cell discharged under CO_2/O_2 compared to the cell discharged under pure O_2. The exact mechanism for the increased discharge capacity is not clear, but is consistent with another

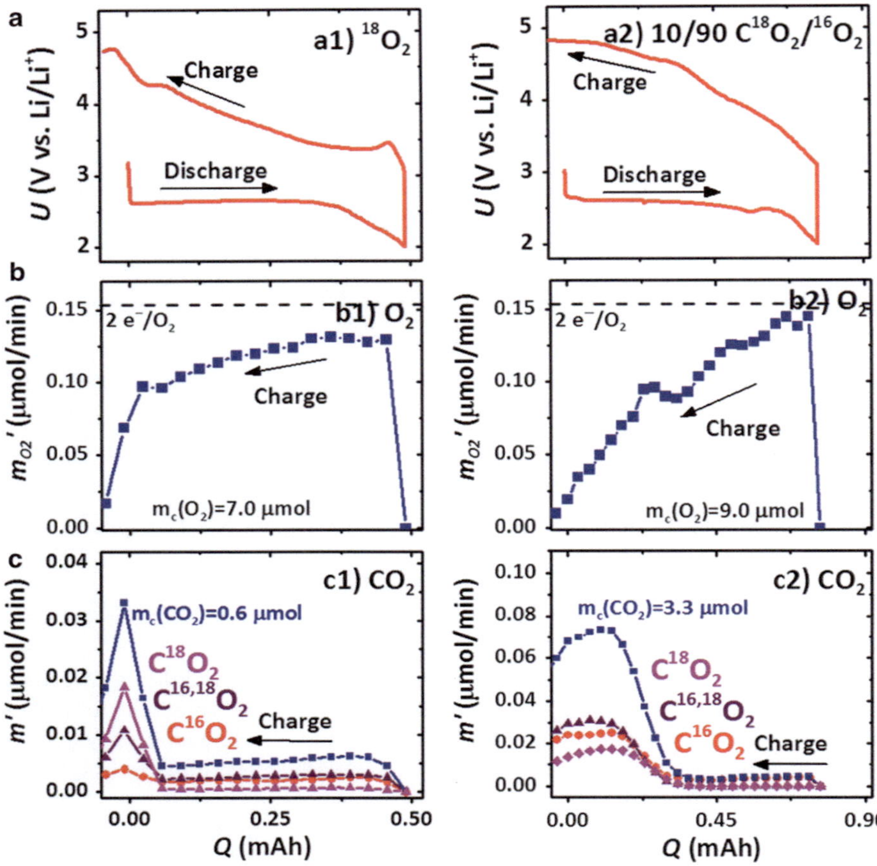

Fig. 3.22 Galvanostatic discharge and charge (0.47 mA/cm^2) of cells discharged under a pure $^{18}O_2$ headspace (**a1–c1**) and a 10:90 $C^{18}O_2/^{16}O_2$ mixture (**a2–c2**). The discharge profiles in the (**a**) are those compared in Fig. 3.21a. The (**b**) O_2 evolution rate (m'_{O_2}) during cell charge. Only a single O_2 isotope, namely, $^{18}O_2$ for **1b** and $^{16}O_2$ for **2b**, was evolved from each cell and corresponded to the O_2 isotope used in the discharge headspace. (**c**) isotopic CO_2 evolution rate (m') during cell charge. Cells were charged under an Ar headspace. From [16] with permission

study that showed that CO_2/O_2 mixtures exhibited higher capacities than under pure O_2 [51]. Perhaps the increase in capacity is related to morphology changes.

The spontaneous reaction between CO_2 and Li_2O_2 also affects the charge behavior of the electrochemical cell. The first galvanostatic discharge and charge cycle of cells discharged under pure O_2 and CO_2/O_2 are compared in Fig. 3.22. The more rapid rise in charging potential with charge capacity for the CO_2/O_2 cell is likely due to increased concentrations of Li_2CO_3 at the Li_2O_2–electrolyte interface (see Sect. 3.2.1.2.4). O_2 evolution at the onset of charge (3–3.5 V) for both cells is ~2.1–2.2 $(e^-/O_2)_{chg}$. Both cells also exhibit a decrease in O_2 generation as the charge proceeds, although the decrease is significantly faster in the CO_2/O_2 cell. CO_2 is also evolved in both cells at $U > 4.2$ V, with significantly more CO_2 evolved from the CO_2/O_2 cell. More CO_2 evolution and faster reduction in the rate of oxygen evolution also indicate the presence of higher Li_2CO_3 concentration compared to Li_2O_2, in the CO_2/O_2 cell cathode than the pure O_2 cell cathode.

In the cell discharged under CO_2/O_2, carbonate deposits are present not only from cathode and electrolyte decomposition but also from the thermal reaction between atmospheric CO_2 and Li_2O_2. By artificially creating a higher Li_2CO_3 concentration in the bulk electrodeposit via the reaction between Li_2O_2 and CO_2, we see that the observed charge overpotential does increase (Fig. 3.22a1 and 3.22a2). These results indicate that any carbonate or insoluble deposit present on the cathode, resulting from atmospheric contamination, will reduce the cell's voltaic cycle efficiency.

3.5.3 Prospects for Purifying Air

Clearly H_2O and CO_2 as well as other minor organic contaminants will be detrimental to a lithium air battery and must be removed from the battery air feed. One method by Zhang et al. envisioned to deliver O_2 or highly O_2-enriched gas from air to a Li–O_2 cell has been an "oxygen-diffusion" membrane that selectively permeates O_2 while retaining other gaseous species [62]. However, the incorporation of such a membrane seems unlikely for high-rate applications, as water vapor and CO_2 are more permeable than O_2 through any membrane, regardless of the membrane's physical properties (including its relative hydrophobicity) [33]. Since water molecules are much smaller than oxygen molecules, preferential O_2 permeability relative to H_2O through porous membranes, where gas separation is controlled by size exclusion mechanisms such as Knudsen diffusion and Poiseuille flow, is not possible. Additionally, gas separation through nonporous polymeric membranes, like those used by Zhang et al., is governed by the solution/diffusion mechanism, where permeability of any gas through a polymer is the product of its ability to dissolve into a polymer (i.e., its solubility) and its ability to diffuse through the polymer (i.e., its diffusion rate) [33]. In other words: $P = DS$, where P is a gas's permeability through a nonporous polymer film, D is its diffusion coefficient, and S is its solubility coefficient. A gas's diffusion coefficient through a polymer film is generally inversely correlated to its molecular size raised to a positive power, and therefore, water will always have a higher diffusion coefficient than O_2 through any membrane. Furthermore, an Arrhenius-type correlation exists between a gas's solubility coefficient and its condensability in any polymer [33]. As a result, water and CO_2 condensabilities are much higher than that of O_2 and will also be much more soluble than O_2 in any membrane. Therefore, water and CO_2 permeabilities through dense polymer films tend to be significantly higher (sometimes many orders of magnitude) than oxygen permeability. It seems

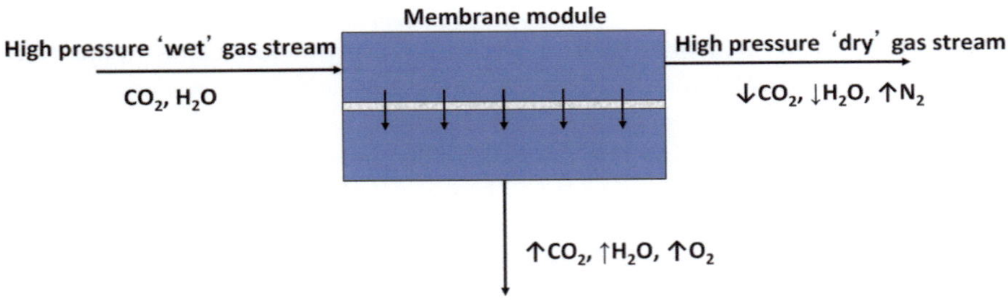

Fig. 3.23 Possible membrane system for removing H_2O and CO_2 for use in a lithium air battery. The *arrows* ↑↓ next to a gas refer to how its concentration changes in the gas stream on passing through the membrane system. The high-pressure "dry" gas stream must then be fed to a polishing step to remove residual H_2O and CO_2

therefore highly unlikely to ever be able to devise an oxygen-selective barrier membrane.

Nonetheless, by preferentially permeating H_2O and CO_2 instead of oxygen, membranes still hold promise to deliver highly oxygen-enriched, dehydrated, and nearly CO_2-free air for EV applications. For example, a possible configuration for air dehydration employs a hydrophilic membrane with a very high H_2O and CO_2 permeability compared to O_2 permeability [1]. As a result, O_2 is retained on the high-pressure side of the membrane as shown in Fig. 3.23. It seems likely that such a scheme could perhaps provide ~99 % O_2/H_2O selectivity and ~90 % O_2/CO_2 selectivity. It will then be necessary to add an additional "polishing" step, e.g., selective adsorption of H_2O, CO_2, etc., on activated zeolites as done routinely in catalytic cleaning processes [8], and it can then be sent to the battery. By combining the two-step cleaning process, it should be possible to obtain H_2O and CO_2 concentrations of ~0.1 ppm.

Of course, the size, weight, and cost of such a system are critical and unknown at this preengineering stage. Thus, although it is technically feasible to eliminate contaminates in an air-breathing system, it is still unclear whether this is practical if/when other technical issues confronting lithium air are solved.

3.6 Fundamental Electrochemistry

3.6.1 Mechanism

The mechanism that was proposed theoretically [18, 19] based on first principles thermodynamics to account for the electrochemistry is

$$2\text{Li} \leftrightarrow 2(\text{Li}^+ + e^-) \quad \text{(anode)} \tag{3.1}$$

$$Li^+ + e^- + O_2^* \leftrightarrow LiO_2^* \quad \text{(cathode)} \tag{3.2}$$

$$Li^+ + e^- + LiO_2^* \leftrightarrow Li_2O_2^* \quad \text{(cathode)} \tag{3.3}$$

where the * refers to a surface-adsorbed species. This mechanism is the Li^+ analogue of the conventional proton-induced charge transfer in the aqueous associative ORR/OER forming H_2O_2 [42, 53]. The fundamental difference to the aqueous case, however, is that the product and intermediate species (Li_2O_2*, LiO_2*) are both formed on the surface by the ion transfers and remain there permanently since these ionic species have essentially no solubility in nonaqueous electrolytes. NMR studies of the solubility of Li_2O_2 in DME show that $[Li^+]$ <1 ppm so that the solubility product $K_{sp} = [Li^+][LiO_2^-] \sim 10^{-10}$. Therefore, at $[Li^+] = 1$ M from the electrolyte, the Li_2O_2 solubility is miniscule. Several variants of the mechanism described by Eqs. (3.1)–(3.3) have also been suggested in the literature to be involved in the electrochemistry. For example, several authors [11, 27] suggest that reaction (Eq. 3.2) is in actuality composed of two time-separated solution steps (Eqs. 3.2a and 3.2b) instead of the concerted ion transfer surface mechanism implied in Eq. (3.2). In this case the formation of LiO_2* is viewed as a precipitation of the insoluble LiO_2. However, to our knowledge, there is no experimental evidence that O_2^- exists even transiently in solution in the presence of high Li^+ concentrations.

$$O_2 + e^- \rightarrow O_2^- \tag{3.2a}$$

$$O_2^- + Li^+ \rightarrow LiO_2^* \tag{3.2b}$$

Laoire et al. [24, 25] and Peng et al. [46] suggest that in addition to or instead of reaction (Eq. 3.3), the chemical disproportionation reaction (Eq. 3.4) occurs during discharge

$$LiO_2^* + LiO_2^* \rightarrow Li_2O_2^* + O_2 \tag{3.4}$$

Since the reverse of this mechanism cannot occur during charge (at least under charging with an Ar atmosphere), it has been suggested [46] [24, 25] that a simultaneous $2e^-$ direct charge reaction occurs by passing the LiO_2* intermediate.

$$Li_2O_2 \rightarrow 2(Li^+ + e^-) + O_2 \tag{3.5}$$

Other authors [27, 63] have suggested that net electrochemical reactions such as

$$4(Li^+ + e^-) + O_2 \rightarrow 2Li_2O^* \tag{3.6}$$

$$2(Li^+ + e^-) + Li_2O_2^* \rightarrow 2Li_2O^* \tag{3.7}$$

also contribute during deep discharge or with reactive metal catalysts.

In order to probe the mechanism (and kinetic overpotentials) in some detail, we have combined $Li–O_2$ discharge and charge experiments in the rapidly stirred bulk

Fig. 3.24 Cyclic voltammogram for Li–O_2 electrochemistry on GC in the bulk electrolysis cell with $P = 1$ bar O_2 and 1 M LiTFSI in DME as the electrolyte. From [38] with permission

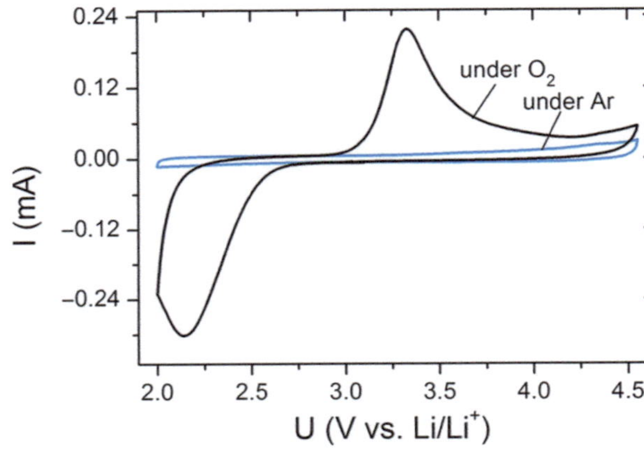

electrolysis cell of Fig. 3.2b with equivalent ones using the DEMS Swagelok™ cells of Fig. 3.2a. The three-electrode electrolysis cell measures the Li–O_2 current–voltage characteristics without any distortions due to cell *iR* drops and concentration polarizations (see Sect. 3.7 for a qualification), but this does not guarantee that all the current comes from the desired Li–O_2 electrochemistry.

Figure 3.24 shows a cyclic voltammogram (CV) for Li–O_2 in the bulk electrolysis cell. If the peaks can be identified as due to Li–O_2 ORR and OER, then the kinetic overpotentials must be quite small, since the onset of currents is quite close to the open-circuit potential of 2.85 V (when the surface is covered with Li_2O_2). The peaks in the CV occur when the rate of charge transport through the growing Li_2O_2 film decreases enough to overcome the kinetic charge transfer rate that is increasing with overpotential, i.e., that the surface becomes electrically passivated (see Sect. 3.7). Note that there are only two peaks, one for discharge and one for charge. Some authors [25] have assigned extra peaks in the CV to various one-electron processes. However, whenever extra peaks were observed in CVs by us, they have all been linked to the presence of impurities, e.g., H_2O, and ultimately disappear with cycling when the electrolyte and GC are sufficiently clean. In addition to the two peaks, there is a significant high-voltage tail in the OER portion of the Li–O_2 CV extending to U ~4.5 V. We have presented evidence that this high-voltage tail results from a roughly monolayer of carbonate forming at the Li_2O_2–electrolyte interface during charging and is not a fundamental kinetic overpotential [38]. The high-voltage tail creates significant problems for reversible cycling of a battery since there is significant electrochemical oxidation of the electrolyte at the higher U. Similar CV were obtained with other moderately stable electrolytes.

Figure 3.25a shows linear sweep voltammetry using the Swagelok™-type DEMS cell. In these experiments, the potential is swept slowly negative (0.5 mV/s) to ~2 V relative to Li^+/Li from the OCV under a ~1.2 bar O_2 headspace. After this cathodic sweep, the headspace is changed to ~1.2 bar Ar and the potential is swept slowly positive (0.5 mV/s) from $U_0 = 2.85$ V. The solid line is the current, while the points indicate either O_2 or CO_2 molar consumption/evolution during ORR/OER. The ratio

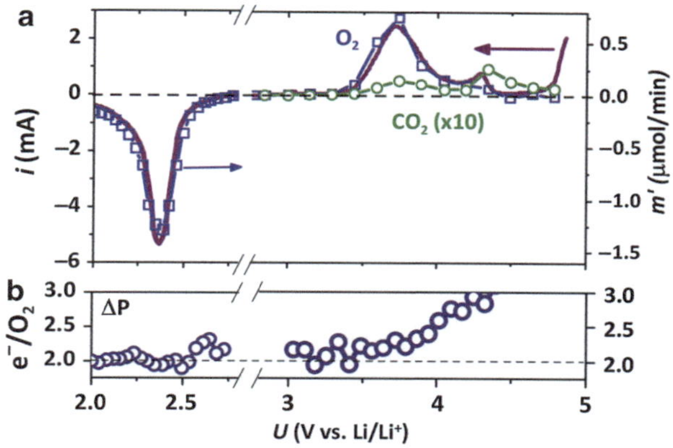

Fig. 3.25 Linear sweep voltammetry at 0.5 mV/s using the DEMS cell (XC72 carbon cathode, 1 M LiTFSI in DME). (**a**) Current is the solid line and the points are the O_2 (and CO_2) measured by the DEMS. There was very little gas evolution (<1 mol% of ORR) during the cathodic sweep. (**b**) e^-/O_2 determined by comparing the current to the moles of O_2 consumed or evolved (by measuring the pressure change ΔP in the isolated cell headspace). The *dashed line* is at the ideal $e^-/O_2 = 2.00$ when only $2Li + O_2 = Li_2O_2$ contributes to the electrochemistry. From [37] with permission

of current to O_2 change is given in Fig. 3.25b for a discharged/charged cell identical to that monitored by the DEMS of Fig. 3.25a. The dominant ORR and OER peaks are clearly related to O_2 consumption and production. Ideally, 2.00 electrons per O_2 would be involved both in ORR and OER if Li_2O_2 is the only species produced/consumed by any of the electrochemical mechanisms discussed above. Discharge does correspond to ~2.0 ± 0.1 e^-/O_2 throughout its potential range and this is consistent with the claim that Li_2O_2 formation is the dominant electrochemical process occurring during discharge. However, this does not by itself guarantee that none of the Li_2O_2 undergoes a thermal chemical reaction with the electrolyte once it is formed, as discussed in Sect. 3.3. Approximately 2.2 ± 0.1 e^-/O_2 is observed for charging up to a potential of ~3.8 V. However, for $U > 3.8$ V, the e^-/O_2 ratio exhibits a continuously increasing deviation from ~2.2. This implies that some electrochemically initiated reaction between the electrolyte and Li_2O_2 (or its OER intermediates) occurs from the outset but increases dramatically at the higher potentials during charging. O_2 evolution continues until ~4.5 V in the DEMS cell, indicating that some Li_2O_2 persists on the surface up to these high potentials. This result is similar to that observed in the bulk electrolysis cell CV, where an oxidative scan to 4.55 V was necessary to produce a fully cyclical CV [38]. Above ~4.8 V, electrolyte oxidation dominates the electrochemistry.

One important conclusion from Fig. 3.25b is that there is absolutely no potential for which a one-electron process dominates the electrochemistry of either discharge or charge within the time scale of the experiment (~10 s). This means that once LiO_2* is initially formed, it transforms to Li_2O_2 by either reaction (Eq. 3.3) or

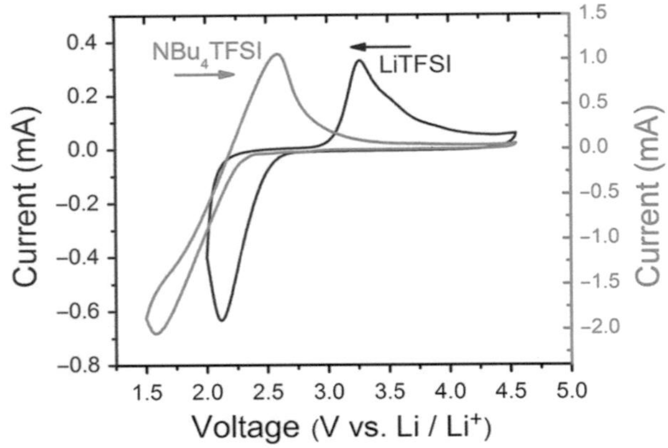

Fig. 3.26 Cyclic voltammogram (current i vs. voltage U) showing O_2 reduction and oxidation in 0.5 M NBu_4TFSI (*green, right ordinate*) in DME as the electrolyte and 1 M LiTFSI (*blue, left ordinate*) in DME as the electrolyte; 100 mV/s scan rate, ~1 bar O_2 pressure in the bulk electrolysis cell. From [37] with permission

(Eq. 3.4) rapidly and therefore can't be a thermodynamically stable species at any potential as suggested earlier [24, 25].

Figure 3.26 shows a CV for oxygen reduction reaction (ORR) and oxygen evolution reaction (OER) for 0.5 M NBu_4TFSI salt in DME compared to an equivalent CV experiment in 1 M LiTFSI in DME. The comparison in Fig. 3.26 is similar to some presented previously [24, 25, 46], but our conclusions from this result are quite different than those of the earlier work. It is well known that a solution-phase O_2^- is produced during ORR in NBu_4TFSI. However, the onset potential for ORR in LiTFSI is ~0.25 V higher than the onset potential for ORR in NBu_4TFSI. This means that it is easier to reduce O_2 in the presence of Li^+ and the thermodynamic stability of the LiO_2* intermediate formed must be greater than that of an O_2^- solution intermediate since $\Delta G = -eU$. We believe the comparison in Fig. 3.26 argues that the ORR in LiTFSI must involve the concerted process of Eq. (3.2) followed by Eq. (3.3) or (3.4) rather than time-separated processes as implied by Eqs. (3.2a) and (3.2b). If it was the latter, then the ORR in LiTFSI should occur at the same potentials as for NBu_4TFSi since Eq. (3.2b) is purely chemical and can't contribute to the potential generated by the electrochemistry. The free energy stabilization for a chemical step $Li^+ + O_2^- \rightarrow LiO_2^*$ (ΔG_{2b}) would have to appear as a kinetic overpotential.

Figure 3.27 shows the DEMS for a linear oxidative scan following discharge with a mixture of 50 % $^{16}O_2$ and 50 % $^{18}O_2$ in the headspace. At most, <2 % $^{16,18}O_2$ is produced during the charge and this confirms that almost no dissociation of the O_2 occurs during the discharge–charge process. Since reactions (Eqs. 3.6 and 3.7) imply dissociation of O_2, we conclude that they are insignificant in normal battery discharge using a C cathode. At most, only a small fraction of Li_2O (<2 %) could be formed during either discharge or charge because of the minimal isotopic scrambling. When reactive metals such as Pt are used as catalysts, however, O_2 dissociation readily occurs on the Pt and this produces decomposition of the electrolyte and possibly also Li_2O [36, 61].

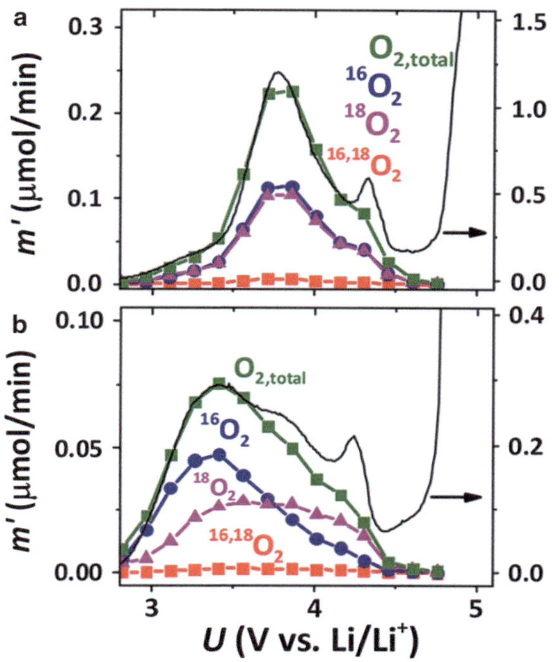

Fig. 3.27 (**a**) DEMS measurements of O_2 isotopes during an oxidative potential scan at 0.5 mV/s following galvanostatic discharge at 0.5 mA to a capacity of 0.41 mAh under a ~50:50 mixture of $^{16}O_2$ and $^{18}O_2$. The current is the *solid line*. (**b**) DEMS measurements of O_2 isotopes during an oxidative potential scan at 0.5 mV/s following discharge of 83 μAh $^{18}O_2$ followed by purging with Ar and then discharge of 83 μAh $^{16}O_2$. The current is the *solid line*. From [37] with permission

We also studied the dependence of the kinetic rate (i.e., current) at constant overpotential on the O_2 partial pressure P at 1 M Li^+ concentration and the Li^+ ion concentration $[Li^+]$ at $P = 1$ bar. Figure 3.28 presents the dependence of the discharge current, i_d, at a constant discharge potential of 2.4 V (vs. Li/Li^+) on the GC cathode for both (a) $[Li^+]$ and (b) an extended range of O_2 partial pressures (0.05 bar to 6 bar). These were obtained from the initial part of the ORR peak in CV scans at 100 mV/s for the various $[Li^+]$ and O_2 partial pressures and a subset of these CV is presented in the insets to Fig. 3.28. Qualitatively similar results were produced at other U in the CV scans, e.g., at 2.3 and 2.5 V.

For the galvanostatic conditions appropriate for battery discharge, steady state must be achieved for the intermediate LiO_2* since Li_2O_2 is the ultimate product of the discharge. This implies that $d[LiO_2^*]/dt = 0$ and thus that the current at constant overpotential $i_d \propto k_2[Li^{+*}][O_2^*]$. $[Li^{+*}]$ and $[O_2^*]$ are the surface concentrations of the two species. This same equation for i_d occurs both when Eq. (3.3) or (3.4) finally produce Li_2O_2 following Eq. (3.2). Figure 3.28a shows that at very low $[Li^+]$ concentrations, the current is roughly $\propto$ $[Li^+]$ but that the rate fully saturates by 0.1 M $[Li^+]$. This implies that $[Li^{+*}] \propto [Li^+]$ prior to saturation at higher $[Li^+]$ bulk concentrations. $[O_2^*]$ is related to the partial pressure of O_2 gas P through a conventional Langmuir isotherm, i.e., $[O_2^*] = bP/(1 + bP)$. b is the Langmuir parameter and depends on the adsorption energy of O_2 on the surface and the vibrational frequencies of the adsorbed species. Since neither of these are known, b is simply treated as an adjustable parameter in the usual fashion for

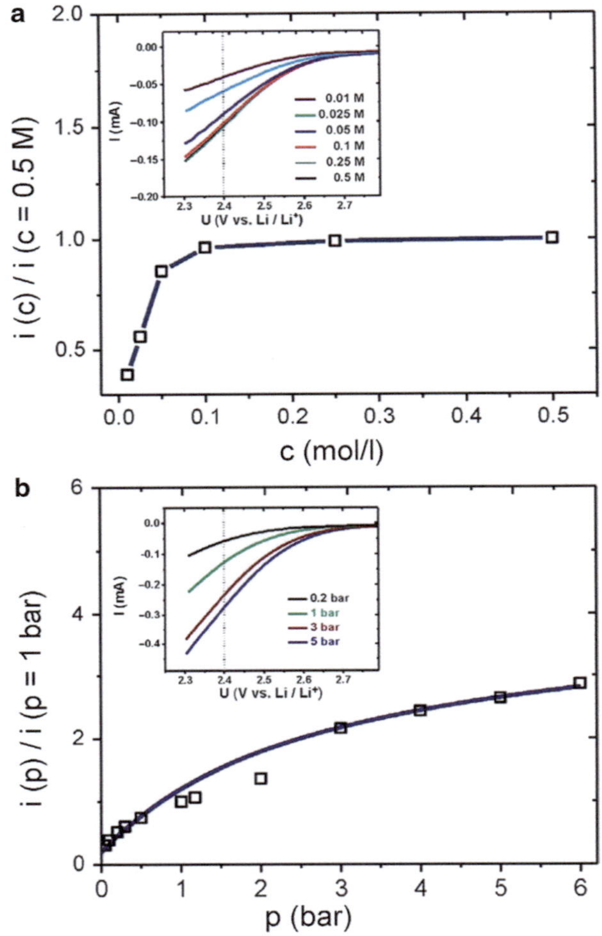

Fig. 3.28 Discharge current at 2.4 V during cyclic voltammograms (100 mV/s, LiTFSI in DME) vs. (**a**) [Li^+] molar concentration c in mol/L and (**b**) the equilibrated O_2 pressure above the solution. The points are from the experiment and in (**a**) the currents are normalized to that at [Li^+] $= c = 0.5$ M and in (**b**) are normalized to the current at $P = 1$ bar. The *insets* show the early discharge portions of the CVs from which the data points were obtained. The line through the points in (**a**) is simply to guide the eye but in (**b**) is a Langmuir fit to the O_2 pressure dependence with the Langmuir parameter $b = 0.35$. From [37] with permission

Langmuir adsorption isotherms. Figure 3.28b shows a fit to the experiments for $b = 0.35$, and this is in good agreement with the observed pressure dependence.

In summary, we find that only cathode reactions (Eq. 3.2) and (Eq. 3.3) or (Eq. 3.4) are consistent with experiments. However, none of the experiments determine whether Eq. (3.3) or (3.4) is dominating for the second step of ORR. Which dominates depends on whether the chemical barrier to disproportionation and O_2 evolution is larger than the Li^+ de-solvation barrier (and overpotential) that defines the electrochemical rate of reaction (Eq. 3.3). The latter is estimated as ~0.7 eV from temperature-dependent measurements of the overpotentials at fixed currents [54], and we believe that this is the dominant second reaction step in ORR. The reverse direction of Eq. (3.3) must dominate during OER since simultaneous two-electron processes have much higher barriers than sequential one-electron processes through an intermediate state [22].

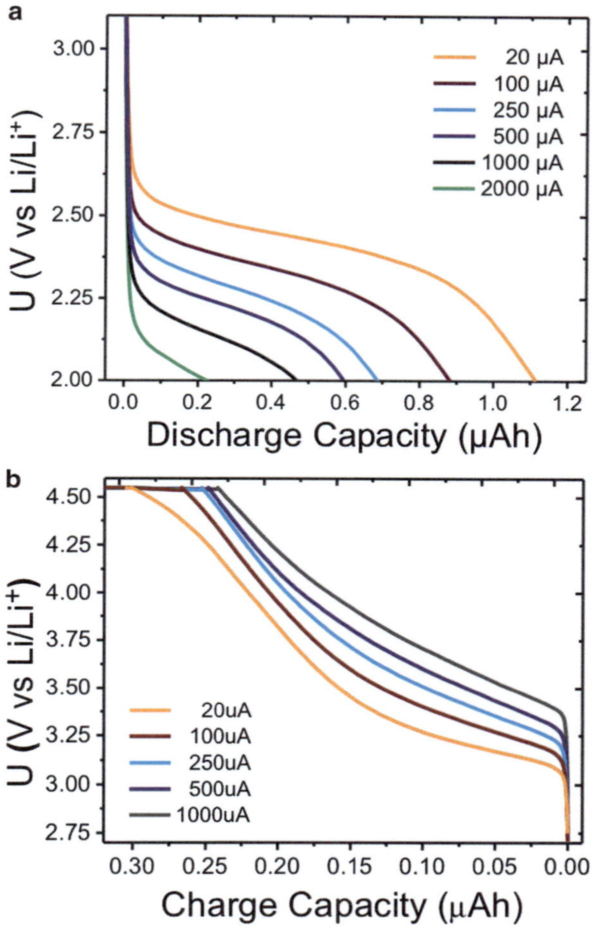

Fig. 3.29 (**a**) Galvanostatic discharges on GC in the electrolysis cell at the currents in the legend. (**b**) Galvanostatic charging at the currents in the legend following discharge of 0.33 μAh discharge on GC in the electrolysis cell. From [55] with permission

3.6.2 Kinetic Overpotentials

Figure 3.5 showed galvanostatic discharge plots in the Swagelok™ cells (*U* vs. Q_{dis}) and Fig. 3.6 a plot of U_{dis} vs. *j* in these cells. As pointed out earlier, these results do not measure the fundamental kinetic overpotentials, but rather are dominated by the cell impedance. Figure 3.29a shows galvanostatic discharge of Li–O_2 at various current densities *j* in the bulk electrolysis cell using the same electrolyte as in Figs. 3.5 and 3.6. In this cell, Li_2O_2 forms a continuous film on the GC surface as a result of discharge [55]. At a given *j*, the discharge plot shows an initial drop in potential due to the kinetic overpotential η_{dis}, a linear decrease in potential with Q_{dis} ascribed to a resistive *jR* drop through the thickening Li_2O_2 film, and a "sudden death" characterized by a rapid decrease in *U*. The latter two phenomena have previously been assigned as charge transport limitations through the Li_2O_2 film [55]. In this section, we discuss only the initial drop in U at $Q_{dis} \approx 0$ and its j dependence. We define $U_{dis}(Q_{dis} = 0)$ as the linear extrapolation of the initial part of the discharge curve (jR drop region) to $Q_{dis} = 0$. This defines the kinetic overpotential η_{dis} (when U_0 is known) in the regime for Li_2O_2 growth on Li_2O_2 that we believe dominates the electrochemistry during battery discharge [19, 54, 55]. Figure 3.29b shows galvanostatic charging (U vs. charging capacity Q_{chg}) for various charging currents, all following a galvanostatic discharge of $Q_{dis} = 0.33$ μAh at a current of 20 μA. This Q_{dis} is ~1/3 of that for sudden death at 20 μA. To maintain the constant charging current, U increases with Q_{chg}. We discussed in Sect. 3.2.1.2 that this increase in potential is *not* due to a fundamental kinetic overpotential, but simply results from the buildup of contamination products at the electrolyte–Li_2O_2 interface [38]. We define the initial U defined by a linear extrapolation to $Q_{chg} = 0$ from the first ~0.05 μAh charging as a measure of initial charging potential for a given discharge capacity, $U_{chg}(Q_{dis})$. Repeating the experiment at other Q_{dis} and extrapolating $U_{chg}(Q_{dis})$ to $Q_{dis} = 0$ gives the fundamental charging overpotential without distortion due to the film jR drop.

Figure 3.30a shows the Tafel plot for discharge and charge, i.e., log (j) vs. $U_{dis}(Q_{dis} = 0)$ and $U_{chg}(Q_{dis} = 0)$ as defined above. Once the surface is covered with Li_2O_2, the open-circuit or equilibrium potential is $U_0 \approx 2.85$ V (vs. Li/Li^+). *This figure shows that the kinetic overpotentials, both for discharge and charge, are very small.* The separation between $U_{dis}(Q_{dis} = 0)$ and $U_{chg}(Q_{dis} = 0)$ at 10 μA/cm^2, a current density equivalent to a ~10 mA lithium air battery with a cathode of ~0.1 m^2 active surface area, is <0.4 V. This implies that an electrical efficiency 1—($U_{chg} - U_{dis})/U_0$ of ~85 % should in principle be achievable in a discharge–charge cycle if the battery were limited only by the kinetic overpotentials. The Tafel plot for discharge is highly nonlinear and that for charge is slightly nonlinear. We offer reasons for this below.

The *iR* potential loss due to the cell impedance in the Swagelok™ cells in Fig. 3.6 is much larger than the kinetic overpotentials measured here. For example, a current of 10 mA in the Swagelok™ cells (XC72 with a total active cathode surface area of ~1,000 cm^2) with $U_0 - U_{dis} \approx 0.9$ V is equivalent to only a current density of ~10 μA/cm^2 in the electrolysis cell (smooth GC cathode) with $\eta_{dis} \approx 0.2$ V. Thus, practical Li–O_2 batteries show considerably larger potential losses from *iR* drops related to the cell impedance than from the kinetic overpotentials. Therefore, we suggest that the central issue in achieving higher power densities in lithium air batteries is not electrocatalysis to minimize kinetic overpotentials, but rather minimizing cell impedances.

3.6.3 First Principles Theoretical Overpotentials

During discharge/charge a continuous crystalline deposit of Li_2O_2 is formed/dissolved. Therefore, a theoretical treatment of the current–voltage response of the Li–O_2 battery requires a detailed understanding of crystal growth and

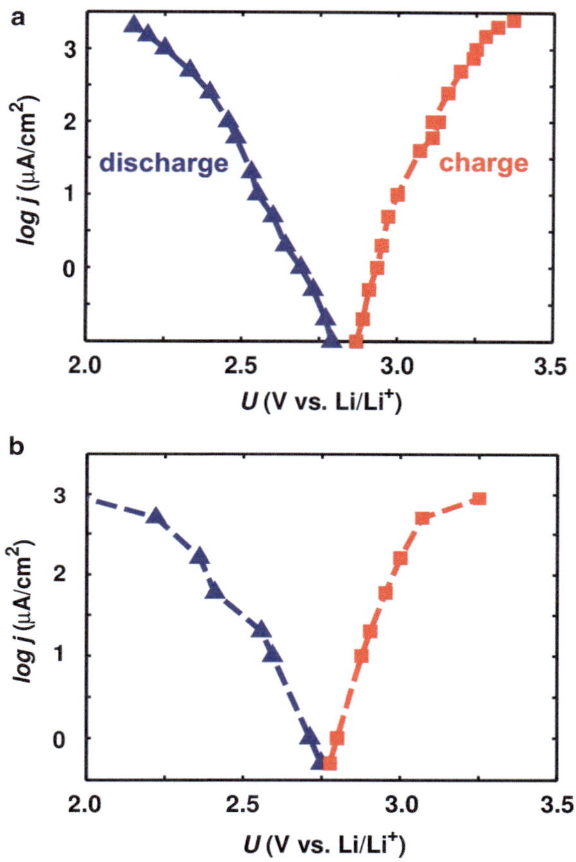

Fig. 3.30 (**a**) Experimental Tafel plots for Li–O_2 discharge (ORR—*blue triangles*) and for charging following discharge (OER—*red squares*). (**b**) First principles theoretical prediction of the Tafel plots for Li_2O_2 crystal growth, the dominant process in discharge (*blue triangles*) and Li_2O_2 dissolution, the dominant process in charge (*red squares*). Adapted from [54] with permission

dissolution of Li_2O_2 for discharge and charge, respectively. While the nucleation/dissolution on the cathode material itself is important at the outset of discharge/end of charge, the growth and dissolution of Li_2O_2 on the Li_2O_2 itself dominate the overall discharge/charge processes. In Sect. 3.6.1 we argued that growth/dissolution proceeds through sequential transfers of ($Li^+ + e^-$), with potential limiting steps of ($Li^+ + e^-$) + O_2* → LiO_2* for crystal growth and Li_2O_2 → LiO_2* + ($Li^+ + e^-$) for dissolution [18, 19, 55]. These charge transfers can occur on different crystal facets, on different terminations on those facets, and at different sites (terrace, step, or kink) and could involve different combinations of nucleation and diffusion. Thus, the overall mechanism for growth/dissolution of crystalline Li_2O_2 can be quite complicated. The surface energies of the most stable facets and terminations are given in Fig. 3.31. While the surface energies of the stoichiometric surfaces are potential independent, the surface energies of nonstoichiometric O-rich surfaces are U dependent.

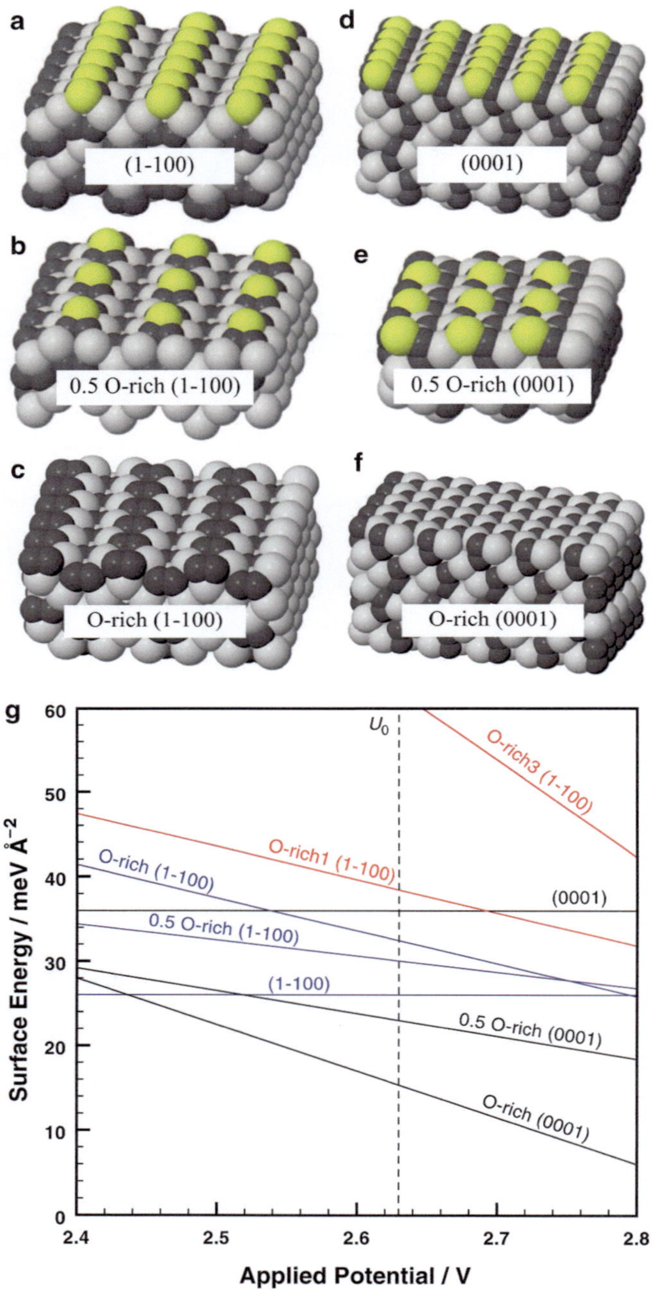

Fig. 3.31 Models of and stabilities of some of the most stable Li_2O_2 surfaces near the equilibrium potential U_0. Nonstoichiometric surfaces show a potential dependent stability. From [19] with permission

A thermodynamic analysis of growth and dissolution on some of the possible paths (different stable surface structures and terminations and the different charge transfer sites) has been presented elsewhere [19]. Using DFT calculations, the theoretical limiting potentials for discharge and charge for a given path are defined as the highest (lowest) potential at which all of the discharge (charge) steps are downhill in free energy. A detailed discussion of how these theoretical overpotentials are obtained from DFT calculations is given in [18] and [19] and is beyond the scope of this chapter. This work focused on some (but not all) paths that have a low overpotential at a given facet/termination/charge transfer site since these will dominate the activity due to the exponential dependence of the rate on the overpotential.

This thermodynamic treatment predicts a lower limit to the overpotentials associated with the discharge and charge for a given path [22]. Although de-solvation barriers could in principal cause higher kinetic overpotentials, modest de-solvation barrier as anticipated here should increase the onset currents minimally. The validity of this thermodynamic treatment in predicting overpotentials in agreement with experiments is well documented for the aqueous oxygen reduction reaction (ORR) [42] and oxygen evolution reaction (OER) [32], where de-solvation barriers are likely of similar magnitude.

Given the multitude of paths for electrochemistry, the theoretical current density j is given as $j = \Sigma_k \theta_k j_k$, where k refers to the possible paths for electrochemistry on Li_2O_2. θ_k is the relative surface concentration of the site responsible for path k and j_k is the current density through path k and is given by a simple Tafel expression. The θ_k are obtained from the relative facet and termination surface energies and the step and kink formation energies, all of which are potential dependent [54]. Certainly the θ_k and overpotentials for all possible paths have not been calculated, but including only those in [19] gives the theoretical Tafel plot of Fig. 3.30b.

There is excellent semiquantitative agreement with the experiments of Fig. 3.30a, including all of the unusual aspects in the experiments. The curvature arises because the overpotentials at the steps and kinks are small and these limited number of sites are the only ones necessary at low j. However, as j increases, higher θ_k terrace sites become necessary to support the current and these have higher overpotentials. The asymmetry in the curvature arises because $\eta_{dis} > \eta_{chg}$ on the terraces. See [54] for details of the experimental and first principles theoretical Tafel analysis.

3.6.4 Electrocatalysis

We suggested in Sect. 3.6.2 that the Li–O_2 kinetic overpotentials are in fact quite small and that the current dependent potential loss in batteries is dominated by the *iR* drop from the cell impedance. Therefore, we believe there is no strong motivation to look for electrocatalysts to lower the kinetic overpotentials. However, the search for electrocatalysts has been a central theme in lithium air research, in part motivated by the high charging overpotentials initially observed with carbonate solvents and the erroneous conclusions about its origin.

Given the mechanism for Li–O_2 growth and dissolution, it is somewhat hard to envision how electrocatalysis could actually occur. Electrocatalysis occurs when an active catalytic site lowers the barrier (or overpotential) to the kinetic limiting step in the mechanism, but the catalytic site itself is not changed during the reaction, i.e., it has a large turnover number. For Li–O_2 ORR and OER, this would imply that both the reactant and product have mobility to and away from the catalytic site. Since Li_2O_2 and likely LiO_2 are completely insoluble in the electrolyte, it is difficult to see how these species could have enough mobility for true catalysis to occur. For example, if some particle is a catalyst for ORR, it would rapidly be covered up by solid Li_2O_2 and the dominant electrochemistry then becomes simply Li–O_2 ORR on Li_2O_2. On the other hand, in OER, how does the Li_2O_2 crystal (or LiO_2 on the surface of the crystal) get to the catalyst particle since they are insoluble? Of course during the initial nucleation/final dissolution phase, when the electrochemistry is dominated by Li–O_2 electrochemistry on some other surface, some electrocatalysis is possible by the nature of the surface itself. However, this represents a minimal part of the total discharge/charge. In general, any "electrocatalyst" for Li–O_2 electrochemistry must necessarily be unreactive (i.e., to breaking the O—O bond) and therefore would be completely unrelated to active catalysts for aqueous ORR/OER since these modify the intermediate surface-adsorbed states in the electrochemistry when the O—O bond is broken [42].

Figure 3.32 shows an attempt to observe electrocatalysis in LiTFSI/DME cells for various added nanoparticles which were suggested to have electrocatalytic properties in Li–O_2, either for discharge or charge [36]. Both the initial potential of charging and the onset of O_2 evolution are identical for all cathodes, regardless of the presence of the nanoparticle "catalysts." In fact, in the presence of Pt nanoparticles, O_2 dissociation and subsequent electrolyte decomposition occur and CO_2 is the dominant gas evolved. We therefore conclude that there is a low efficacy for electrocatalysis for the Li–O_2 electrochemistry. On the other hand, when carbonate solvents are used instead of DME, the charging potential is strongly dependent upon the nature of the added nanoparticles, as observed previously by others. However, in all cases CO_2 is the dominant gas evolved, so that the electrocatalytic process observed during charging is related to carbonate oxidation (presumably due to a solution intermediate) rather than OER. This once again points out the danger of trying to explain electrochemical processes without measuring the chemistry quantitatively.

While we believe there is no true electrocatalysis in the same way as observed for aqueous ORR/OER where soluble products are formed, this does not rule out that added nanoparticles could have some beneficial effects on morphology and charge transport at higher currents and capacities. However, we have not yet looked carefully into these possibilities.

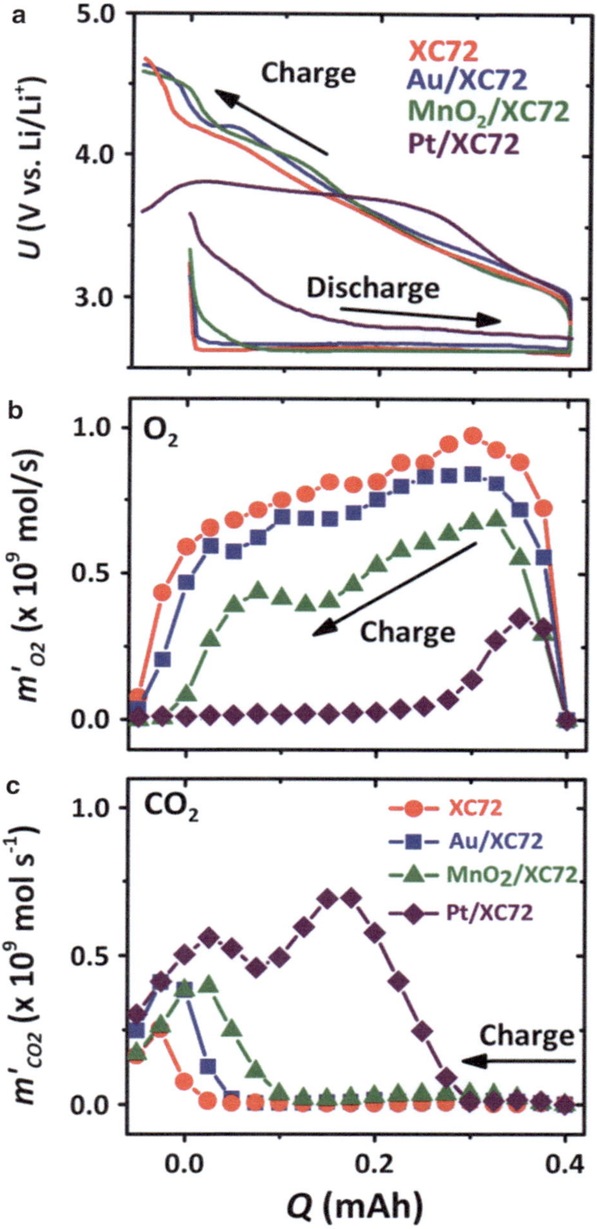

Fig. 3.32 Gas evolution from cells employing 1 M LiTFSI/DME as the electrolyte and different nanoparticle "catalysts" on XC72 carbon particles. (**a**) Discharge–charge voltage curves, corresponding O_2 (**b**) and CO_2 (**c**) evolution during charging of cells using the various cathode catalysts. From [36] with permission

3.7 Charge Transport Limitations

It is well accepted that during discharge, electrical passivation of the cathode occurs. This is because the Li_2O_2 produced during discharge is completely insoluble in the nonaqueous electrolyte and builds up as a deposit on the cathode surface. Since Li_2O_2 is a wide band-gap insulator [18, 47], electrical passivation occurs after some Li_2O_2 deposits and this manifests itself as a "sudden death" in the discharge at some discharge capacity, Q_{max}. A variety of relaxation-type experiments prove that under most discharge conditions both Li^+ and O_2 arrivals at the surface are sufficiently fast so that this "sudden death" is in fact related to "concentration polarization" of the other reactant, i.e., charge, reaching the electrolyte–Li_2O_2 interface where the electrochemistry occurs. What is particularly disturbing is that this "sudden death" occurs at lower and lower Q_{max} as the current increases, as shown in Figs. 3.5 and 3.29a. Therefore, although a high specific energy (high Q_{max}) is currently achievable at low currents (~0.05 mA/cm^2), the specific energy is more than an order of magnitude less at practical current densities (~5 mA/cm^2). We believe these issues are entirely due to charge transport limitations rather than Li^+ or O_2 transport limitations (concentration polarizations) as relaxation experiments at sudden death show no recovery in discharge U following extended waits at open circuit. The basic problem is outlined in Fig. 3.33. Although Li^+ and O_2 arrive through the solution at the Li_2O_2–electrolyte interface where electrochemistry occurs, the charge must travel through (or around) the Li_2O_2 insulator to reach the active interface.

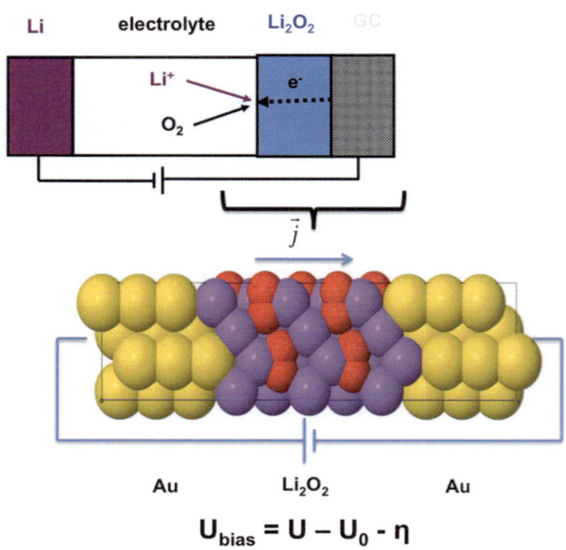

Fig. 3.33 Upper part is a schematic of a typical Li–O_2 battery showing the necessary reactant transport. Bottom part is a metal-insulator-metal (MIM) configuration used to theoretically calculate charge transport through a Li_2O_2 film as a function of bias applied across the film. From [55] with permission

We believe that in Li–O_2 batteries, charge transport occurs both along the surface of Li_2O_2 nanoparticles and through the bulk of the Li_2O_2 nanoparticles [47]. The surface charge transport is much more facile than charge transport through the bulk, but cannot support high currents. However, at higher *i* bulk charge transport dominates to maintain the galvanostatic discharge and we believe that this is the important regime for practical lithium air batteries. However, even at these higher *i* where charge transport through the bulk dominates, there is still a ~5× decrease in Q_{max} with *i* (from ~0.1 to 5 mA/cm^2).

3.7.1 Experimental Studies of Li_2O_2 Charge Transport

To study the more important bulk charge transport in detail, galvanostatic discharges were performed on GC in the bulk electrolysis cell of Fig. 3.2b at various *j*. We discussed earlier that the Li_2O_2 grows as a thickening continuous film on the GC surface so that there is only bulk charge transport after the nucleation phase and Q_{dis} is an approximate measure of film thickness. To prove that charge transport limitations were responsible for the sudden death, an outer-sphere reversible redox couple Ferrocene/Ferrocenium (Fc/Fc$^+$) was added to the electrolyte in the bulk electrolysis cell. Since the oxidation/reduction of the Fc/Fc$^+$ couple only depends upon electron (or hole) transport through the Li_2O_2 film to the solution, the current of this reaction provides a direct measurement of charge transport through the film. This current was followed at various stages of discharge corresponding to different average thickness *d* of the Li_2O_2 film. The extent of charge transport through the film was measured by electrochemical impedance spectroscopy using the Fc/Fc$^+$ couple as the reversible reaction. Results are shown in Fig. 3.34 and compared to a typical galvanostatic discharge. It is clear that the exchange current for the facile Fc/Fc$^+$ redox reaction precipitously dies at the same thickness as Q_{max}. We call this a "critical thickness" for charge transport.

Figure 3.29a showed galvanostatic Li–O_2 discharges (U vs. Q_{dis}) on planar glassy carbon (GC) electrodes in a bulk electrolysis cell with 1 M LiTFSI in DME at various current densities. In Sect. 3.6.2, we discussed the immediate drop in U due to η_{dis}. In this section, we discuss the linear decrease in U with Q_{dis} and the final "sudden death." Both of these have previously been attributed to charge transfer limitations [55], with the linear decrease as simply some unspecified *jR* drop through the film and the sudden death as due to the finite tunneling length through the film and this will be discussed in more detail in Sect. 3.7.2. We will report elsewhere extensive studies similar to Fig. 3.29a on the interrelated current, temperature, [Li$^+$], and *P* dependences of galvanostatic discharges [31].

The shapes of galvanostatic discharges in the electrolysis cell of Fig. 3.29a are qualitatively similar, but quantitatively different, to galvanostatic discharges in the Swagelok™ cell of Fig. 3.5. In Fig. 3.29a, U_{dis} vs. Q_{dis} exhibits a *jR* drop in addition to a sudden death. The *jR* drop is absent in Fig. 3.5 and only an equivalent sudden death is apparent. The *jR* drop in Fig. 3.29a increases somewhat with *j*. We believe

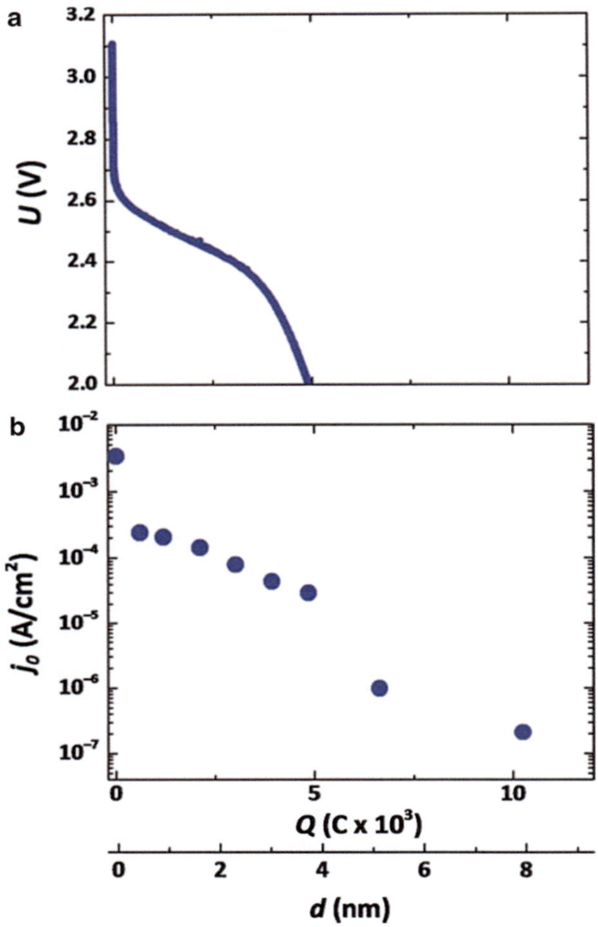

Fig. 3.34 (**a**) Galvanostatic discharge of a Li–O_2 cell at ~1 μA/cm^2 discharge current density on GC in the electrolysis cell. The average Li_2O_2 film thickness d is calculated from the discharge capacity Q and the density of Li_2O_2. (**b**) Exchange current density j_O for the Fc/Fc$^+$ reaction from EIS measurements under similar discharge conditions as in (**a**) but periodically stopped to run the EIS. From [55] with permission

that its absence in Fig. 3.5 is both due to the ~10^3 lower current densities in the Swagelok™ cell (that minimizes the jR drop) and to the continual nucleation of new nanocrystals as Q_{dis} proceeds. We will present elsewhere a detailed comparison of charge transport in the two configurations of Li_2O_2 crystal growth: nanocrystals and continuous films.

3.7.2 *Theoretical Treatment of Charge Transport*

The charge transport through the electrochemical deposit during cell discharge resembles that of a metal-insulator-metal composite (MIM), with the Fermi level of the C cathode represented by an equivalent metal, the Li_2O_2 film as the insulator, and the electrolyte Li–O_2 redox level also represented by an equivalent metal.

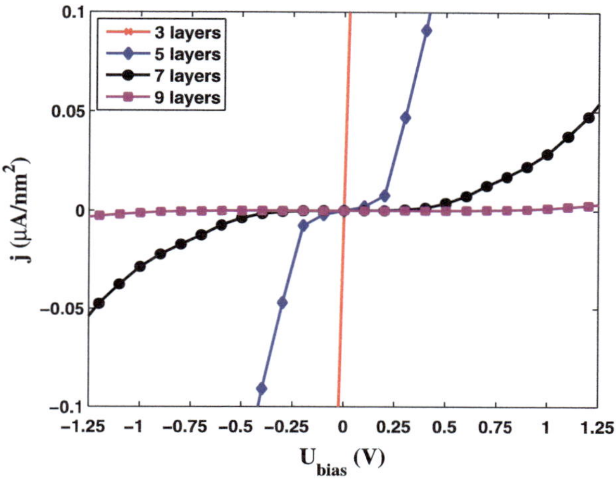

Fig. 3.35 Charge transport current through Li_2O_2 as a function of the thickness of the Li_2O_2 film (number of layers of Li_2O_2) and bias across the film. From [55] with permission

We used this theoretical MIM configuration consisting of Au|Li_2O_2|Au layers to investigate the charge transport through Li_2O_2 films of different thickness d [55]. This is a reasonable model for pure charge transport through the Li_2O_2 film during battery operation if the Fermi levels of the two metal leads approximate both the Li–O_2 redox level in the electrolyte and the Fermi energy of the GC cathode. The work function of Au ($\Phi_{Au} \approx 5.1$ eV) is quite similar to that of GC (~5 V) so that the alignment of its Fermi level is similar to that of the GC cathode. The Li–O_2 redox potential relative to vacuum zero is ~4.6 V ±0.3 (overpotentials). Again, this is moderately close to $\Phi_{Au,}$ so that the Au Fermi energy is close to the Li–O_2 redox level. Thus, both leads in the MIM configuration have reasonable band alignments relative to the true electrochemical cell during discharge/charge. Using Au in place of the electrochemical discharge reaction is equivalent to using a perfectly reversible redox couple in the electrochemistry. Therefore, the current calculated through the MIM structure depends only on the conductivity of the peroxide film. We then add the electrochemical requirements for galvanostatic discharge as constraints on this charge transport.

Figure 3.33 shows the MIM model and its relationship to the electrochemical charge transport problem. Coherent charge transport calculations for this MIM configuration were performed using a nonequilibrium Green's function method using DFT to represent the electron structure of the total MIM structure and each part separately. The left (L) and right (R) Au leads are kept at different chemical potentials μ to simulate an applied bias voltage of $U_{bias} = (\mu_L - \mu_R)/e$ across the central region. Figure 3.35 shows the calculated current-bias response curve for Li_2O_2 structures of varying thickness (number of layers of Li_2O_2). The current for a fixed thickness exhibits a roughly linear + exponential dependence with bias characteristic of MIM structures. From the density of states for the combined Au|Li_2O_2|Au system, the Fermi energy for this coupled system is ~0.5 eV above the valence

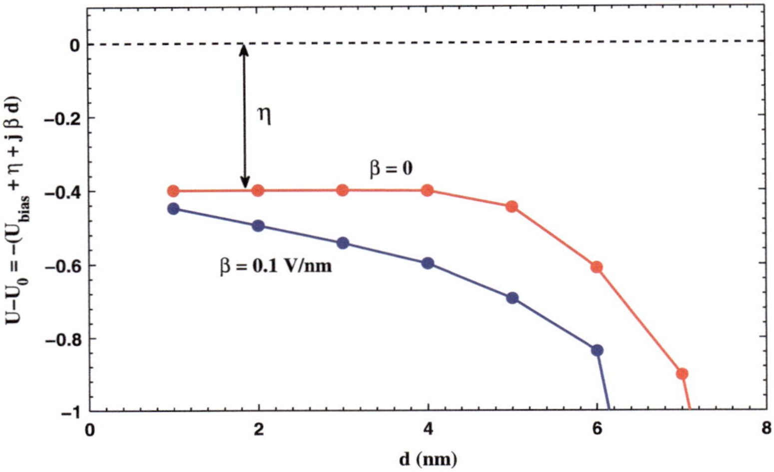

Fig. 3.36 Theoretical calculation of electrochemical galvanostatic discharge at 1 μA/cm^2 subject to charge transport limitations. From [55] with permission

band maximum. Therefore the coherent charge transport through the film must be dominated by tunneling of holes. We discuss briefly in Sect. 3.7.3 why we believe that this hole tunneling dominates the charge transport relative to other possible charge transport processes, e.g., through small polarons or defects.

The electrochemical current for galvanostatic discharge is simply the electrochemical reaction rate (or current j) subject to the conditional constraint that the rate of charge transport through the Li_2O_2 film is sufficient to drive the required electrochemical current. As the film thickens, U_{bias} must increase in order to maintain enough charge transport to keep the electrochemical rate constant at j (as required for galvanostatic discharge). U_{bias} simply represents a potential drop across the Li_2O_2 film so that it is not available as output at the cathode. Therefore, as long as charge transport is adequate to maintain the conditional constraint, $U_{\text{bias}} \approx 0$, but when not, U_{bias} increases nearly exponentially with d or Q_{dis} in an attempt to keep the charge transport equal to the galvanostatic current. This behavior is shown in Fig. 3.36 and is nearly identical to the phenomenon of sudden death observed in the experiments, even to the magnitude of d at sudden death. Thus, Q_{max} and the critical thickness are simply defined as a tunneling length. The MIM model is based on a perfect crystalline Li_2O_2 lattice at $T = 0$ K so that there is no mechanism for resistivity in the film. However, there is also a jR drop through the film prior to reaching the critical thickness because of the gradual linear drop in U with Q_{dis}. We will show elsewhere that this is most likely due to scattering from charged Li vacancies and thermally populated hole polarons in the film [56], but at this stage simply add a phenomenological resistance to the tunneling transport ($\beta = 0.1$ V/nm). This gives excellent agreement with the experimental result in Fig. 3.33a.

This simple MIM model coupled to electrochemistry does not explain the important lowering of the onset of sudden death with current, i.e., $Q_{max}(i)$ apparent in Figs. 3.5 and 3.29a. Since the tunneling current is nearly exponential with the tunneling length (Li_2O_2 thickness d), this model predicts very small changes in Q_{max} with i. However, simple modifications to this model to more clearly reflect how the band alignments are affected by the electrochemistry give excellent agreement with this aspect as well. This generalization of the transport model and its comparison to experiments will be reported elsewhere [31].

3.7.3 Charge Transport via Polarons

Theoretical calculations have shown that when adding an extra electron to Li_2O_2, it is more favorable to have that electron in a local O_2^{3-} state than delocalized in the conduction band (CB) by ~2.3 eV [14, 21]. This local O_2^{3-} is called a small electron polaron. The electron associated with this small polaron could then migrate through the Li_2O_2 by a hopping mechanism from one O_2^{2-} to another. However, the barrier for this motion is ~0.5–1 eV, too large for this to be a viable charge transport mechanism at room temperature. However, in a similar manner, if an electron is removed from Li_2O_2, it was predicted to be more favorable to have the hole localized in a single O_2^{-} than in the valence band (VB) by ~0.12 eV, thus forming a small hole polaron [43]. The barrier for this hole polaron migration was calculated as only ~0.1–0.2 eV. These barriers were predicted to be low enough that charge transport was theoretically possible at room temperature via a hole hopping mechanism if the hole polarons are present. In addition, these theoretical calculations predicted that the hole polarons could trap at Li vacancies if these exist during discharge and the authors suggested that this could also provide a mechanism for conductivity (although it was not detailed in any way). However, other calculations of the hole polaron [14] suggest significantly higher hole polaron stabilities and barriers to migration (0.5–1 eV), which would make this a less likely mechanism for bulk charge transport.

The one-dimensional (x) charge transport of small polarons will be governed by the Nernst-Planck equation which describes the combined diffusion and migration of the polarons:

$$j = j_D + j_M = eD\frac{dn}{dx} + e\mu n(x)\frac{dU_{bias}}{dx}$$

where j_D is the diffusion term and j_M is the migration term. D is the diffusion constant of the polaron and depends strongly on the barrier to hopping in an Arrhenius fashion; the ion mobility $\mu = eD/k_BT$ and $n(x)$ is the polaron density at x. The diffusion term is maximum when $n(x = 0) = n_0$ and $n(x = d) = 0$ with $x = 0$ as the Li_2O_2–electrolyte interface and $x = d$ the Li_2O_2–C interface. When migration dominates, $n(x) = n_0$ (a constant). U_{bias} is the potential across the film (our bias in the MIM description). Since $j_M/j_D \sim eU/k_BT$, migration should dominate for any reasonable bias potential across the film >25 mV. Assuming n_0 is constant for a given current, migration then predicts an approximate linear decrease in U with d for galvanostatic discharges since U_{bias} must increase linearly to keep j constant. While this could qualitatively rationalize the linear decrease with U observed in the initial discharge region, there is no way we see that polarons could ever give the dominant phenomenon of a sudden death. It is this aspect that is the important hallmark for bulk charge transport limitations in Li–O_2 batteries and why we believe that tunneling rather than hole polaron migration is the dominant bulk charge transport mechanism in Li_2O_2 formed in Li–O_2 batteries at these currents. These arguments have been quantified in other publications and both mechanisms compared to a wide variety of other experimental results [31, 54, 56].

3.7.4 Is It Possible to Avoid the Li_2O_2 Charge Transport Issues?

One way to avoid the charge transport issues that affect current-capacity trade-offs in Li–O_2 is to devise an alternative higher conductivity mechanism for charge transport through Li_2O_2, either by doping or some other mechanisms of band-gap engineering. However, at present no mechanism for this has yet been devised and it is still unclear if this is at all physically possible. For example, even if dopants can be found that introduce new states near the Fermi level in Li_2O_2 that could enhance charge transport, it is difficult to see how they could be reversibly electrochemically deposited/removed at the same potential as the Li–O_2 electrochemistry.

An alternative is to simply avoid building up deposits of Li_2O_2 greater than a few nm thick on the electrochemically active cathode surface. Certainly, one way to minimize the deposit thickness is to have a very high surface area nano-engineered cathode surface, although this approach will have its limits. Another possibility would be to solubilize Li_2O_2 so that it is transported away from the active cathode surface (and possibly precipitates elsewhere in the cell). Xie et al. [58] have shown that the anion receptor complex tris(pentafluorophenyl) borane (TPB) as an additive induces significant solubility of Li_2O_2 in nonaqueous electrolytes (and therefore results in larger discharge capacities Q_{max}). We have used quantitative DEMS in order to probe whether this enhanced capacity is also compatible with rechargeable Li–O_2 electrochemistry. Figure 3.37 shows DEMS results for O_2 consumption and O_2 evolution during linear sweep voltammetry using 1 N LiTFSI in DME with 0.33 N TPB as an additive. Comparing with the equivalent experiment in pure DME shown in Fig. 3.25, it is apparent that O_2 evolution is not the dominant electrochemical process occurring during charging, although initial formation of Li_2O_2 still appears to dominate discharge. Therefore, adding TPB (and probably other anion receptor complexes) to induce some solubility of Li_2O_2 appears incompatible with rechargeable Li–O_2 nonaqueous electrochemistry.

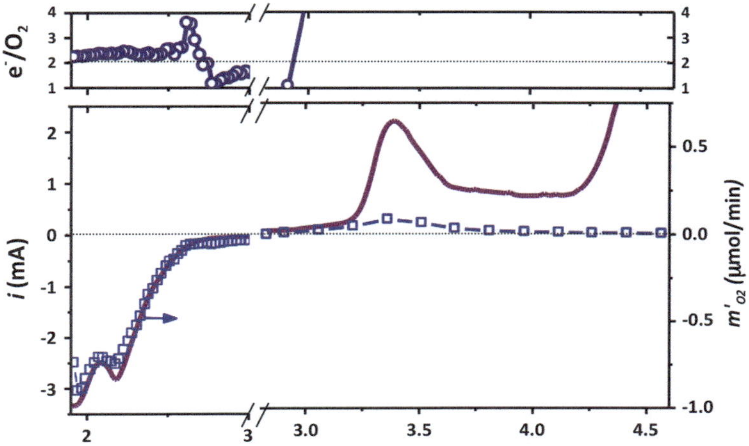

Fig. 3.37 *Bottom*—linear sweep voltammetry at 0.5 mV/s for Li–O_2 with 1 N LiTFSI in DME with 0.33 N TPB added. *Line* is the current and *blue points* are O_2 consumption and evolution. *Top*—(e^-/O_2) derived by dividing the current by the moles of O_2 consumed/evolved. The *dotted line* corresponds to the ideal ratio of 2.00. Note that very little O_2 is evolved during charge so (e^-/O_2) $\gg$ 4

3.8 Summary and Outlook

In this chapter we have tried to summarize our view of the electrochemistry occurring at the cathode in nonaqueous lithium air batteries. Of course, since this is a young and fast-moving field, none of the perspectives presented here are yet etched in stone. At present, there is not uniform agreement on all of the issues discussed, so that other authors may have different perspectives on some aspects presented here. Therefore readers should exercise their own judgment on arguments presented throughout this book.

One aspect that we have tried to highlight throughout this chapter is that to study the electrochemistry of lithium air batteries, you need to study *both* the Coulometry and the quantitative chemistry occurring at the same time. There are far too many studies in the literature making extravagant claims of cyclability based solely on Coulometric studies, all of which we think are wrong. Even qualitative spectroscopic probes are not sufficient to fully evaluate the chemistry. We have extensively used quantitative DEMS in this regard. However, even this probe does not allow us to quantitatively separate chemical problems in discharge from electrochemical ones during charging.

It is well known that there are many challenges in trying to develop practical lithium air batteries, many of which are discussed in other chapters in this book. One overriding question that has been the focus of much of our work is the following: If one could solve all the practical challenges, is the fundamental cathode electrochemistry "good enough" to form the basis for a rechargeable high-energy and high-power battery?

In answer to this question, we argue that the fundamental kinetic overpotentials for the Li–O_2 ORR and OER are incredibly small, probably because you do not dissociate/associate the O—O bond during the electrochemical cycle as you do in aqueous ORR and OER. If this was the only issue in Li–O_2 batteries, then a discharge–charge cycle electrical efficiency of >85 % should be possible (although potential losses due to *iR* drops from the cell impedance are larger). There is, however, a significant increase in the potential required for galvanostatic charging (U_{chg}) with the charging capacity (Q_{chg}) that we believe is due to electrolyte decomposition products depositing at the Li_2O_2–electrolyte interface. Unfortunately, only a few monolayers of such deposition can cause U_{chg} to rise to very high potentials (>4 V), and this causes severe electrochemical stability issues for all electrolytes in the presence of Li_2O_2. We have studied many electrolytes with quantitative DEMS and find that none are capable of giving chemical rechargeability (defined as OER/ORR) >90 %. Therefore, we believe that at present no electrolyte is stable enough to give cyclability, regardless of the cathode used. Stability issues are worse when using C as a cathode substrate since it also reacts with Li_2O_2 to form Li_2CO_3. Finding a suitable electrolyte and cathode material that is both chemically and electrochemically stable enough to avoid the buildup of a few monolayers of contaminants at the Li_2O_2–electrolyte interface is indeed a challenge!

Another significant cathode electrochemistry challenge is the limited electrical conductivity of the Li_2O_2 deposit since this causes electrical passivation of the active cathode. The practical result is that this limits the deposits to thin films (or small nanocrystals at low currents). This causes an adverse capacity-current trade-off that will be unacceptable for a combined high-energy and high-power battery. It seems unlikely that cathode surface area can be increased enough to compensate for this adverse trade-off. Because of these limitations, we have studied the mechanism and properties of bulk charge transport in Li_2O_2 in some detail. We believe that the dominant conductivity mechanism is via tunneling of holes and that a tunneling length defines the critical length for charge transport. This short-length scale is responsible for limitations of batteries at useable current densities. We believe that another charge transport mechanism must ultimately be engineered to avoid this charge transport bottleneck.

In summary, while the fundamental Li–O_2 electrochemistry itself is certainly an excellent basis for a rechargeable high-energy and high-power battery, there are still electrolyte and cathode stability issues that must be resolved and a way to engineer better charge transport through the Li_2O_2 must be found. While these are significant challenges, they by no means seem insurmountable.

Note added in proof

An iodometric titration for $n(Li_2O_2)$ combined with the DEMS has now allowed full separation of the parasitic chemistry occurring during discharge from the parasitic electrochemistry occurring during charge. Both contribute to electrolyte stability issues, with the thermal chemistry during discharge and its consequences during charging being the more dominant cause for the loss in rechargeability. This work is reported in McCloskey BD, Valery A, Luntz AC et al. (2013) Combining Accurate O_2 and Li_2O_2 Assays to Separate Discharge and Charge Stability Limitations in Nonaqueous Li-O_2 Batteries. J Phys Chem Lett 4:2989–2993

Acknowledgements The authors wish to thank their many collaborators, both at IBM and SLAC/Stanford, for their contributions to the work presented here. These include at IBM, Girish Gopalakrishnan, Angela Speidel (VW), Rouven Scheffler (VW), and Greg Wallraff and at SLAC/ Stanford, Jens Hummelshøj and Jens Nørskov. In addition, we especially thank Winfried Wilcke for initiating the program in lithium air at IBM and for stimulating interest in this field throughout the world. V. V. also acknowledges support from the US Department of Energy, Basic Energy Sciences, through the SUNCAT Center for Interface Science and Catalysis and for a UTRC fellowship.

References

1. Baker R (2001) Future directions of membrane gas-separation technology. Membr Technol 2001(138):5–10
2. Beattie SD, Manolescu DM, Blair SL (2009) High-capacity lithium-air cathodes. J Electrochem Soc 156(1):A44–A47
3. Black R, Oh SH, Lee J-H et al (2012) Screening for superoxide reactivity in Li-O_2 batteries: effect on Li_2O_2/LiOH crystallization. J Am Chem Soc 134(6):2902–2905
4. Bryantsev VS, Blanco M (2011) Computational study of the mechanisms of superoxide-induced decomposition of organic carbonate-based electrolytes. J Phys Chem Lett 2(5):379–383
5. Bryantsev VS, Giordani V, Walker W et al (2011) Predicting solvent stability in aprotic electrolyte Li–air batteries: nucleophilic substitution by the superoxide anion radical ($O_2\bullet^-$). J Phys Chem A 115(44):12399–12409
6. Chen Y, Freunberger SA, Peng Z et al (2012) Li–O_2 battery with a dimethylformamide electrolyte. J Am Chem Soc 134(18):7952–7957
7. Christensen J, Albertus P, Sanchez-Carrera RS et al (2012) A critical review of Li/air batteries. J Electrochem Soc 159(2):R1–R30
8. Dahl S (2012) Gas cleaning—perspectives from catalytic processes. In: Beyond lithium ion V: symposium on scalable energy storage, Berkeley, CA
9. Débart A, Bao J, Armstrong G et al (2007) An O_2 cathode for rechargeable lithium batteries: the effect of a catalyst. J Power Sources 174(2):1177–1182
10. Fan W, Cui Z, Guo X (2013) Tracking formation and decomposition of abacus-ball-shaped lithium peroxides in Li–O_2 cells. J Phys Chem C 117(6):2623–2627
11. Freunberger SA, Chen Y, Drewett NE et al (2011) The lithium–oxygen battery with ether-based electrolytes. Angew Chem Int Ed 50(37):8609–8613
12. Freunberger SA, Chen Y, Peng Z et al (2011) Reactions in the rechargeable lithium-O_2 battery with alkyl carbonate electrolytes. J Am Chem Soc 133(20):8040–8047

13. Gallant BM, Mitchell RR, Kwabi DG et al (2012) Chemical and morphological changes of Li–O_2 battery electrodes upon cycling. J Phys Chem C 116(39):20800–20805
14. García-Lastra JM, Myrdal JSG, Christensen R et al (2013) DFT + U study of polaronic conduction in Li_2O_2 and Li_2CO_3: implications for Li-air batteries. J Phys Chem C 117:5568–5577
15. Girishkumar G, McCloskey B, Luntz AC et al (2010) Lithium–air battery: promise and challenges. J Phys Chem Lett 1(14):2193–2203
16. Gowda SR, Brunet A, Wallraff GM et al (2012) Implications of CO_2 contamination in rechargeable nonaqueous Li–O_2 batteries. J Phys Chem Lett 4:276–279
17. Hartmann P, Bender C, Vračar M et al (2012) A rechargeable room-temperature sodium superoxide (NaO_2) battery. Nat Mater 12:228–232
18. Hummelshoj JS, Blomqvist J, Datta S et al (2010) Communications: elementary oxygen electrode reactions in the aprotic Li-air battery. J Chem Phys 132(7):071101
19. Hummelshoj JS, Luntz AC, Norskov JK (2013) Theoretical evidence for low kinetic overpotentials in Li-O_2 electrochemistry. J Chem Phys 138(3):034703
20. Jung H-G, Hassoun J, Park J-B et al (2012) An improved high-performance lithium-air battery. Nat Chem 4(7):579–585
21. Kang J, Jung YS, Wei S-H et al (2012) Implications of the formation of small polarons in Li_2O_2 for Li-air batteries. Phys Rev B 85(3):035210
22. Koper MTM (2011) Thermodynamic theory of multi-electron transfer reactions: Implications for electrocatalysis. J Electroanal Chem 660(2):254–260
23. Laino T, Curioni A (2012) A new piece in the puzzle of lithium/air batteries: computational study on the chemical stability of propylene carbonate in the presence of lithium peroxide. Chemistry 18(12):3510–3520
24. Laoire CO, Mukerjee S, Abraham KM et al (2009) Elucidating the mechanism of oxygen reduction for lithium-air battery applications. J Phys Chem C 113(46):20127–20134
25. Laoire CO, Mukerjee S, Abraham KM et al (2010) Influence of nonaqueous solvents on the electrochemistry of oxygen in the rechargeable lithium-air battery. J Phys Chem C 114 (19):9178–9186
26. Lim H, Yilmaz E, Byon HR (2012) Real-time XRD studies of Li–O_2 electrochemical reaction in nonaqueous lithium–oxygen battery. J Phys Chem Lett 3(21):3210–3215
27. Lu Y-C, Gasteiger HA, Crumlin E et al (2010) Electrocatalytic activity studies of select metal surfaces and implications in Li-air batteries. J Electrochem Soc 157(9):A1016–A1025
28. Lu Y-C, Gasteiger HA, Parent MC et al (2010) The influence of catalysts on discharge and charge voltages of rechargeable Li-oxygen batteries. Electrochem Solid State Lett 13(6):A69–A72
29. Lu Y-C, Shao-Horn Y (2012) Probing the reaction kinetics of the charge reactions of nonaqueous Li–O_2 batteries. J Phys Chem Lett 4(1):93–99
30. Lu Y-C, Xu Z, Gasteiger HA et al (2010) Platinum-gold nanoparticles: a highly active bifunctional electrocatalyst for rechargeable lithium-air batteries. J Am Chem Soc 132(35):12170–12171
31. Luntz AC, Viswanathan V, Voss J et al (2013) Tunneling and polaron charge transport through Li_2O_2 in Li–O_2 batteries. J Phys Chem Lett 4:3494–3499
32. Man IC, Su H-Y, Calle-Vallejo F et al (2011) Universality in oxygen evolution electrocatalysis on oxide surfaces. ChemCatChem 3(7):1159–1165
33. Matteucci S, Yampolskii Y, Freeman BD et al (2006) Transport of gases and vapors in glassy and rubbery polymers. In: Yampolskii Y, Pinnau I, Freeman BD (eds) Materials science of membranes. Wiley, New York, NY
34. McCloskey BD, Bethune DS, Shelby RM et al (2011) Solvents' critical role in nonaqueous lithium–oxygen battery electrochemistry. J Phys Chem Lett 2(10):1161–1166
35. McCloskey BD, Bethune DS, Shelby RM et al (2012) Limitations in rechargeability of Li-O_2 batteries and possible origins. J Phys Chem Lett 3:3043–3047

36. McCloskey BD, Scheffler R, Speidel A et al (2011) On the efficacy of electrocatalysis in nonaqueous Li–O_2 batteries. J Am Chem Soc 133(45):18038–18041
37. McCloskey BD, Scheffler R, Speidel A et al (2012) On the mechanism of nonaqueous Li–O_2 electrochemistry on C and its kinetic overpotentials: some implications for Li–air batteries. J Phys Chem C 116(45):23897–23905
38. McCloskey BD, Speidel A, Scheffler R et al (2012) Twin problems of interfacial carbonate formation in nonaqueous Li–O_2 batteries. J Phys Chem Lett 3(8):997–1001
39. Meini S, Piana M, Beyer H et al (2012) Effect of carbon surface area on first discharge capacity of Li-O_2 cathodes and cycle-life behavior in ether-based electrolytes. J Electrochem Soc 159 (12):A2135–A2142
40. Meini S, Piana M, Tsiouvaras N et al (2012) The effect of water on the discharge capacity of a non-catalyzed carbon cathode for Li-O_2 batteries. Electrochem Solid State Lett 15(4): A45–A48
41. Mizuno F, Takechi K, Higashi S et al (2013) Cathode reaction mechanism of non-aqueous Li–O_2 batteries with highly oxygen radical stable electrolyte solvent. J Power Sources 228:47–56
42. Norskov JK, Rossmeisl J, Logadottir A et al (2004) Origin of the overpotential for oxygen reduction at a fuel-cell cathode. J Phys Chem B 108(46):17886–17892
43. Ong SP, Mo Y, Ceder G (2012) Low hole polaron migration barrier in lithium peroxide. Phys Rev B 85(8):081105
44. Ottakam Thotiyl MM, Freunberger SA, Peng Z et al (2013) The carbon electrode in non-aqueous Li-O_2 cells. J Am Chem Soc 135:494–500
45. Peng Z, Freunberger SA, Chen Y et al (2012) A reversible and higher-rate Li-O_2 battery. Science 337(6094):563–566
46. Peng Z, Freunberger SA, Hardwick LJ et al (2011) Oxygen reactions in a non-aqueous Li(+) electrolyte. Angew Chem Int Ed 50(28):6351–6355
47. Radin MD, Rodriguez JF, Tian F et al (2012) Lithium peroxide surfaces are metallic, while lithium oxide surfaces are not. J Am Chem Soc 134(2):1093–1103
48. Read J (2002) Characterization of the lithium/oxygen organic electrolyte battery. J Electrochem Soc 149(9):A1190–A1195
49. Ryan KR, Trahey L, Ingram BJ et al (2012) Limited stability of ether-based solvents in lithium–oxygen batteries. J Phys Chem C 116(37):19724–19728
50. Takechi K, Higashi S, Mizuno F et al (2012) Stability of solvents against superoxide radical species for the electrolyte of lithium-air battery. ECS Electrochem Lett 1(1):A27–A29
51. Takechi K, Shiga T, Asaoka T (2011) A Li-O_2/CO_2 battery. Chem Commun (Cambridge, UK) 47(12):3463–3465
52. Veith GM, Nanda J, Delmau LH et al (2012) Influence of lithium salts on the discharge chemistry of Li-air cells. J Phys Chem Lett 3(10):1242–1247
53. Viswanathan V, Hansen HA, Rossmeisl J et al (2012) Unifying the 2e– and 4e– reduction of oxygen on metal surfaces. J Phys Chem Lett 3(20):2948–2951
54. Viswanathan V, Nørskov JK, Speidel A et al (2013) Li–O_2 kinetic overpotentials: Tafel plots from experiment and first-principles theory. J Phys Chem Lett 4(4):556–560
55. Viswanathan V, Thygesen KS, Hummelshoj JS et al (2011) Electrical conductivity in Li_2O_2 and its role in determining capacity limitations in non-aqueous Li-O_2 batteries. J Chem Phys 135(21):214704
56. Varley JB, Viswanathan V, Nørskov JK, Luntz AC (2014) Lithium and oxygen vacancies and their role in Li_2O_2 charge transport in Li–O_2 batteries. Energy Environ Sci 7:720–727
57. Wu X, Viswanathan VV, Deyu W et al (2011) Investigation on the charging process of Li_2O_2-based air electrodes in Li-O_2 batteries with organic carbonate electrolytes. J Power Sources 196(8):3894–3899
58. Xie B, Lee HS, Li H et al (2008) New electrolytes using Li_2O or Li_2O_2 oxides and tris (pentafluorophenyl) borane as boron based anion receptor for lithium batteries. Electrochem Commun 10(8):1195–1197

59. Xu W, Hu J, Engelhard MH et al (2012) The stability of organic solvents and carbon electrode in nonaqueous Li-O_2 batteries. J Power Sources 215:240–247
60. Xu W, Xu K, Viswanathan VV et al (2011) Reaction mechanisms for the limited reversibility of Li-O_2 chemistry in organic carbonate electrolytes. J Power Sources 196(22):9631–9639
61. Xu Y, Shelton WA (2010) O_2 reduction by lithium on Au(111) and Pt(111). J Chem Phys 133(2):024703
62. Zhang D, Li R, Huang T et al (2010) Novel composite polymer electrolyte for lithium air batteries. J Power Sources 195(4):1202–1206
63. Zhang SS, Foster D, Read J (2010) Discharge characteristic of a non-aqueous electrolyte Li/O_2 battery. J Power Sources 195(4):1235–1240
64. Zhang ZC, Lu J, Assary RS et al (2011) Increased stability toward oxygen reduction products for lithium-air batteries with oligoether-functionalized silane electrolytes. J Phys Chem C 115(51):25535–25542

Chapter 4
The Kinetics and Product Characteristics of Oxygen Reduction and Evolution in LiO_2 Batteries

Betar M. Gallant, Yi-Chun Lu, Robert R. Mitchell, David G. Kwabi, Thomas J. Carney, Carl V. Thompson, and Yang Shao-Horn

4.1 Current Challenges and Demonstrated Energy and Power of Rechargeable $Li–O_2$ Batteries

Lithium–oxygen ($Li–O_2$) batteries are attractive for future energy storage devices because of their projected several-fold increase in gravimetric energy compared to lithium-ion batteries [1, 2]. Combining operational elements of a fuel cell (e.g., reduction of gaseous O_2 from the environment during discharge) with that of a battery (storage of electrons and Li^+ within a solid phase, Li_2O_2), $Li–O_2$ systems represent a departure from conventional constraints of Li-ion positive electrode development. Specifically, the necessity of using intercalation structures with heavy transition metals is removed, with the theoretical gravimetric or volumetric energy of the $Li–O_2$ electrode limited not by properties of an intercalation compound but by the ability to efficiently fill the electrode void volume with the newly formed phase, Li_2O_2. It is from this fundamental difference that $Li–O_2$ cells are largely expected to derive their gravimetric energy advantage.

$Li–O_2$ batteries face substantial challenges that currently limit their practical applications, including poor stability of the carbon electrode and electrolyte solvent

---

B.M. Gallant • Y.-C. Lu • D.G. Kwabi
Department of Mechanical Engineering, Massachusetts Institute of Technology, Cambridge, MA 02139, USA

R.R. Mitchell • T.J. Carney • C.V. Thompson
Department of Materials Science and Engineering, Massachusetts Institute of Technology, Cambridge, MA 02139, USA

Y. Shao-Horn (✉)
Department of Materials Science and Engineering, Massachusetts Institute of Technology, Cambridge, MA 02139, USA

Departments of Mechanical Engineering and of Materials Science and Engineering, Massachusetts Institute of Technology, Cambridge, MA 02139, USA
e-mail: shaohorn@mit.edu

N. Imanishi et al. (eds.), *The Lithium Air Battery: Fundamentals*,
DOI 10.1007/978-1-4899-8062-5_4,

on discharge and charge [3–8], large voltage hysteresis and low round-trip efficiency [1, 9], and short cycle life (typically below 100 cycles) [1, 10]. Research efforts have focused on addressing the chemical challenges that limit the utility of Li–O_2 batteries, for example, by searching for more stable electrolytes such as ethers (forming Li_2O_2 on the first discharge) [3, 4, 6, 11, 12] rather than the widely used carbonate electrolytes (which form Li_2CO_3 and not Li_2O_2) [5, 13].

The formation and growth mechanisms of bulk Li_2O_2; its electronic structure, chemistry, and morphology; and its relation to parasitic products such as Li_2CO_3 in the electrode have received less attention from researchers [4, 12, 14–16], although they are likely of critical importance for cell-level performance. Understanding these processes could provide insight into the origin of electrode characteristics such as the voltage profile during discharge, the rate capability, the discharge capacity, the charging voltage, and the ability to cycle electrodes reversibly.

Despite these challenges, developments in electrode design have enabled the demonstration of extraordinarily high electrode gravimetric energies. Figure 4.1a compares the discharged gravimetric energy and power of various Li–O_2 positive electrode materials reported to date [4, 10, 12, 17, 18] with two conventional Li-ion positive electrodes, $LiCoO_2$ [19] and $LiNi_{0.5}Mn_{0.5}O_2$ [20]. The Li–O_2 electrodes achieved a significant increase (2–5× at 50 W/$kg_{positive}$) in gravimetric energy at the laboratory scale, with maximum electrode energies ranging from 1,000 to 3,000 Wh/$kg_{positive}$ compared to 500–600 Wh/$kg_{positive}$ for Li-ion electrodes. The energies for Li–O_2 electrodes shown in Fig. 4.1a include the weight of the reaction product, Li_2O_2, which is formed within the positive electrode during discharge. Unfortunately, there has been a tendency in the literature to neglect this additional weight, leading to a higher and unrealistic projected energy gain (~20× compared to Li-ion) [21, 22]. The gravimetric power capability of reported Li–O_2 electrodes (Fig. 4.1a), which is a critical metric for the viability of Li–O_2 batteries for electric vehicles, compares well with that of Li-ion electrodes (20–4,000 W/$kg_{positive}$).

Several of the discharged gravimetric energies shown in Fig. 4.1a begin to approach the theoretical gravimetric energy associated with formation of monolithic Li_2O_2 ($3,212\ \mathrm{Wh/kg_{Li_2O_2}}$; a practical discharge voltage of 2.75 V vs. Li/Li^+ is assumed in the calculation) [2, 12]. In this context, "theoretical" refers to the gravimetric energy stored in an electrode consisting entirely of Li_2O_2 and no substrate (e.g., no carbon). Therefore, discharged electrode gravimetric energies that approach the theoretical value of Li_2O_2 correspond to electrodes where the weight of the substrate (e.g., carbon) is insignificant compared to the weight of Li_2O_2 formed during discharge. However, high gravimetric energy approaching the theoretical value does not necessarily translate into high volumetric energy, as the gravimetric energy does not include information about the distribution of Li_2O_2 in the electrode or how it fills the available space. For example, the CNF electrodes shown in Fig. 4.1a have maximum volumetric energies of ~2,400–3,000 Wh/$L_{electrode}$ (accounting for uncertainty in electrode thickness), which is only ~30–40 % of the theoretical volumetric energy of $7,388\ \mathrm{Wh/L_{Li_2O_2}}$ (assuming a Li_2O_2 density of 2.3 g/cm^3).

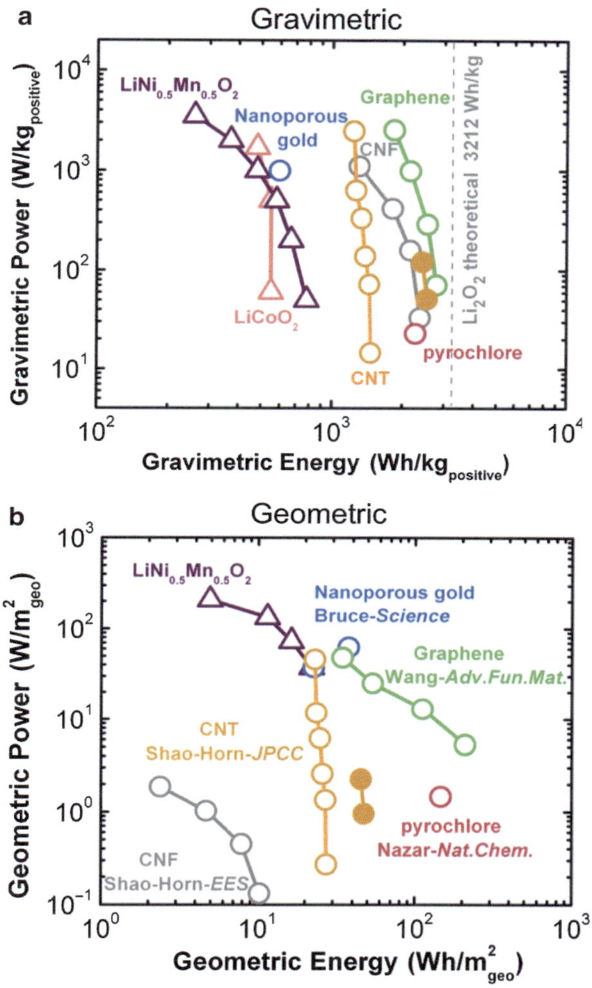

Fig. 4.1 Discharged (**a**) gravimetric and (**b**) geometric Ragone plots (power vs. energy) of reported Li–O_2 electrodes: carbon nanotubes (CNTs) [4], nanoporous gold [10], carbon nanofibers (CNFs) [12], pyrochlore [17], and freestanding hierarchically porous carbon ("graphene") [18]. Conventional Li-ion positive electrodes, $LiCoO_2$ [19] and $LiNi_{0.5}Mn_{0.5}O_2$ [20], are shown for comparison. *Open orange circles* are for CNTs discharged to a limited capacity of 1,000 mAh/g_C [4], while *closed orange circles* are from CNT electrodes discharged to 5,000 mAh/g_C. The theoretical gravimetric energy of Li_2O_2 is calculated assuming a practical discharge voltage of 2.75 V and $1,168\ \mathrm{mAh/g_{Li_2O_2}}$

Figure 4.1b shows the geometric energy and power of electrodes in Fig. 4.1a. In some cases (e.g., for nanoporous gold [10], pyrochlore [17], and graphene [18] electrodes), high gravimetric energy translates well to high geometric energy, and these electrodes exhibit comparable or higher energy (30–300 Wh/m^2_{geo}) compared to Li-ion positive electrodes such as $LiNi_{0.5}Mn_{0.5}O_2$ (<30 Wh/m^2_{geo}) [20]. In other cases (e.g., carbon nanotube (CNT) [4] and carbon nanofiber (CNF) [12] electrodes) with extremely high void volumes (>90 %) and relatively low thicknesses (several to hundreds of micrometers), there is no or only moderate geometric advantage even though the gravimetric energy advantage compared to Li-ion is significant (Fig. 4.1a). It is important to note that the Li–O_2 electrodes in Fig. 4.1a, b were not necessarily optimized for gravimetric or geometric energy, and further improvements, particularly in geometric energy, should be possible.

The corresponding geometric powers, which are critical for the development of compact devices, are currently significantly lower than those achievable by Li-ion electrodes. For example, typical Li–O_2 electrode geometric powers in Fig. 4.1b range from ~0.1 to 70 W/m^2_{geo}, whereas those for Li-ion electrodes are much higher, e.g., ~200 W/m^2_{geo} for $LiNi_{0.5}Mn_{0.5}O_2$ [20]. However, the rate capability trend of certain electrodes such as CNTs (Fig. 4.1b) shows negligible change in attainable energy with increasing power, and higher geometric powers should be achievable with optimization. For instance, the discharge capacity of CNTs (open orange symbols) in Fig. 4.1b was limited to 1,000 mAh/g_C [4] and any changes in energy with increasing power originate from changes in the discharge voltage rather than in the capacity. As the Li_2O_2 morphology at 1,000 mAh/g_C consists of small particles [4], which poorly utilize the available void volume, tailoring of the electrode to minimize void space by enabling higher Li_2O_2 filling could yield improvements.

At the fundamental level, the intrinsic activity of various electrode surfaces is a critical parameter that has received significant attention in the literature in the search for active [11, 14, 17, 23] and stable [4, 7, 10, 24, 25] surfaces for the oxygen reduction reaction (ORR). We therefore compare the maximum geometric and true surface area-specific power capability of select electrodes from Fig. 4.1 in Fig. 4.2 to show the intrinsic power capability of some of today's high-performing electrode materials [4, 10, 18] that are known to be moderately stable in non-carbonate electrolytes. The true surface area powers are similar, with a maximum achieved for gold (25 mW/m^2_{true}, Fig. 4.2b) that is only slightly higher than that achieved using CNT electrodes (23 mW/m^2_{true}) and graphene electrodes (18 mW/m^2_{true}). It is of further interest in Fig. 4.2 that high true surface area power, for example, as obtained from CNTs (Fig. 4.2b), does not directly translate into the highest geometric power (Fig. 4.2a), which underscores that the design of the electrode is critical to realize the intrinsic benefits of different material surfaces for ORR.

Despite high geometric energies and powers achievable at the electrode scale, packaged Li–O_2 cell prototypes have not been widely developed, and the true gravimetric energy advantage of packaged devices remains unknown. Furthermore, predicting the performance of fully packaged devices based on electrode performance alone is not trivial. For typical lithium-ion batteries, the gravimetric energy of the packaged product is approximately three to five times lower than that of the electrode once the weight of current collectors, separators, electrolyte, and packaging has been included in the mass [26]. However, such scaling assumptions are

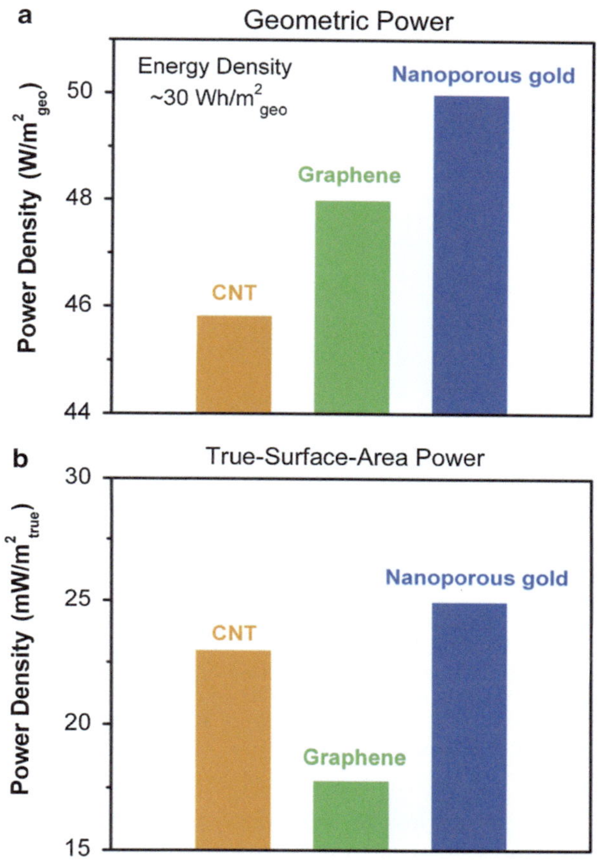

Fig. 4.2 Maximum geometric and true surface area-specific power comparison of CNT, graphene, and nanoporous gold electrodes [4, 10, 18]

only valid for electrodes with thicknesses on the order of several hundreds of micrometers [26]. As many of the most promising Li–O_2 battery materials to date (Fig. 4.1) are based on nanoscale materials that are not compatible with existing manufacturing and packaging techniques, the added weight of non-active cell components is difficult to project. We therefore attempt to quantify the gravimetric energy of Li–O_2 vs. lithium-ion batteries at the active material level, i.e., considering the negative and positive electrode weights but neglecting the electrolyte, separator, current collector, packaging, and other contributions which are not easily estimated for a packaged Li–O_2 device. Such a projection represents an upper bound on the gravimetric energy of Li–O_2 batteries as the losses due to packaging, even in highly optimized systems, will probably only approach that of lithium-ion batteries and will likely be higher due to system complexity.

Figure 4.3 shows two estimates of the theoretical gravimetric energy of Li–O_2 cells assuming Li_2O_2 to be the final reaction product and either a Li or graphite (C_6) negative electrode [2]. The Li negative electrode is assumed to contain a twofold excess of lithium beyond the capacity required to lithiate the positive electrode due

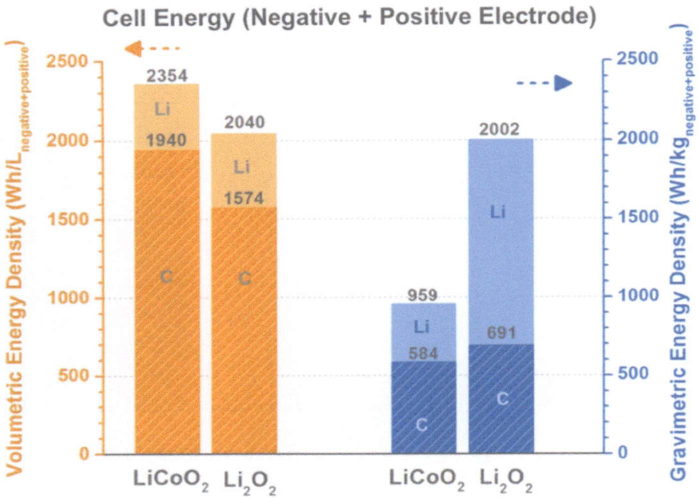

Fig. 4.3 Estimated cell-level volumetric (*orange*) and gravimetric (*blue*) energy of Li–O_2 batteries and comparison with lithium-ion batteries. O_2 (Li_2O_2) or $LiCoO_2$ is used as the positive electrode and carbon (C_6, *dark blue* or *orange-shaded bars* labeled "C"), or Li (*light blue* or *orange-shaded bars* labeled "Li") is used as the negative electrode. The cell voltages used for C–$LiCoO_2$, C–Li_2O_2, Li–$LiCoO_2$, and Li–Li_2O_2 are 3.7, 2.45, 4.0, and 2.75 V, respectively. Neither catalyst (for O_2 cells), carbon, electrolyte, nor packaging was included in the calculations. The positive electrode ($LiCoO_2$ or Li_2O_2) is assumed to be lithiated, and an additional two times excess Li is used as the Li negative electrode [2]. Reproduced by permission in print and electronic media by The Royal Society of Chemistry

to the gradual loss of active Li from imperfect plating during cycling. The C_6 negative electrode represents a more practical alternative from a safety and lifetime perspective. For positive electrodes, a well-designed structure is assumed wherein the mass of the carbon is negligible compared to the mass of the Li_2O_2 formed during discharge; therefore, the electrode weight was considered to be entirely that of Li_2O_2. We also assume a practical discharge voltage of Li_2O_2 of 2.75 V. With these assumptions, the projected gravimetric energy densities are 691 and 2,002 $Wh/kg_{electrodes}$ for a full Li–O_2 cell using C_6 or Li negative electrodes, respectively (Fig. 4.3). Upon comparison with today's lithium-ion batteries (C_6–$LiCoO_2$, 584 $Wh/kg_{electrodes}$), the Li–Li_2O_2 (2,002 $Wh/kg_{electrodes}$) system represents roughly a threefold improvement, but the C_6–Li_2O_2 system (691 $Wh/kg_{electrodes}$) has only a fractional advantage. On a volumetric basis the projected Li–O_2 cell has no advantage. We again emphasize the gravimetric energy of a practical device will be further reduced from the numbers presented here due to added weight associated with making a full device. Polyplus has predicted that it is possible to construct a rechargeable Li–O_2 cell with a gravimetric energy of 700–800 Wh/kg_{device} using a protected Li anode, which is in reasonably good agreement with our calculations [27, 28].

4.2 The Kinetics and Reaction Product Characteristics of Li–O_2 Batteries upon Discharge

4.2.1 Oxygen Reduction Reaction Kinetics on Carbon

Understanding the intrinsic oxygen reduction reaction (ORR) kinetics and limitations associated with Li_2O_2 formation is critical for the development of O_2 electrodes with high rate capability. To probe these kinetics, we have investigated the ORR on CNT electrodes under both galvanostatic and potentiostatic discharge conditions [29]. Under potentiostatic discharge (Fig. 4.4a), the average current

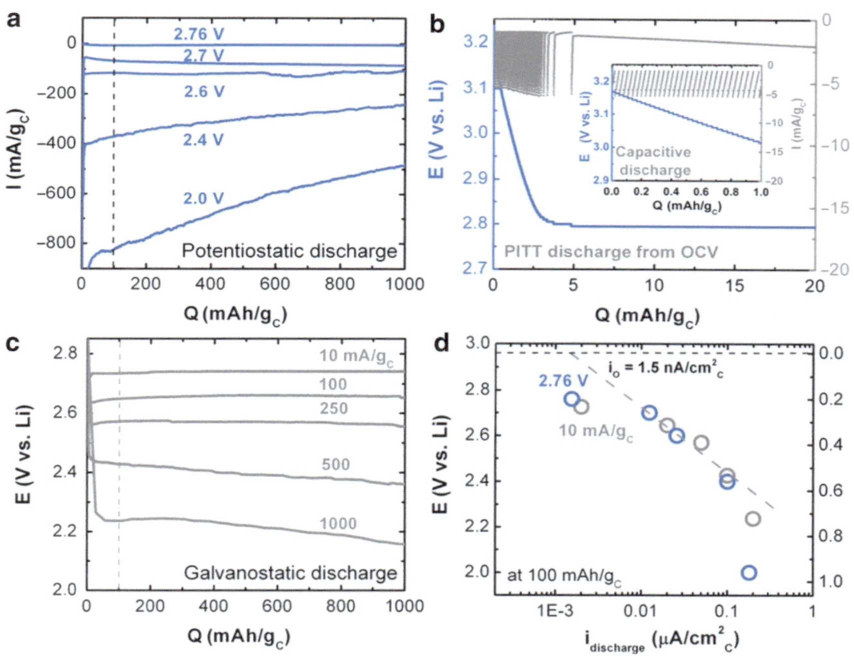

Fig. 4.4 (**a**) Current vs. capacity of CNT electrodes discharged potentiostatically over a range of potentials between 2.0 and 2.76 V. (**b**) Potentiostatic intermittent titration technique (PITT) discharge of CNT electrodes from open circuit (~3.15 V) with a voltage step of 5 mV and a current cutoff per step of 1 mA/g_C. (*Inset*) Zoom-in of the PITT response in the capacity range below 1 mAh/g_C where the current corresponds to only capacitive discharge. The initiation of Li_2O_2 nucleation and growth on carbon at ~5 mAh/g_C occurs at 2.78 V. (**c**) Voltage vs. capacity of CNT electrodes discharged galvanostatically between 10 and 1,000 mA/g_C. (**d**) Tafel plot of voltage vs. carbon true surface area-normalized current for electrodes discharged potentiostatically (*blue circles*) from (**a**) and galvanostatically (*gray circles*) from (**c**). The thermodynamic potential of Li_2O_2 (2.96 V) is indicated as a *dashed horizontal line*, and the right-hand axis shows the ORR overpotential referenced to this potential. The *dashed gray line* is a linear fit to the data points between 0.01 and 0.1 μA/cm^2_C where linear Tafel-like behavior was observed [29]. Reproduced by permission in print and electronic media by The Royal Society of Chemistry

from 0 to 1,000 mAh/g_C increased in magnitude with decreasing voltage from 2.76 (average 6 mA/g_C) to 2.0 V (660 mA/g_C) vs. Li/Li^+ (all voltages in this chapter are referenced to Li/Li^+ unless otherwise noted; we use simply "V" going forward). We note that 2.76 V corresponds to the minimum overpotential (200 mV) required to observe significant ($i_{discharge} > 1$ mA/g_C) ORR currents, as probed by potentiostatic intermittent titration technique (PITT) discharge from open circuit (Fig. 4.4b). 2.76 V is therefore the maximum potential at which nucleation and growth of Li_2O_2 on CNTs occur. At 2.76 and 2.6 V, the current was approximately constant with time during sustained polarization to 1,000 mAh/g_C. Upon lowering the potential to 2.4 and 2.0 V, the current magnitude was found to be at a maximum upon initial polarization and to decrease monotonically with continued discharge (Fig. 4.4a), which could be related to complex nucleation and growth processes occurring on CNTs at high overpotentials [16].

The currents and voltages measured under galvanostatic (Fig. 4.4c) discharge were found to agree well with those obtained potentiostatically (Fig. 4.4a) at low discharge capacities such as 100 mAh/g_C (corresponding to an average thickness of ~0.1 nm, less than one monolayer coverage of Li_2O_2 on CNTs), as shown in a Tafel plot of discharge voltage vs. current normalized to CNT surface area in Fig. 4.4d [29]. At average discharge currents between 0.01 and 0.1 μA/cm_C^2 (corresponding to $i_{galvano}$ of 100–500 mA/g_C or $E_{potentio}$ of 2.7–2.4 V), the discharge voltage exhibited a linear relationship with the log of the average discharge current, consistent with a kinetically controlled ORR process [9, 30]. The corresponding Tafel slope of ~290 mV/decade obtained from the individual galvanostatic and potentiostatic discharge experiments (Fig. 4.4d) agreed well with that obtained from cyclic voltammetry of CNT electrodes in Li–O_2 cells [29] but was higher than that reported previously (~150 mV/decade) [9] at comparable average Li_2O_2 thickness (~0.1 nm) on glassy carbon (GC) and Vulcan carbon electrodes in tetraethylene glycol dimethyl ether (TEGDME) electrolyte. Although the physical origin is not currently understood, the differences among Tafel slopes at low Li_2O_2 coverage could reflect surface-dependent nucleation kinetics of Li_2O_2 on carbon [31]. By extrapolating the Tafel plot to the thermodynamic potential of $E_{Li_2O_2} = 2.96\,V$ [32], the ORR exchange current density was measured to be 1.5 nA/cm_C^2 on CNTs (Fig. 4.4d), which agrees well with that measured on Vulcan carbon (~ 1.5 nA/cm_{GC}^2) [9] in Li–O_2 cells.

4.2.2 Discharge Reaction Product Characteristics at Low Rates and Overpotentials

Several studies have reported that upon discharge to high gravimetric capacities (several thousand [11, 12, 33] to ~10,000 mAh/g_C [15, 18, 34, 35]) in non-carbonate electrolytes, Li_2O_2 forms discrete particles with a characteristic toroid-like shape, which can have very large particle sizes from several hundred nanometers to ~1 μm

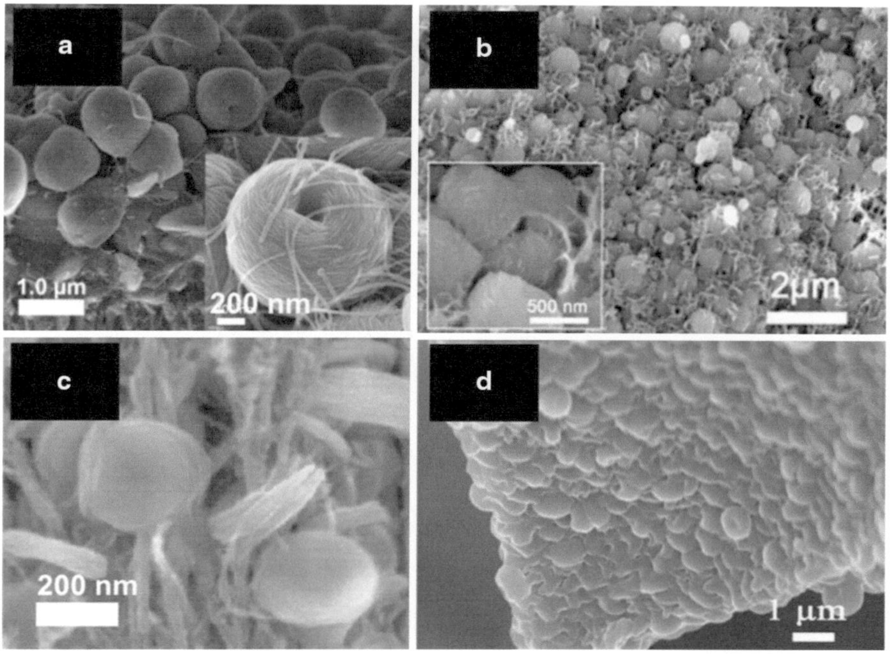

Fig. 4.5 (**a**) Morphology of Li_2O_2 in CNF electrodes discharged to 2.0 V at 46 mA/g_C (4,240 mAh/g_C) and (*inset*) 100 mA/g_C (3,900 mAh/g_C) [12]. (**b**) Morphology of a Ketjen black electrode in 1 M LiTFSI in dimethyl sulfoxide (DMSO) after discharge at 0.05 mA/cm^2 to ~9,400 mAh/g_C [35]. (**c**) Li_2O_2 discs formed in CNT electrodes upon discharge at 90 mA/g_C to ~13,400 mAh/g_C. The electrolyte in (**a**) and (**c**) was 0.1 M $LiClO_4$ in 1,2-dimethoxyethane (DME). (**d**) Morphology of a freestanding hierarchically porous carbon (graphene) electrode after discharge at 0.2 mA/cm^2 to 11,000 mAh/g_C in 1 M LiTFSI in DME [18]. Reproduced by permission in print and electronic media by The Royal Society of Chemistry

(Fig. 4.5) as revealed from ex situ SEM [11, 12, 15, 18, 33–35]. Toroid-like particles have been reported in a broad range of non-carbonate electrolyte solvents (ethers and dimethyl sulfoxide (DMSO)) and carbon/(catalyst) substrates (Fig. 4.5) [11, 12, 15, 18, 33–35], indicating that the toroid-like structure appears to be broadly characteristic of electrochemically formed Li_2O_2 rather than being electrolyte or substrate specific.

Our early work on CNF electrodes [12] in a 1,2-dimethoxyethane (DME)-based electrolyte investigated the different stages of growth of toroid-like particles at low rates. We found that their formation during discharge begins with the nucleation of small particles on the sides of the CNFs at low capacities (350 mAh/g_C upon discharge at 68 mA/g_C, with an average discharge voltage of ~2.6 V) and that the toroid-like shape emerges upon continued discharge to higher capacities (1,880 mAh/g_C at 64 mA/g_C, ~2.7 V). Similar to what was observed on CNFs, at low rates (10 mA/g_C or 2 nA/cm^2_C) on CNTs, sparse particles nucleated on CNT sidewalls as toroid-shaped particles at gravimetric capacities as low as 200 mAh/g_C (Fig. 4.6a).

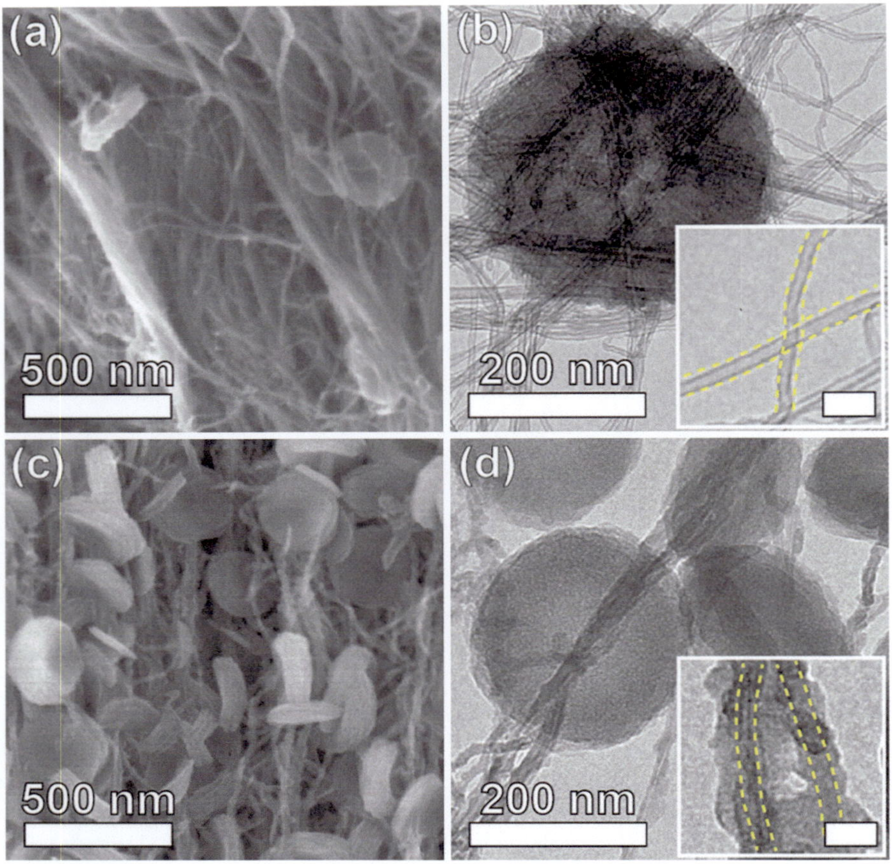

Fig. 4.6 SEM and TEM micrographs of Li_2O_2 particles on electrodes consisting of freestanding CNT carpets discharged at low and intermediate gravimetric rates. (**a**) SEM and (**b**) TEM micrographs of an electrode discharged at 10 mA/g_C to 200 mAh/g_C, with disc-shaped particles and bare CNT sidewalls. (**c**) SEM and (**d**) TEM micrographs of an electrode discharged at 90 mA/g_C to 13,000 mAh/g_C, with a high density of disc particles and a thin coating of discharge product present on the sidewalls of the CNTs. *Insets*: higher magnification TEM images of the CNT sidewalls, indicated with a *dashed yellow line*, showing the (**b**) absence and (**d**) presence of small particles (scale bar = 20 nm). Reprinted with permission in all media from Mitchell RR, Gallant BM, Shao-Horn Y et al (2013), J Phys Chem Lett 4:1060-1064. Copyright (2013) American Chemical Society

As evident from TEM imaging, the CNTs in electrodes discharged at low rates (Fig. 4.6b) are mostly bare, and the Li_2O_2 particle sizes are similar, suggesting simultaneous nucleation at relatively few sites early in the discharge process. The higher surface area of CNTs (500 m^2/g_C), relative to CNFs (100 m^2/g_C), combined with lower gravimetric current densities of 10 mA/g_C (2 nA/cm^2_C) (compared to the lowest rate of 68 mA/g_C (68 nA/cm^2_C) used previously on

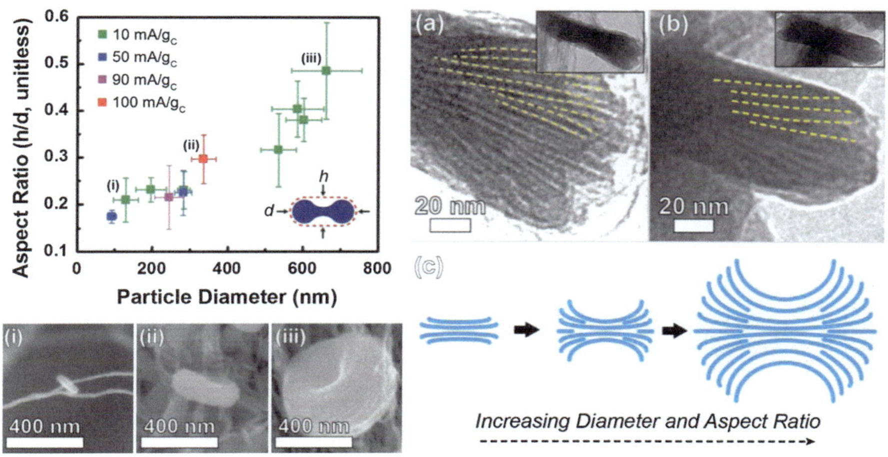

Fig. 4.7 (*Left*) Evolution of the particle aspect ratio as a function of particle diameter for several gravimetric rates. Discs and toroids represent a continuum of shapes for Li_2O_2 particles. As seen in both the plot and the SEM micrographs (i, 10 mA/g$_C$ to 500 mAh/g$_C$; ii, 100 mA/g$_C$ to 14,000 mAh/g$_C$; iii, 10 mA/g$_C$ to 1,100 mAh/g$_C$), as particles grow larger in size, their aspect ratio, defined as the ratio of particle height, *h*, to particle diameter, *d*, increases. (*Right*) Side-profile TEM images of disc particles discharged at (**a**) 50 mA/g$_C$ to 1,000 mAh/g$_C$ and (**b**) 90 mA/g$_C$ to 13,000 mAh/g$_C$. *Yellow lines* highlight the location of some of the discrete plates that compose the microscale particle; *insets* show a lower magnification image. (**c**) Schematic illustration of the microscale shape evolution of Li_2O_2 particles indicating the role of plate splaying and additional plate nucleation in the transition from disc to toroid particles. Adapted with permission in all media from Mitchell RR, Gallant BM, Shao-Horn Y et al (2013), J Phys Chem Lett 4:1060-1064. Copyright (2013) American Chemical Society

CNFs) allows us to better study the structure of Li_2O_2 formed nearer to equilibrium conditions.

Analysis of the Li_2O_2 particle geometry provides insight into how these structures form and grow during discharge. Ex situ SEM imaging was conducted on CNTs discharged galvanostatically to different capacities and showed that the aspect ratio of the toroid-like particles (defined as the thickness *h* divided by the diameter *d*) increased as a function of particle diameter, as seen in Fig. 4.7 (left). As shown in representative SEM images (Fig. 4.7, left, (i)–(iii)), the corresponding morphological evolution at 10 mA/g$_C$ evolves from a small disc-shaped particle (aspect ratio of ~0.17; (i)) at a capacity of 500 mAh/g$_C$ to the more characteristic toroid-like Li_2O_2 particle with an aspect ratio of ~0.5 at 1,100 mAh/g$_C$ (iii). The relationship between discs and toroids was further confirmed by TEM studies of the internal structure of toroid-like particles. As shown in Fig.4.7a, b, the structural units comprising the discs/toroid-like particles are arranged as arrays of plate-like Li_2O_2 crystallites (yellow dashed lines highlight the plate locations). These Li_2O_2 plate-like crystallites appear to stack nearly epitaxially on top of each other along the *c*-axis of Li_2O_2 (P6$_3$/mmc) [36], as revealed by electron diffraction analysis summarized in Fig. 4.8. SEM and TEM images of representative particles can be

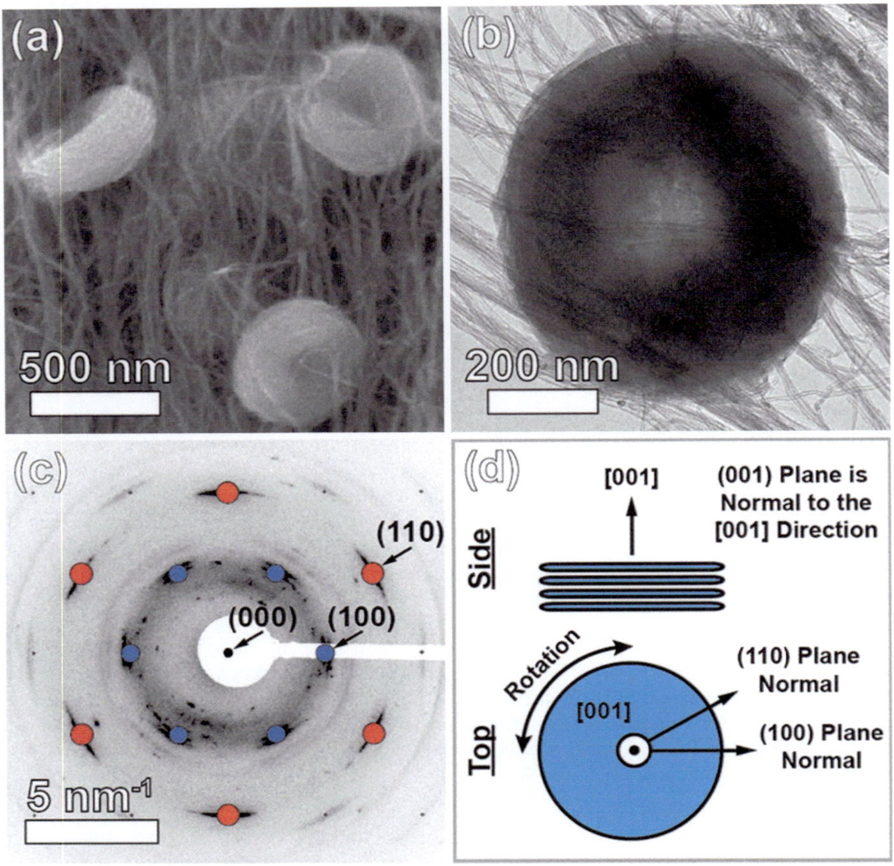

Fig. 4.8 Electron diffraction of individual Li_2O_2 particles. (**a**) SEM and (**b**) bright-field TEM images of toroid particles formed electrochemically at 10 mA/g_C to a gravimetric capacity of 1,400 mAh/g_C. (**c**) Simulated Li_2O_2 [001] zone axis (*red* and *blue dots*) superimposed over an experimental diffraction pattern for the particle pictured in (**b**). (**d**) Side and top view schematics of a stack of crystallite plates which compose the disc and toroid particles. The plate rotation axis, which is related to the arcs in the experimental diffraction patterns, and relevant crystal planes and directions for Li_2O_2 are noted. Reprinted with permission in all media from Mitchell RR, Gallant BM, Shao-Horn Y et al (2013), J Phys Chem Lett 4:1060-1064. Copyright (2013) American Chemical Society

seen in Fig. 4.8a, b, with an accompanying diffraction pattern in Fig. 4.8c. The particles exhibit nearly single crystalline diffraction patterns having a series of broad arcs, which can be attributed to reflections from an array of stacked plate crystallites having a slight rotational misalignment along the *c*-axis of Li_2O_2. In addition to rotational misalignment, the initially parallel plate-like crystallites in the discs appear to splay apart with increasing disc diameter, and additional plates appear to nucleate in the empty space between the splayed plates (Fig. 4.7a, b) as the disc diameter increases. The collective result of these splaying and inter-plate nucleation processes is the evolution of a rim around the disc circumference and the resulting transition from the disc-like to the toroid-like macroscopic shape, as illustrated in Fig 4.7c. While the origin of the plate splaying is not known, one possibility is that internal stress within the plates, arising from either a polar structure or surface stress variations on opposing plate faces, gives rise to spontaneous bending, as has been observed in several polar crystal systems [37–39]. The plates observed in this study had thicknesses on the order of 10 nm, which is on the order of measured crystal sizes obtained from X-ray diffraction [4, 5, 34]. Further, the observation of a faceted crystal with a large (0001) face is consistent with the theoretically calculated Wulff shapes for Li_2O_2 [40, 41].

The toroid-like shape of Li_2O_2 observed in our studies is similar to particles formed chemically in other material systems such as $CaCO_3$ [42–44], ZnO [45], Co_3O_4 [46], and $LiFePO_4$ [47], wherein crystal growth is proposed to occur via a mesocrystallization assembly process. If an analogous process occurs in Li_2O_2, microscale disc and toroid particles can be hypothesized to form through the aggregation of soluble primary units (soluble LiO_2, Li_2O_2 clusters, or other discharge intermediates), wherein the microscale shape is dictated by the interaction between the nanoscale crystallites that ultimately comprise the discs. However, in the absence of a soluble species, it seems likely that Li_2O_2 discs and toroids form by direct electrocrystallization of Li_2O_2 on existing Li_2O_2 surfaces. Further studies are needed to examine these possible mechanisms in detail.

4.2.3 Discharge Reaction Product Characteristics at High Rates and Overpotentials

The morphologies of Li_2O_2 formed at lower discharge potentials were found to be different from those formed at high potential (Fig. 4.9) [16, 29]. In contrast to discharge at high voltages ($E_{potentio} > 2.7$ V) which yielded disc-like Li_2O_2 (Fig. 4.9a), discharge at intermediate to low voltages ($E_{potentio} < 2.7$ V) yielded small Li_2O_2 particles with characteristic sizes of 10–20 nm anchored along the sides of the CNTs (Fig. 4.9b–d). Different from discs, which largely involve Li_2O_2 growth on Li_2O_2, the ORR kinetics at $E < 2.7$ V ($i_{discharge} \geq 0.01$ $\mu A/cm^2_{true}$ in Fig. 4.4d) correspond largely to growth of Li_2O_2 on carbon. We believe that the abrupt transition between disc-like particles above 2.7 V and equiaxed particles below 2.7 V reflects a nucleation barrier for Li_2O_2 on carbon that permits Li_2O_2 growth on carbon to compete with growth on Li_2O_2 only below ~2.7 V, which can be dependent upon the nature of the carbon surface used [31].

The small particles formed upon potentiostatic discharge are similar to those formed under galvanostatic conditions [4, 16], as illustrated in Fig. 4.10a, b for electrodes discharged at 250 mA/g_C to 1,000 and 4,200 mAh/g_C. Upon continued discharge to even higher capacities of 9,800 mAh/g_C (Fig. 4.10c), disc-/toroidal-like Li_2O_2 was observed among the small particles, which was similarly observed

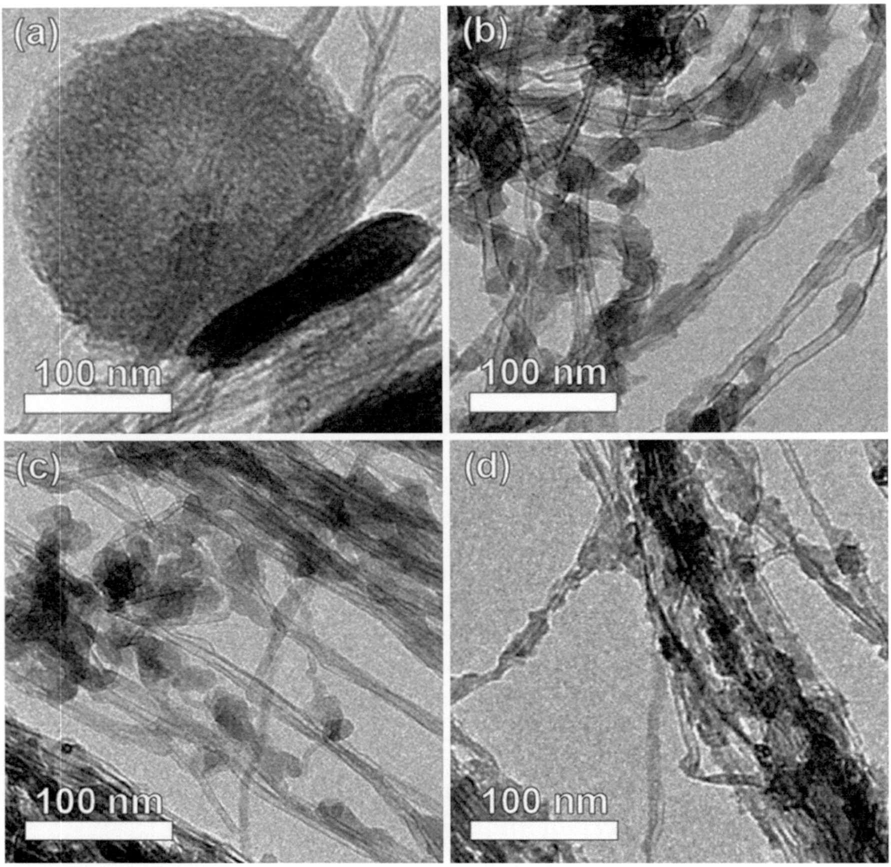

Fig. 4.9 TEM images of potentiostatically discharged CNT electrodes at (**a**) $E > 2.74$ V to 2,000 mAh/g$_C$ (PITT discharge), (**b**) 2.6 V to 5,000 mAh/g$_C$, (**c**) 2.4 V to 5,000 mAh/g$_C$, and (**d**) 2.0 V to 4,500 mAh/g$_C$ [29]. Reproduced by permission in print and electronic media by The Royal Society of Chemistry

upon discharge at 90 mA/g$_C$ to a very high capacity of 13,000 mAh/g$_C$ (Fig. 4.6c, d) [16]. SEM and TEM images in Fig. 4.6c, d reveal a much higher density of particles with nonuniform shapes compared to the toroids formed at lower rates (10 mA/g$_C$, Fig. 4.6a, b), and the CNT sidewalls appear to be coated by numerous small particles (Fig. 4.6d). These observations show that small particles can precede growth of discs/toroids with a transition point that is likely rate dependent. The true surface area-normalized average capacity, 2.0 μAh/cm^2_C, at which the discs were found to first appear at 250 mA/g$_C$ in CNTs (corresponding to 9,800 mAh/g$_C$) is similar to that observed previously in our CNF studies [12], with a transition from small particles to toroids occurring at an average loading of 1.9 μAh/cm^2_C.

Fig. 4.10 TEM images of electrodes discharged at 250 mA/g_C to increasing depth of discharge: (**a**) 1,000, (**b**) 4,200, and (**c**) 9,800 mAh/g_C. (**a**) and (**b**) show increasing coverage of equiaxed Li_2O_2 particles that precede the growth of disc particles as seen in (**c**). Reprinted with permission in all media from Mitchell RR, Gallant BM, Shao-Horn Y et al (2013), J Phys Chem Lett 4:1060-1064. Copyright (2013) American Chemical Society

Thus, the capacity-dependent morphologies agree well when differences in surface area are taken into account. We speculate that the high rates under which particles form might produce high local concentrations of soluble intermediates such as LiO_2 [15, 48, 49] near the CNT surface. With very high local concentrations, disproportionation of LiO_2 to Li_2O_2 could occur before soluble intermediates can diffuse to find the most energetically favored Li_2O_2 surfaces on which to deposit, leading to less preferential growth of Li_2O_2 on Li_2O_2 and to a more distributed coating of Li_2O_2 on the CNTs. Additional studies are needed to provide experimental support for this proposed mechanism.

4.3 Surface Chemistry of Discharged Oxygen Electrodes in Li–O_2 Batteries

4.3.1 Morphology-Dependent Surface Chemistry of Li_2O_2

X-ray absorption near-edge structure (XANES), which has been used to gain insight into the hybridization, phase identification, and electronic structure of elements in traditional Li-ion battery materials such as $LiCoO_2$ [50], $LiFePO_4$ [51], and transition metal oxides [52, 53], is becoming increasingly applicable for Li–O_2 studies [4, 11, 29]. In XANES, the density of unoccupied states is probed by measuring sample absorption as a function of X-ray excitation energy. X-ray absorption may be measured by transmission, by the detection of directly ejected/Auger electrons in the total electron yield (TEY) mode with an escape depth in the tens of nanometers, or by photons created from the annihilation of core holes by higher energy/valence electrons in the fluorescence yield (FY) mode with an escape depth in hundreds of nanometers.

Recently, we have used XANES to develop unique insights into the morphology-dependent chemical differences between the bulk and surfaces of Li_2O_2 [29]. This work builds upon earlier XANES work, which has revealed qualitative differences between commercial and electrochemically formed Li_2O_2 in Li–O_2 cells [11, 54], suggesting the presence of nonstoichiometry and structural defects in the latter [9, 11]. Using the FY mode, the bulk structures of both discs and particles formed in CNT electrodes exhibited features characteristic of stoichiometric commercial Li_2O_2, which is in agreement with XRD data showing that both discs and particles are largely composed of crystalline Li_2O_2 [4, 29]. However, the TEY mode (Fig. 4.11a) revealed that the surfaces of the discs were markedly different from the interior of the discs and also from bulk Li_2O_2.

O K-edge spectral features of Li_2O_2 [55] and LiO_2 [56] have previously been rationalized using molecular orbital principles, with each near-edge peak representing a transition to either σ^* or π^* anti-bonding orbitals of the O–O anionic species. In Fig. 4.11a, the enhanced second peak at 534.9 eV in the TEY mode for discs, which matches well with O 1$s \rightarrow \sigma^*$ transitions in reference LiO_2 reported in the literature [56], was found to be higher than in commercial Li_2O_2 (Fig. 4.11a). This observation suggests that the oxygen anions on the (0001) surface of discs are in a LiO_2-like electronic configuration, which is in agreement with previous TEM data showing that the surfaces of discs are dominated by (0001) facets [16], and with DFT calculations predicting that the equilibrium surfaces of Li_2O_2 at near-equilibrium potentials contain largely O-rich (0001) surfaces [40, 41, 57, 58].

The surface chemical nature of small particles was found to be markedly different from that of the discs (Fig. 4.11a, b) [29]. Specifically, in the O K-edge TEY mode, the surfaces of the small particles were found to have features largely consistent with that of the commercial Li_2O_2 but with an enhanced peak intensity at 532.1 eV and a decreased peak intensity at 534.9 eV. The enhanced first peak in the

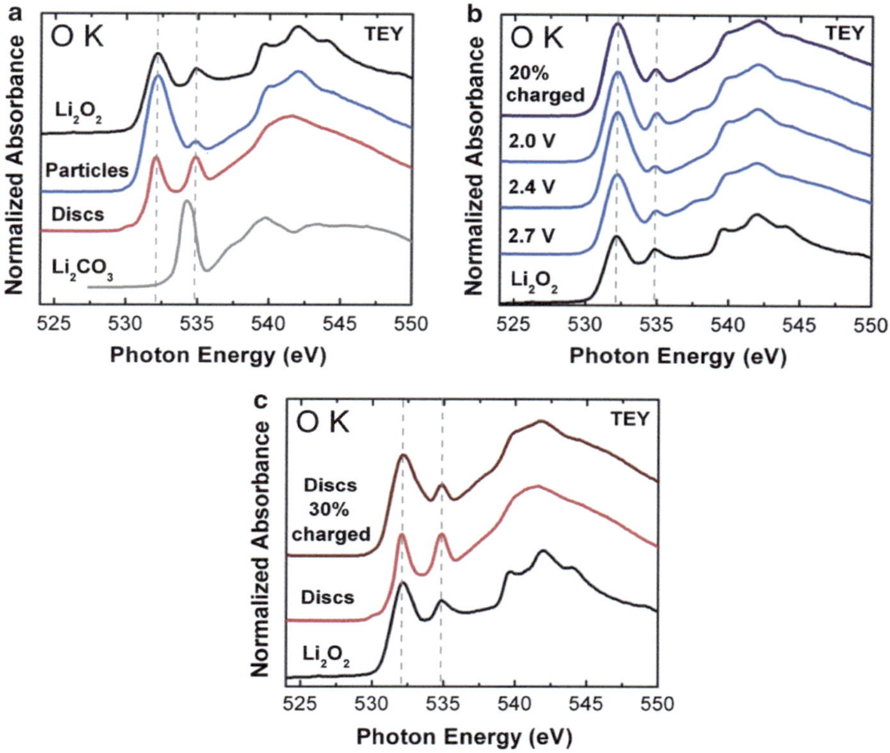

Fig. 4.11 XANES O K-edge data of discharged and partially charged discs and particles and comparison with commercial Li_2O_2 (90 %, Sigma-Aldrich) and Li_2CO_3 (99 %, Alfa Aesar). (**a**) TEY spectra of discharged discs formed galvanostatically at 10 mA/g_C to 4,000 mAh/g_C and of particles formed potentiostatically at 2.4 V to 5,000 mAh/g_C. (**b**) TEY spectra of particles formed potentiostatically at 2.7, 2.4, or 2.0 V to 6,200, 5,000, and 5,200 mAh/g_C, respectively, and comparison with a 20 % partially charged electrode containing particle morphologies. The 2.4 V trace is the same as that in (**a**). (**c**) Comparison of discharged discs from (**a**) with a 30 % partially charged disc electrode terminated at the onset of the plateau region [29]. Reproduced by permission in print and electronic media by The Royal Society of Chemistry

particles indicates an increased number of transitions at ~532.1 eV above that of the O 1*s* → σ* transition in reference Li_2O_2. One possible origin of the enhanced peak at ~532.1 eV could be trapped holes (polarons) in the π* anti-bonding state of Li_2O_2, which, according to recent computations [59], lower the energy of Li_2O_2 relative to non-polaronic Li_2O_2. Although the surface facets of small particles are not known, Hummelshoj et al. [57] have predicted that with increasing overpotential, the thermodynamically stable surfaces of Li_2O_2 change from predominantly O-rich (0001) surfaces to increasingly stoichiometric and O-rich (1-100) surfaces, which are possible candidates for surface facets of small particles. Additional studies are needed to directly probe the surface electronic structure and elucidate the electronic differences between discs and particles.

4.3.2 Chemical Reactivity Between Li_2O_2 and Carbon

To understand the origins of poor ORR/OER reversibility and rapid capacity fading upon cycling in Li–O_2 cells, it is critical to investigate the chemical stability of Li_2O_2 and other cell constituents. We have employed XANES to help understand the formation of parasitic Li_2CO_3 species in the electrode during discharge and their evolution upon cycling. Upon discharge to low capacities (1,000 mAh/g_C) corresponding to electrodes with sparse small particle morphologies, the TEY and FY spectra contained some Li_2O_2 features but were largely dominated by Li_2CO_3 (Fig. 4.12a, b). However, consistent with the findings in Fig. 4.11 at higher capacities, the discharge product surface and bulk became predominantly Li_2O_2-like with further discharge to 4,700 mAh/g_C (Fig. 4.12a, b), corresponding to increased coverage of Li_2O_2 particles on CNTs. This finding provides evidence that Li_2CO_3 forms on the CNT surface upon initial discharge and is covered by Li_2O_2 with continued discharge to higher capacity, as depicted schematically in Fig. 4.12d. Additionally, as the TEY (surface) signal at 4,700 mAh/g_C was predominantly that of Li_2O_2 with negligible Li_2CO_3 detected (Fig. 4.12b), the surface of Li_2O_2 was found to be stable in contact with the DME electrolyte. Taken together, these findings revealed that the carbon surface is the main site of parasitic species growth during discharge with CNTs [4, 6], which we proposed could result from reactivity between carbon and the initial monolayers of Li_2O_2 formed during discharge or between carbon and other discharge intermediates (e.g., LiO_2 or decomposition products in the electrolyte). The proposed reactivity between Li–O_2 intermediates or Li_2O_2 and the carbon electrode was further supported by XANES data in the C K-edge spectrum (Fig. 4.12c), where the peak at 290.2 eV, which was attributed to bonding between carbon and O in Li_2O_2 [4], was found to increase with increasing capacity from 1,000 to 4,700 mAh/g_C.

Upon charging, Li_2CO_3 was nearly completely removed at voltages below 4.5 V [4]. However, Li_2CO_3 became increasingly difficult to oxidize with continued cycling and aggregated in the electrode, ultimately dominating the discharge product chemistry after just a few cycles as shown in Fig. 4.12e, f. The corresponding charging voltage profile evolved to reflect these changes in electrode chemistry. Within a few cycles, an increasing fraction of the charge capacity occurred at higher potentials (~4.4 V) corresponding to Li_2CO_3 oxidation, while a decreasing fraction of charge capacity occurred at lower potentials (~4.0 V) corresponding to Li_2O_2 oxidation.

In a recent study, Thotiyl et al. [7] used chemical titration and mass spectroscopy of evolved gases from solid products formed at different capacities, which provides unique data on the amount of parasitic species that form during discharge. The authors showed that the Li_2CO_3 formed during discharge contains ^{12}C predominantly originating from the electrolyte (DMSO and TEGDME) rather than from the isotopically labeled ^{13}C carbon electrode (Fig. 4.13c). We note that such an apparent discrepancy regarding the source of C in Li_2CO_3 formed during discharge,

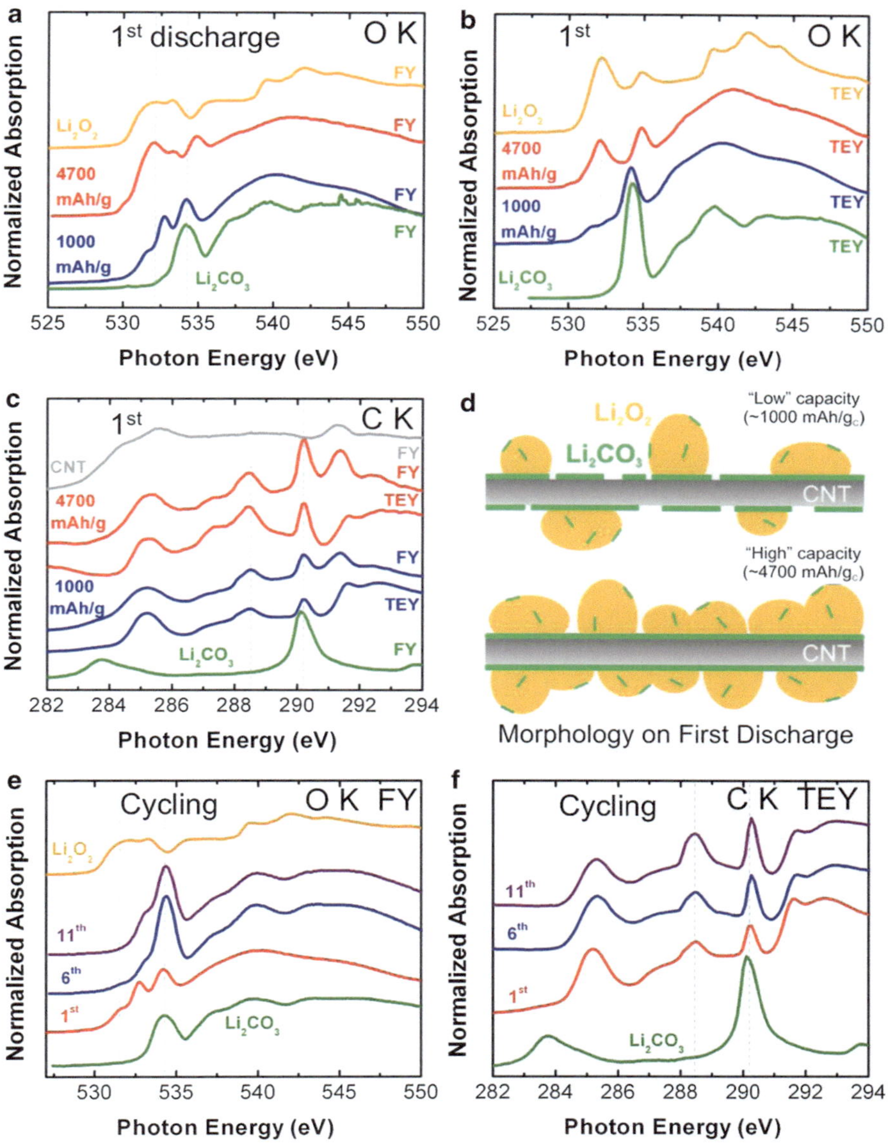

Fig. 4.12 (**a**) O K-edge FY, (**b**) O K-edge TEY, and (**c**) C K-edge TEY/FY XANES spectra of electrodes discharged to 1,000 (at 250 mA/g_C) and 4,700 mAh/g_C (at 100 mA/g_C) on the first discharge. The reference spectra of commercial Li_2O_2, commercial Li_2CO_3, and pristine CNTs (for C K-edge) are included. (**d**) Schematic of discharge products formed at low and high capacity on CNTs on the first discharge. (**e**) O K-edge FY and (**f**) C K-edge TEY spectra of cycled electrodes ending on the 1st, 6th, or 11th discharge at 250 mA/g_C to 1,000 mAh/g_C. Adapted with permission in all media from Gallant BM, Mitchell RR, Kwabi DG et al (2012), Journal of Physical Chemistry C 116:20800–20805. Copyright (2012) American Chemical Society

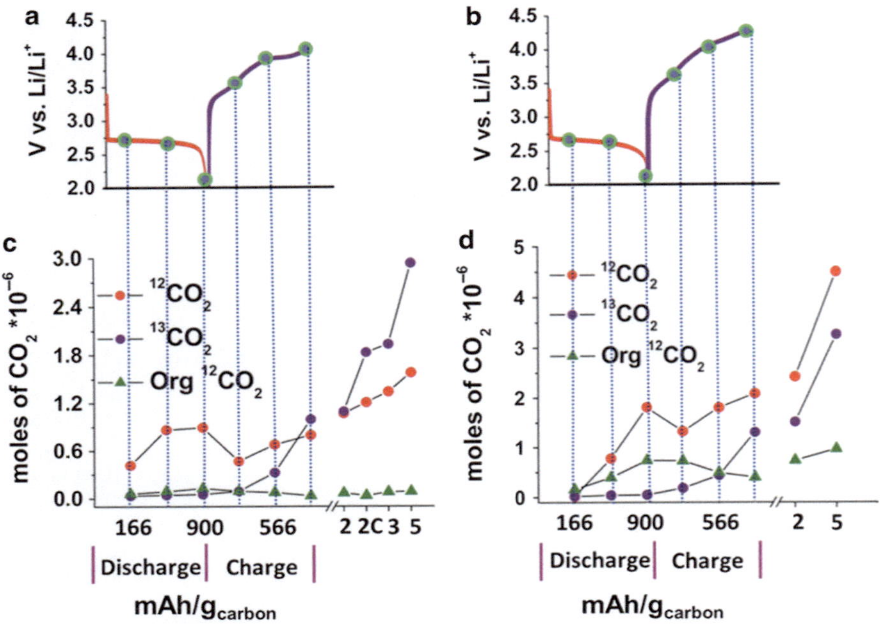

Fig. 4.13 (**a**, **b**) Discharge–charge curves on the first cycle for (**a**) DMSO- and (**b**) TEGDME-based electrolytes with a carbon cathode at 70 mA/g_C. (**c**, **d**) Moles of CO_2 evolved from the carbon cathodes, removed from the cells at the states of discharge and charge indicated by the *green dots* in (**a**) and (**b**), and then treated with acid to decompose $Li_2{}^{12}CO_3$ and $Li_2{}^{13}CO_3$ and Fenton's reagent to decompose the lithium carboxylates (or organic carbonates labeled Org $^{12}CO_2$). The values on the *x*-axis indicate the capacities at which the cathodes were sampled. The numbers 2, 3, and 5 in (**c**) and 2 and 5 in (**d**) correspond to the electrodes analyzed at the end of discharge on those cycles, and 2C in (**c**) corresponds to the analysis of the carbon electrode at the end of the second charge. Reprinted with permission in all media from Thotiyl MMO, Freunberger SA, Peng Z et al (2013), J Am Chem Soc 135:494-500. Copyright (2013), American Chemical Society

whether from the carbon electrode [4, 6] or from the electrolyte [7], could be partly explained if decomposition products originating from the electrolyte deposit directly onto and cover the surface of carbon rather than by direct reaction between Li_2O_2 and carbon. However, such a view cannot easily explain the apparent Li–O–C (different from that in Li_2CO_3) bonding that we detected by XANES in the C K-edge spectra (Fig. 4.12c), which seems to suggest that Li_2O_2 reacts at least partly with the carbon substrate. The reactivity of the carbon surface is further supported by the fact that a pure gold electrode [10] was found to significantly reduce the formation of carbonate compared to carbon electrodes. Further studies are needed to determine the extent to which the carbon from the electrode participates parasitically during discharge, which may be dependent upon the carbon or electrolyte used [7].

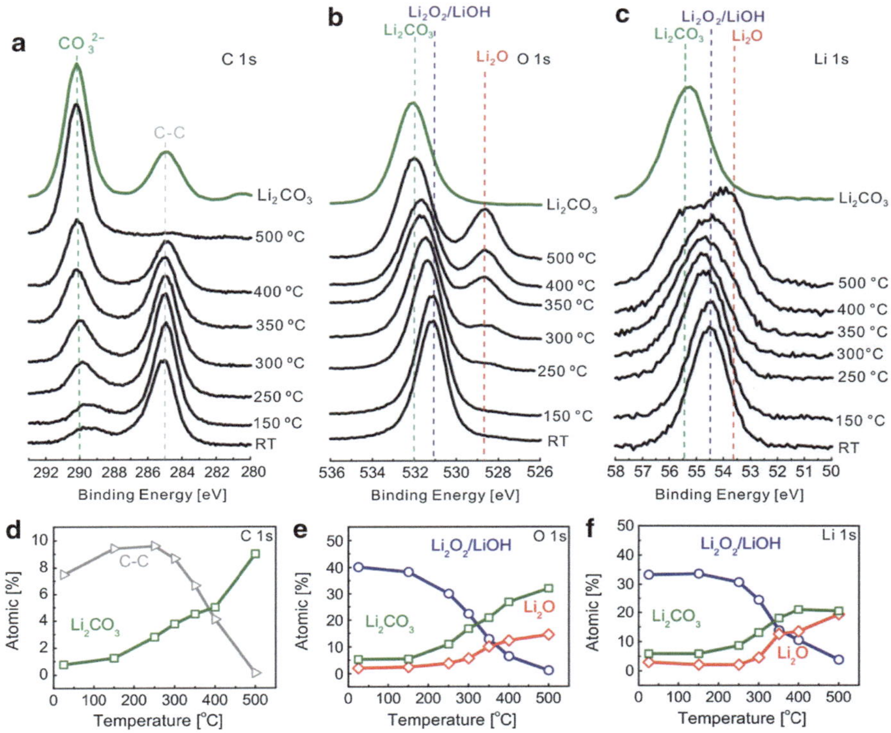

Fig. 4.14 (**a**) Temperature-dependent in situ XPS spectra of a Li_2O_2 pellet in the (**a**) C 1*s*, (**b**) O 1*s*, and (**c**) Li 1*s* regions. (**d–f**) Atomic percentage of C, O, and Li in Li_2O_2, Li_2CO_3, and Li_2O obtained from quantitative component analysis of the (**d**) C 1*s*, (**e**) O 1*s*, and (**f**) Li 1*s* spectra. Reprinted with permission in all media from Yao KPC, Kwabi DG, Quinlan RA et al (2013), J Electrochem Soc 160:A824-A831. Copyright 2013, The Electrochemical Society

The chemical reactivity between Li_2O_2 and carbon-containing species was further highlighted by our recent findings from in situ XPS [60] where we studied heating of commercial Li_2O_2 powder from room temperature to 500 °C in an XPS chamber (Fig. 4.14). Our results show that Li_2O_2 begins to decompose to Li_2O beginning at 250 °C, as seen in the O 1*s* (Fig. 4.14b) and L 1*s* (Fig. 4.14c) spectra. The atomic percentage of O or Li in Li_2O increases with continued heating to 500 °C while the percentage of O or Li in Li_2O_2 decreases (Fig. 4.14e, f). XPS also revealed the appearance of Li_2CO_3 at temperatures as low as 150 °C (Fig. 4.14a, d), which is attributed to the reaction between surface Li_2O_2 and hydrocarbon species likely present as contamination in the XPS chamber. This finding suggests that it is important to consider the kinetics of Li_2O_2 reactivity with carbon-based Li–O_2 cell components, including trace contaminants.

4.3.3 Chemical Reactivity Between Li_2O_2 and Ether-Based Electrolytes

While the bulk of the discharge product is Li_2O_2 in ether electrolytes, recent studies have shown that Li_2O_2 can react with ether-based electrolyte salts and solvents. For example, ex situ XPS has been employed by Younesi et al. [61] to study species formed during discharge of electrodes (super P + α-MnO_2 + PVdF) in 0.5 M LiB$(CN)_4$ in polyethylene glycol dimethyl ether (PEGDME) and TEGDME electrolytes. XPS spectra (C 1*s*, O 1*s*, B 1*s*, F 1*s*) of cycled positive electrodes in comparison with those of pristine and solvent-soaked positive electrodes are shown in Fig. 4.15. The C 1*s* and F 1*s* spectra shown in Fig. 4.15a, d indicate the growth of $–CH_2$ and $–CF_2$ groups in PVdF (at 286.6 and 291.0 eV in C 1*s*), as well as the appearance of LiF (from peak at 685.0 eV in F 1*s*) formed from the degradation of binder (which is the only F source). The cycled electrodes were also found to consist of products formed as a result of salt decomposition. The B 1*s* spectra (Fig. 4.15c) show that new B-containing compounds in addition to the salt have developed after cycling, where additional B 1*s* components at 190.1 and 192.7 eV (B–F or B–O bonds, e.g., Li_3BO_5 [62] or B_2O_3 [63]) were noted after cycling in PEGDME and TEGDME, respectively. This suggests that some of the bonds between B and C in the salt are broken during cycling. Additionally, two O 1*s* components centered around 531.5–532.0 and 533.5 eV were found to increase after cycling (Fig. 4.15b). Considering the O 1*s* binding energy (e.g., Li_2O_2 531.3 eV, carboxylate/carbonate (C=O) 532.0 eV, ethers (C–O) 533.6 eV) [64] and the C 1*s* binding energy (e.g., carboxylate 289 eV, carbonate 290 eV, ethers 287 eV) [64] of various reference compounds, the newly developed oxygenated species can be related to Li_2O_2 and carboxylates such as Li-formate/Li-acetate. This work clearly demonstrates that both the electrolyte salt and solvent can participate in Li–O_2 parasitic reactions and is in agreement with other studies suggesting a high degree of chemical reactivity between Li–O_2 reaction products/intermediates and electrolyte salt [15, 64, 65]. Veith et al. [65] have also shown that reaction products in binder-free Li–O_2 electrodes consisting of graphite foam and discharged in LiTFSI, $LiBF_4$, $LiPF_6$, and $LiClO_4$-containing electrolytes contain lithium halide species, likely formed from the chemical reaction of oxygen radical intermediates (such as O_2^-) with the salt. These studies point to the need for chemically stable electrode and electrolyte components for practical Li–O_2 battery chemistries to be realized.

4.3.4 In Situ APXPS Studies of Rechargeable Li–O_2 Solid-State Cells

Probing the Li–O_2 chemistry during battery operation in the absence of chemical instabilities is critical to provide further insights into the intrinsic behavior of Li–O_2 redox couples. Recently, we utilized in situ ambient pressure XPS (APXPS) to

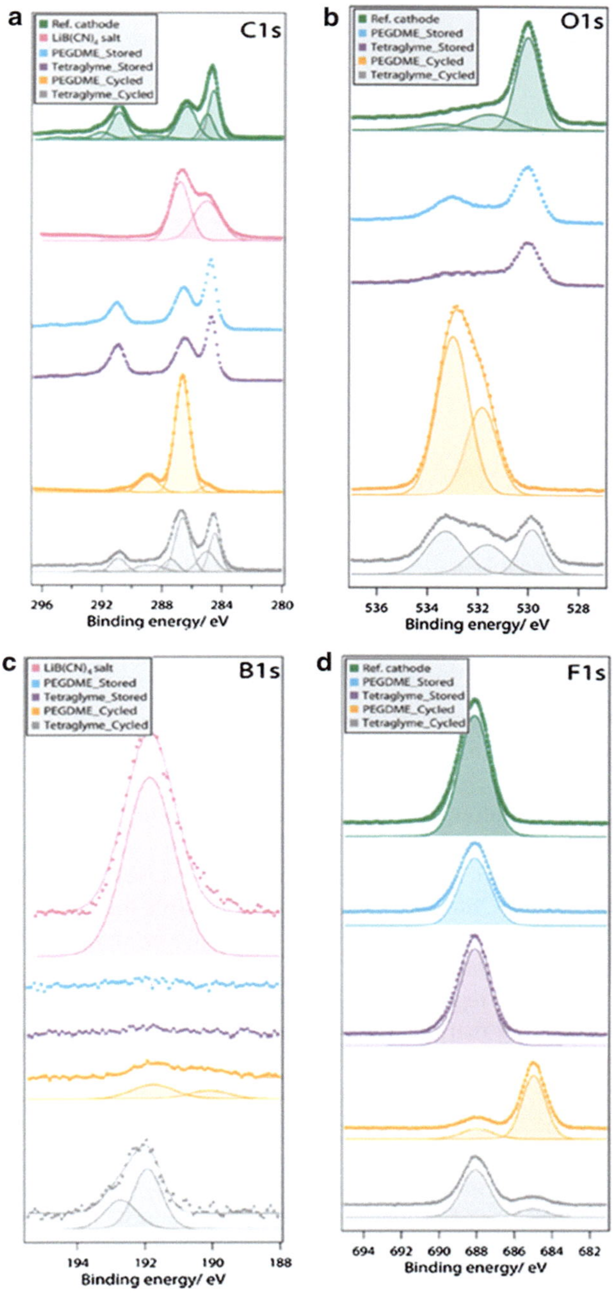

Fig. 4.15 C 1*s* (**a**), O 1*s* (**b**), B 1*s* (**c**), and F 1*s* (**d**) XPS spectra of reference and cycled cathodes using $LiB(CN)_4$ in PEGDME or TEGDME. Reprinted with permission in all media from Younesi R, Hahlin M, Treskow M et al (2012), J. Phys. Chem. C 116:18597-18604. Copyright (2012), American Chemical Society

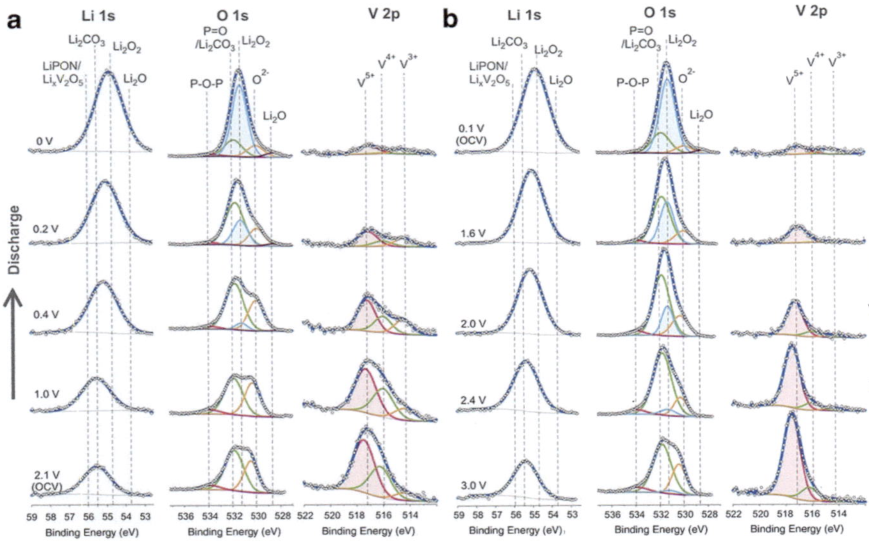

Fig. 4.16 In situ APXPS data of the Li 1*s*, O 1*s*, and V 2*p* regions collected (**a**) during discharge and (**b**) during charge in $p(O_2) = 5 \times 10^{-4}$ atm with the LLTO/LiPON/$Li_xV_2O_5$ cell. Reproduced with permission in print and electronic media from [67]

examine Li–O_2 electrochemistry on a lithium vanadium oxide ($Li_xV_2O_5$) surface using a solid-state electrolyte (lithium phosphorous oxynitride, LiPON) and a non-lithium negative electrode (lithiated lithium titanium oxide, LLTO) [66]. This system intentionally used a carbon-free electrode and electrolyte to investigate the intrinsic Li–O_2 (electro)chemistry.

Examining the surface chemistry of the $Li_xV_2O_5$ electrode with applied bias under $p(O_2) = 5 \times 10^{-4}$ atm, we observed the reversible formation and removal of Li_2O_2 and Li_2O on the surface of $Li_xV_2O_5$. Figure 4.16a shows high-resolution in situ APXPS spectra of the Li 1*s*, O 1*s*, and V 2*p* regions during discharge. The intensity of the Li 1*s* region increased while that of V 2*p* region decreased significantly, indicating that newly formed Li–O_2 discharge products covered the $Li_xV_2O_5$ surface. The Li 1*s* binding energy shifts from 55.5 to 54.8 eV, which is consistent with the binding energy of Li_2O_2 [60]. This observation is further substantiated by the appearance and growth of a new peak centered at 531.3 eV in the O 1*s* region, which became dominant at the end of discharge (at 0 V_{cell}) (Fig. 4.16a). This new species can be assigned to Li_2O_2 based on the binding energies of both the O 1*s* peaks of a Li_2O_2 reference sample (O 1*s* 531.3 eV) [60].

Interestingly, we note that the C 1*s* carbonate component at ~290.0 eV increased about 15 % upon discharge in $p(O_2) = 5 \times 10^{-4}$ atm (Fig. 4.17b) while no changes were observed upon discharge in UHV (Fig. 4.17a). This might result from reaction between Li_2O_2 and surface hydrocarbon species, which is consistent with findings from XPS studies under in situ heating detailed above (Fig. 4.14).

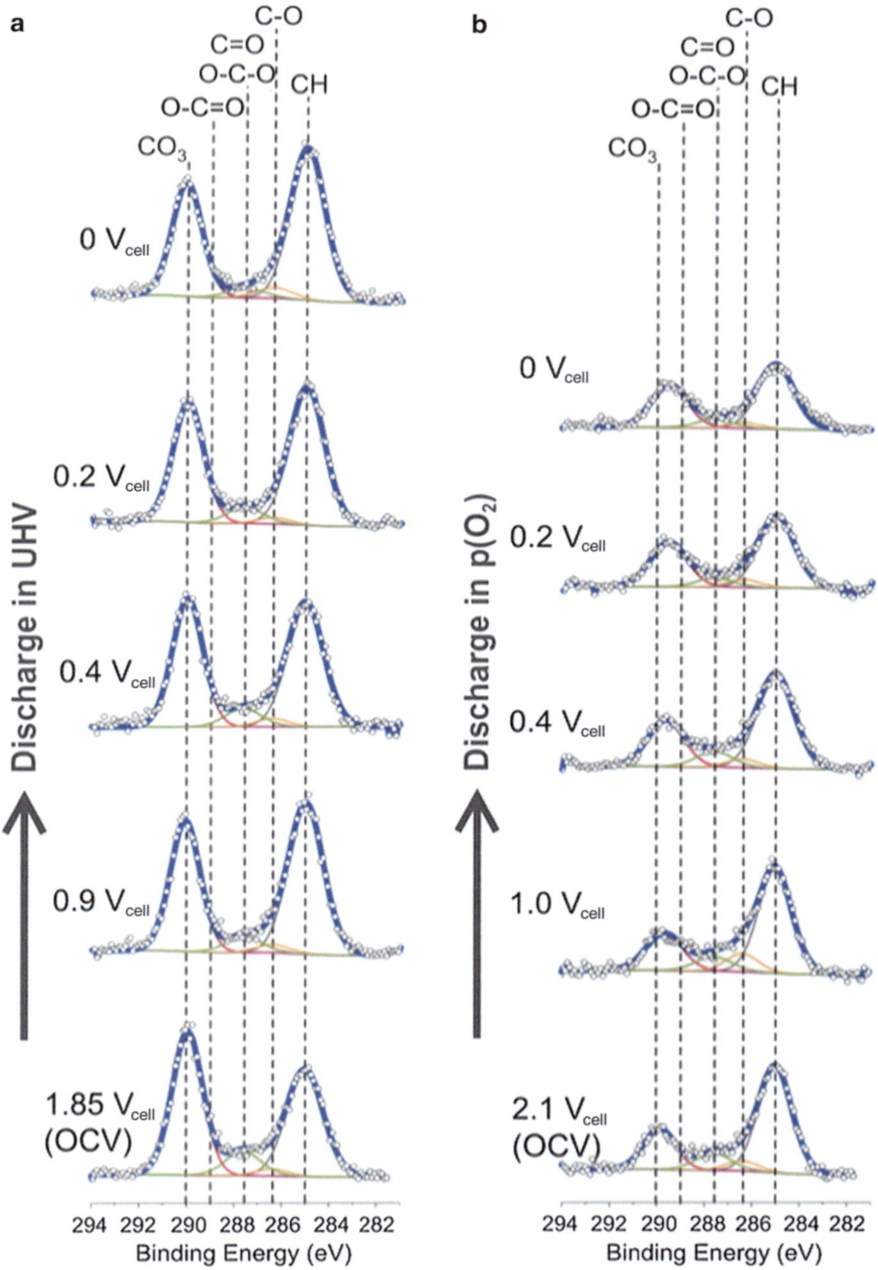

Fig. 4.17 In situ APXPS data of the C 1*s* region collected (**a**) during discharge in UHV and (**b**) during discharge in $p(O_2) = 5 \times 10^{-4}$ atm with the LLTO/LiPON/$Li_xV_2O_5$ cell. Reproduced with permission in print and electronic media from [67]

Upon charge, the oxidation of Li_2O_2 was found to initiate at a low potential of ~3.2 V. To explain this low overpotential, several differences between solid-state and conventional Li–O_2 cells are noted. First, Li_2O_2 formed on the surface of $Li_xV_2O_5$ is approximately 0.7 nm in average thickness assuming uniform coverage, which is significantly thinner than typical deposits in high-surface area electrodes (typically up to tens or hundreds of nanometer average thickness depending on the morphology, Fig. 4.9). The low overpotential required to charge very thin Li_2O_2 in solid-state cells agrees well with the typical low charging voltages for the first sub-nanometer thicknesses in conventional O_2 cells, as discussed in more detail in Sect. 4.4 and as seen in Fig. 4.19. Second, the surfaces of Li_2O_2 could be less reactive toward the electrolyte on charge in solid-state cells than in carbon cells, which could potentially influence the charging process. Lastly, the Li_2O_2 formed in a solid-state cell may have different structures, e.g., stoichiometry and electronic properties, compared to that formed in a conventional liquid-based cell. Further studies are needed to understand the morphology and its link to charging behavior in solid-state cells. Overall, these results highlight that understanding the surface chemistry of Li_2O_2 is critical to enable improvements in stable electrode design. In situ XPS uniquely allows researchers to determine if these conditions are satisfied at the sub-nanometer scale and in real time.

4.4 Influence of Discharge Product Morphology and Surface Chemistry on Li_2O_2 Oxidation

Understanding the relationship between discharge product morphology and charging behavior can provide essential insight into the physical properties of Li_2O_2 that influence the intrinsic OER kinetics. We have systematically investigated this link through the charging of controlled discharge morphologies of particle and disc-like Li_2O_2. These two distinct morphologies lead to notably different charging potential profiles and rate capabilities, as shown in Fig. 4.18 [29]. Charging of small particles over a wide range of currents (5–500 mA/g_C) exhibited a sloping voltage profile (Fig. 4.18a), where charging voltages are largely lower than 4 V at rates equal to or lower than 100 mA/g_C. The high-voltage tail at the end of charge (corresponding to the last 100–150 mAh/g_C of charge capacity) is likely related to oxidation of limited amounts of Li_2CO_3 formed parasitically on the carbon surface [4, 6]. The rate dependence of the OER overpotential of small particles is shown in Fig. 4.18c, where the potential after the first 200 mAh/g_C of charge from Fig. 4.18a (corresponding to the first 20 % of charge) is plotted against the carbon true surface area-normalized current in a Tafel graph. The Tafel slope related to charging of small particles was ~340 mV/decade, which is comparable to that reported previously with electrochemically discharged [9] and commercial Li_2O_2 pre-loaded [67] Vulcan carbon electrodes, although the physical origin of such a slope requires further investigation. The exchange current density of the OER kinetics of small

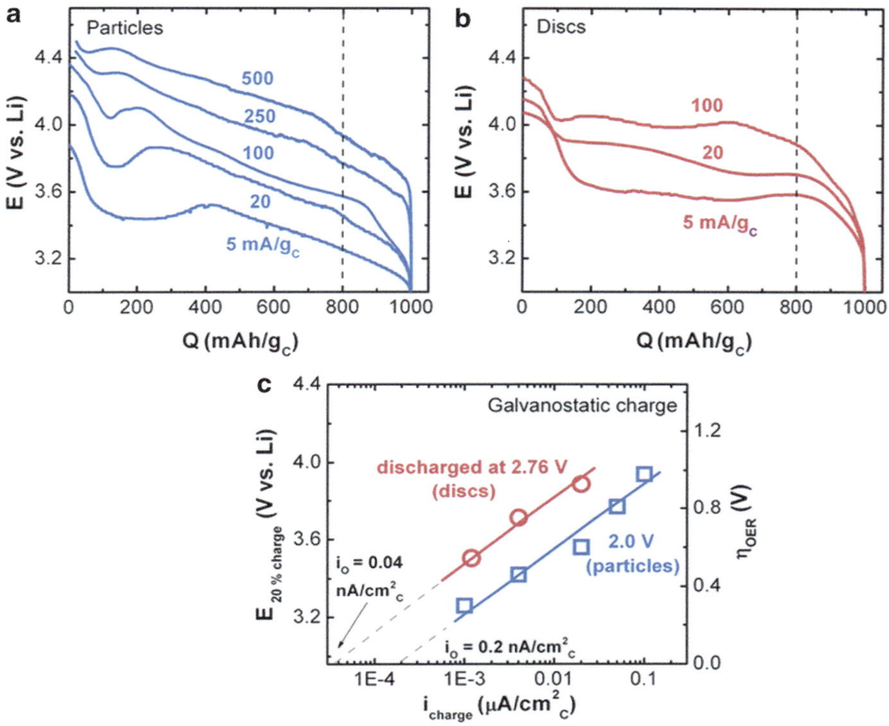

Fig. 4.18 Galvanostatic rate capability charging of (**a**) Li_2O_2 discharged at 2.0 V to 1,000 mAh/g$_C$ with particle morphologies and (**b**) Li_2O_2 discharged at 2.76 V to 1,000 mAh/g$_C$ with disc morphologies. (**c**) Tafel plot of galvanostatic charge voltage vs. carbon true surface area-normalized current for electrodes discharged under potentiostatic conditions at 2.76 or 2.0 V to 1,000 mAh/g$_C$. The current and voltage data were obtained from (**a**) and (**b**) at a fixed capacity of 800 mAh/g$_C$ (corresponding to 20 % charge). The right-hand axis shows the OER overpotential referenced to the thermodynamic potential of Li_2O_2, $E_{Li_2O_2} = 2.96\,V$ [29]. Reproduced by permission in print and electronic media of The Royal Society of Chemistry

Li_2O_2 particles was 0.2 nA/cm^2_C, which is ten times smaller than 1.5 nA/cm^2_C noted on discharge (Fig. 4.4d).

In contrast, charging of discs was found to exhibit two distinct regions on charge (Fig. 4.18b): (1) a sloping profile was observed during the first ~200 mAh/g$_C$ of charging, where the charging potentials in the sloping region were higher for discs than for the small particles; (2) disc morphologies exhibited a voltage plateau upon continued charging beyond the first ~200 mAh/g$_C$ (Fig. 4.18b). As evident from the Tafel plot in Fig. 4.18c, Li_2O_2 discs, which exhibited a similar Tafel slope as the particles, require significantly higher (~200 mV) overpotentials upon charge compared to particles. In addition, discs exhibited a significantly lower exchange current density of 0.04 nA/cm^2_C than particles (1.5 nA/cm^2_C).

It is interesting to compare the observed charging profiles obtained with disc and particle morphologies to that of Li_2O_2 reported elsewhere in the literature, which typically follows one of two distinct trends. On the one hand, thin films (<1 nm) [6, 30, 68] of Li_2O_2 formed at relatively high true surface area currents (> 1 μA/cm^2_C) on glassy carbon electrodes have been reported to charge at relatively low overpotentials (majority of charge occurring below 4.0 V) and to exhibit a sloping charging profile, with the charging onset potential close to the thermodynamic potential of 2.96 V [6, 30]. Such observations have formed the basis for arguments that the OER overpotential for Li_2O_2 is intrinsically low. On the other hand, charging of Li_2O_2 in more conventional Li–O_2 cells employing high-surface area positive electrodes (e.g., Vulcan carbon [9, 11] or CNTs [4, 12]) and with higher average thicknesses of discharge product (~1–10 nm) [4, 9] has been observed to exhibit the two major characteristic features, namely an initially sloping profile followed by a higher-voltage plateau, similar to what is observed with Li_2O_2 discs. Our group proposed that the sloping region in the initial charging stages is related to delithiation of the first sub-nanometer thickness (< 0.05 μAh/cm^2_{true}) of Li_2O_2 or surface nonstoichiometric $Li_{2-x}O_2$ [9], which dominates the charging profile for very thin films (<1 nm) of Li_2O_2 but represents a relatively small charging capacity for thicker (1–10 nm) deposits. As shown in Fig. 4.19a, this first sub-nanometer thickness of Li_2O_2 was found across many studies to charge at consistently low potentials of ~3.2–3.6 V regardless of the total Li_2O_2 loading in the electrode. We found that discharged CNTs with small Li_2O_2 particle morphologies agree well with this trend, whereas Li_2O_2 discs require higher overpotentials (Fig. 4.19a). In contrast, the charging plateau region, which is rate-sensitive (Fig. 4.19b) [9], was attributed to a two-phase charging process, for example, bulk oxidation of Li_2O_2 by a direct two-electron or a disproportionation process. We believe such a two-stage mechanism to be representative of the intrinsic kinetics of thicker, bulk-like deposits of Li_2O_2 that form in high-surface area carbon electrodes in practical cells. However, the nature (e.g., morphology and composition) of the Li_2O_2 formed electrochemically in many earlier studies [9, 30, 68], which can potentially influence not only the kinetics but also electronic transport during charge, is not well understood.

The minimum overpotentials required to drive Li_2O_2 oxidation of particles and discs were probed by PITT [9, 69–71], which can provide further insights into electrochemical processes occurring in different regions of the charging voltage profiles [29]. The voltage responses of Li_2O_2 discs under PITT charging (Fig. 4.20a) exhibit a sloping voltage profile during the first 200 mAh/g_C of charge followed by a voltage plateau, which agrees well with the profiles obtained from galvanostatic charging (Fig. 4.18b). The sloping voltage profile upon charge reflects initial Li^+ removal from Li_2O_2 discs via a solid-solution process such as that observed in $LiCoO_2$ [72] and in the solid-solution regime of $LiFePO_4$ [73]. This hypothesis is further supported by the fact that the current response in the sloping region (Fig. 4.20b) revealed a monotonic decay, suggesting a diffusion-limited process such as Li^+ de-intercalation from $LiCoO_2$ [72, 74], $LiFePO_4$ [70], Li_yFeSO_4F [69], and graphite [75, 76].

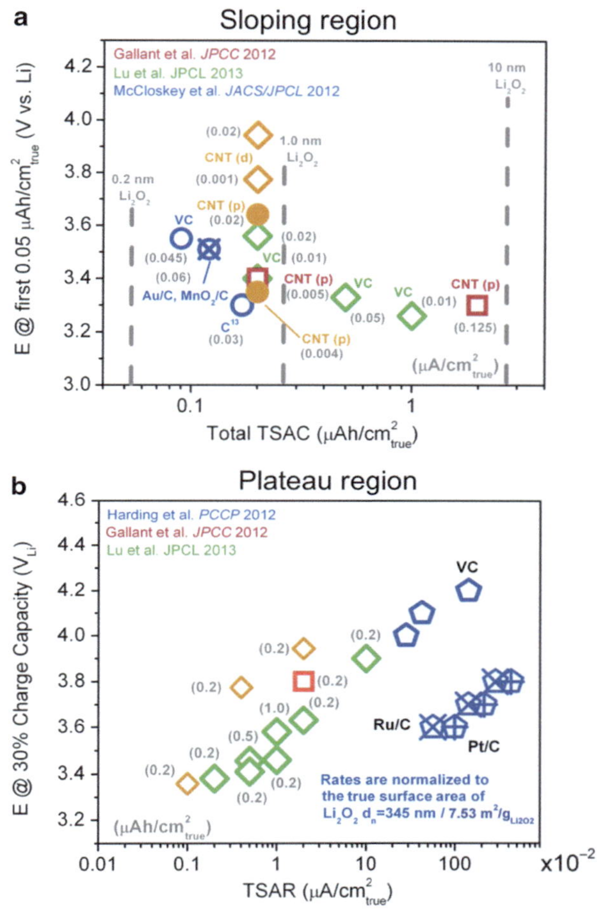

Fig. 4.19 Charge potentials (at the first 0.05 μAh/cm^2_{true} or ~0.2 nm of Li_2O_2 average thickness) of select studies as a function of true surface area-normalized capacity (TSAC) (μAh/cm^2_{true}) at the end of discharge. The *gray dashed lines* denote the Li_2O_2 thickness estimated from the total TSAC assuming a uniform distribution of Li_2O_2 on the electrode surface. The number in the parentheses denotes the true surface area-normalized rate (TSAR) during charging in μA/cm^2_{true}. *Orange-filled* and *open symbols* correspond to Li_2O_2 particles (labeled "p") and discs (labeled "d"), respectively, in CNT electrodes from Fig. 4.18. (**b**) *Green*, *red*, and *orange symbols*: charge potentials (at the first 30 % of charge capacity) of select studies as a function of TSAR (μA/cm^2_{true}). *Open orange symbols* correspond to Li_2O_2 discs shown in Fig. 4.18b. The number in the parentheses denotes the TSAC in μAh/cm^2_{true}. For *blue symbols*, the specific surface area of Li_2O_2 is estimated to be 7.53 $m^2/g_{Li_2O_2}$ assuming spherical particles of ~345 nm diameter. Adapted with permission in all media from Lu Y-C and Shao-Horn Y (2013), J Phys Chem Lett 4:93-99. Copyright 2013, American Chemical Society

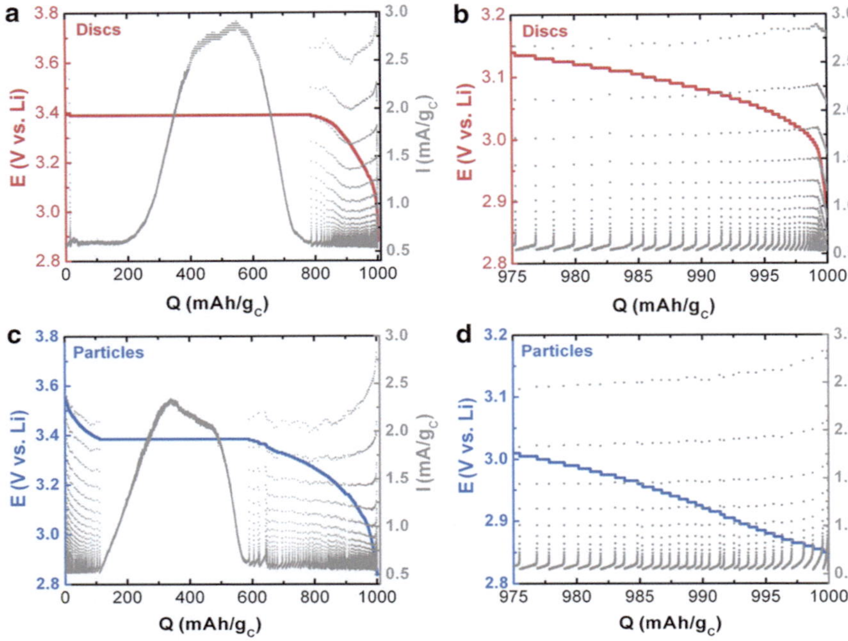

Fig. 4.20 PITT charging of CNT electrodes discharged to 1,000 mAh/g_C at (**a**, **b**) 2.76 V with Li_2O_2 disc morphologies and (**c**, **d**) 2.0 V with particle morphologies. The voltage change per step was 5 mV, and the current cutoff was 0.5 mA/g_C. (**b**) and (**d**) show the initial charging stages (1,000–975 mAh/g_C) of the data in (**a**) and (**c**), respectively [29]. Reproduced by permission in print and electronic media of The Royal Society of Chemistry

Following the initial sloping process, a voltage plateau at ~3.4 V was observed for the rest of charge, where the current exhibited a significant rise to a maximum value at ~2–3 mA/g_C and a subsequent decay (Fig. 4.20a) [29]. Such a voltage response is indicative of a two-phase reaction, which is supported by the observed current response that is characteristic of a kinetically limited nucleation and growth [9, 67, 70, 77, 78] process. Such a two-phase process with corresponding nucleation and growth-type current response is similar to that observed recently during PITT charging of electrochemically formed Li_2O_2 in Vulcan carbon electrodes [9] and also to that observed during potentiostatic charging of Vulcan carbon electrodes pre-loaded with Li_2O_2 [67]. Additionally, a comparable PITT current response has been observed in Li-ion battery compounds exhibiting known two-phase de-intercalation kinetics related to spinodal decomposition such as $LiFePO_4$ [70]. The minimum charging potential for the two-phase reaction, ~3.4 V, is consistent with that observed previously [9] upon PITT charging in Vulcan carbon electrodes from lower discharge capacities (200 mAh/g_C) and in a different electrolyte solvent (TEGDME) than what is used here and appears to be characteristic of electrochemically formed Li_2O_2.

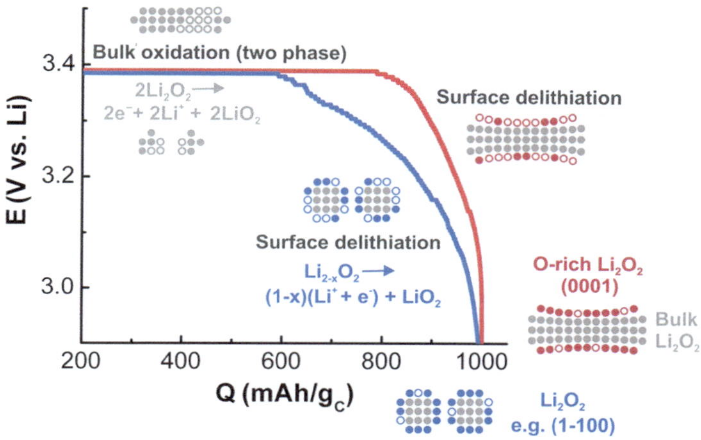

Fig. 4.21 Proposed charging processes for Li_2O_2 with disc (*gray/red*, typical dimensions ~50–200 nm) or particle (*gray/blue*, typical dimensions < 20 nm) morphologies overlaid on PITT data from Fig. 4.20. Discs surfaces are largely O-rich (0001) with LiO_2-like surface species, while particles consist predominantly of Li_2O_2-like surfaces. During the initial stage of charging (to ~800 mAh/g_C for discs and to ~600 mAh/g_C for particles), both discs and particles exhibit a sloping voltage profile attributed to solid-solution-like delithiation from surfaces. Upon further charge, discs and particles exhibit a voltage plateau corresponding to bulk oxidation via a two-phase process with a resulting voltage plateau at ~3.4 V [29]. Reproduced by permission in print and electronic media of The Royal Society of Chemistry

Consistent with the discs, PITT data of Li_2O_2 particles (Fig. 4.20c, d) also exhibit the two stages on charge with the voltage plateau occurring at ~3.4 V. However, the sloping region corresponding to solid-solution Li removal from small Li_2O_2 particles was found to extend over a significantly wider capacity (to 40 % of the charge capacity compared to 20 % of the charge capacity for discs). While a charging plateau is clearly observed under PITT conditions for the Li_2O_2 particles, such a voltage plateau is not evident even at very low galvanostatic rates such as 5–20 mA/g_C in Fig. 4.18a, suggesting that the solid-solution pathway may be more kinetically facile than the two-phase pathway in the particles during charge.

We propose that the differences in charging profiles of Li_2O_2 discs and particles can be attributed to these morphological and surface compositional and electronic differences, as shown schematically in Fig. 4.21. Upon initial charging, Li_2O_2 discs consisting predominantly of O-rich (0001) surfaces [16] with LiO_2-like electronic states (Fig. 4.11a) exhibit limited capacity (~200 mAh/g_C) in the sloping region attributed to solid-solution-like delithiation. In contrast, particles consisting of less O-rich and/or stoichiometric surfaces (Fig. 4.11a, b), for example, the (1-100) facet [57], exhibit a significantly enhanced (~400 mAh/g_C) sloping region with lower charging overpotentials for a given amount of Li extracted. The lower charging overpotentials of particles with proposed (1-100) surfaces compared to discs with (0001) surfaces are consistent with the fact that lower kinetic overpotentials (0.15–0.2 V) are predicted on charge for (1-100) facets compared to the oxygen-rich

(0001) facets (0.36 V) as computed recently from DFT calculations by Hummelshoj et al. [57]. It is also important to note that the small particles (<20 nm) have a higher surface area to volume fraction than the discs (50–200 nm) and any surface-related phenomena influencing the charging pathways will likely be enhanced for the particles. Charging in the sloping region (Fig. 4.21) therefore corresponds to delithiation and subsequent removal (e.g., by disproportionation of the remaining delithiated phase LiO_2) of the different surface regions of discs or particles. This is supported by the fact that XANES O K-edge data of 30 % charged discs (charged just beyond the end of the sloping region to the onset of the plateau region) showed the disappearance of LiO_2-like features (peak occurring at 534.9 eV, Fig. 4.11c) and instead exhibited only Li_2O_2-like features. In contrast, the XANES data of particles remained largely unchanged (i.e., still retained the enhanced 532.1 eV peak) after partial charging (Fig. 4.11b) within the sloping regime, consistent with the fact that there is less of a distinction between surface and bulk in the small particles (i.e., small particles are largely "surface-like") compared to discs with distinct surface and bulk regions.

Following complete delithiation and removal of the surfaces of discs and particles, which exposes the interior bulk Li_2O_2, charging then proceeds via a two-phase process for both the particles and the discs (Fig. 4.21). We propose that this two-phase process involves bulk oxidation of Li_2O_2 involving the coexistence of phases, e.g., stoichiometric Li_2O_2 and O-rich (e.g., LiO_2) species within the particles and discs, where the latter can further disproportionate to generate O_2 on charge. Further detailed studies are needed to investigate this two-phase mechanism and the associated kinetics in discs and particles.

4.5 Electronic Charging Limitations at High Overpotentials Probed by In Situ TEM

Additionally, we have reported [79] the first in situ TEM study of electrochemical oxidation of Li_2O_2 disc-/toroid-like particles, where electron transport limits the oxidation kinetics of Li_2O_2 at very high overpotentials (having charge transfer kinetics faster than transport). To investigate the electrochemical oxidation process, a solid-state in situ microbattery, consisting of an MWCNT/Li_2O_2 positive electrode (pre-discharged galvanostatically to a condition forming discs) and a Si nanowire (delithiated) negative electrode coated with $LiAlSiO_x$ solid electrolyte (SE) [80], was assembled and then placed inside a TEM. Figure 4.22a is a schematic representation of the assembled in situ cell. The contacted Li_2O_2 particle was oxidized potentiostatically through the application of a potential between the external leads of the microbattery. Typically, very large applied potentials in the range of 8–10 V are needed to induce oxidation of Li_2O_2 particles within the TEM experiment timeframe (typically 30–60 min, compared to tens or hundreds of hours in Li–O_2 cells). The applied potentials required for the electrochemical oxidation of

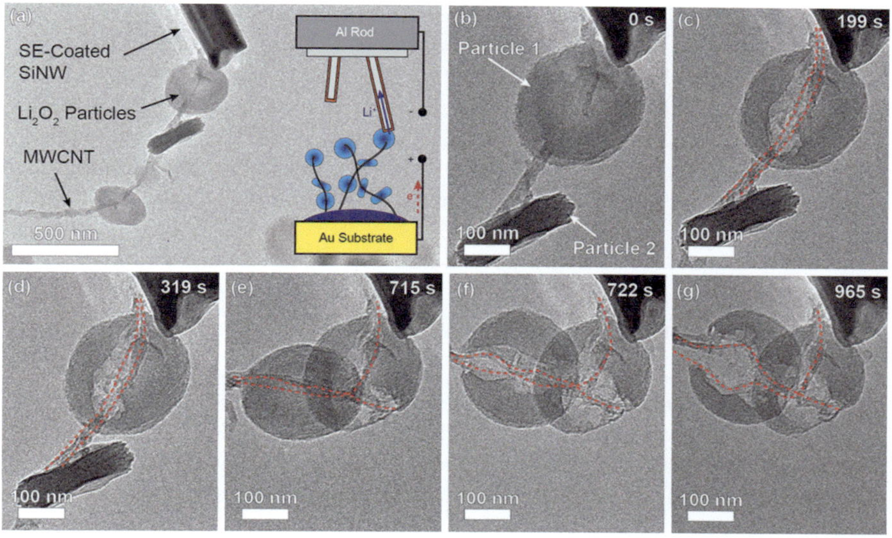

Fig. 4.22 In situ TEM oxidation of Li_2O_2 particles. (**a**) Schematic illustration of the in situ TEM microbattery superimposed over a low magnification TEM image of a solid electrolyte-coated Si nanowire contacting a single Li_2O_2 particle. (**b**) Higher magnification TEM image of the particles in (**a**), showing a MWCNT bundle contacting two physically separated Li_2O_2 particles labeled as Particle 1 and Particle 2, respectively. (**c–g**) Oxidation of Particles 1 and 2 during application of a 10 V_{cell} potential to the MWCNT/Li_2O_2 positive electrode against the Si NW negative electrode. Adapted with permission in all media from Liu Y, Mitchell RR, Zhong L et al (2013) Nano Letters 13:2209-2214.

Li_2O_2 in the TEM were therefore significantly greater than those observed in macroscale Li–O_2 cells [4, 16] and provide useful insights into processes occurring at very high overpotentials.

In situ TEM imaging revealed that the oxidation of individual Li_2O_2 particles at very high overpotentials initiated preferentially at the MWCNT/Li_2O_2 interface, as evidenced by the development of a light-contrast stripe beginning along the CNT axes (highlighted by red dashed lines) of the MWCNT bundle contacting the Li_2O_2 particle (Particle 1 in Fig. 4.22c). The formation of the light-contrast stripe in TEM can be explained by thinning of Li_2O_2 particles at the Li_2O_2/MWCNT interface associated with electrochemical oxidation of Li_2O_2 ($Li_2O_2 \rightarrow Li^+ + O_2 + 2e^-$) and the concurrent release of O_2, where lithium ions migrate through the SE into the Si NW (Fig. 4.22a) and electrons flow through MWCNTs to the Au substrate. The preferential oxidation at the MWCNT/Li_2O_2 interface, but not at the interface between Li_2O_2/SE, suggests that the electrochemical oxidation of Li_2O_2 is electron-transport-limited instead of Li^+-transport-limited at very high overpotentials.

This work revealed processes occurring under limiting high rate charging conditions and suggested that electronic, and not Li^+, transport would govern the rate capability of rechargeable Li–O_2 batteries operating at the high rates required for electrical vehicle applications [21]. Future experiments at lower overpotentials and longer times will be useful to study the kinetics of Li_2O_2 oxidation at intermediate charge rates, where both transport processes and surface electron transfer kinetics influence the oxidation rate. Our findings suggest that electrodes with large specific surface areas, which maximize the Li_2O_2/electrode interfacial area, and high electronic conductivity, providing facile electron transport to the reaction sites, are desirable for the design of high rate rechargeable Li–O_2 batteries.

4.6 Conclusions and Perspective

Significant opportunities exist to improve fundamental understanding of discharge and charge kinetics and processes occurring in Li–O_2 batteries. As detailed in this chapter, the link between physical properties (morphology and surface vs. bulk chemistry) of Li_2O_2 and the corresponding kinetics of Li_2O_2 formation and oxidation is still not well understood. Important topics warranting further investigation in this area include the intrinsic kinetics of Li_2O_2 formation and decomposition in the absence of contaminants (substrate, electrolyte, and ambient), the role of the surface (e.g., carbon, Au, or oxide) on the discharge kinetics and morphology, the detailed formation processes of Li_2O_2 at high and low overpotentials, the electronic nature (surface and bulk) of electrochemically formed Li_2O_2, the Li_2O_2 intrinsic oxidation pathways, and the thermodynamic and kinetic barriers influencing the overpotentials observed on charge.

In particular, improved morphological, chemical, and kinetic understanding in non-carbon electrodes will become increasingly relevant for the development of practical O_2 electrodes. Although a significant portion of our morphological studies on CNTs focused on the first discharge and charge, which corresponds predominantly to formation and decomposition of Li_2O_2 with limited formation of Li_2CO_3, studies in the absence of parasitic reactions are required to elucidate reaction mechanisms. In situ XPS presents an exciting capability for probing the chemical stability and kinetics of Li_2O_2 oxidation by removing the influence of carbon present in most electrolytes and substrates. Further studies that integrate both chemical and morphological understanding in carbon-free cells are needed to examine the implications of these findings for practical devices. Overall, increased fundamental understanding of the areas outlined in this chapter will be critical to enable improved, rational design of Li–O_2 electrodes that can combine high rate capability on discharge, high round-trip efficiency on charge, and reversible formation and oxidation of Li_2O_2 for improved cycle life.

Acknowledgments This work was supported in part by the MRSEC Program of the National Science Foundation under award number DMR-0819762; by the Ford-MIT Alliance and the Assistant Secretary for Energy Efficiency and Renewable Energy, Office of FreedomCAR and Vehicle Technologies of the DOE (DE-AC03-76SF00098 with LBNL); and by the US Department of Energy's US-China Clean Energy Research Center for Clean Vehicles (Grant DE-PI0000012).

This work was performed in part at the Center for Nanoscale Systems (CNS) at Harvard University, a member of the National Nanotechnology Infrastructure Network (NNIN), which is supported by the National Science Foundation under NSF award number ECS-0335765. B.M.G. acknowledges a National Science Foundation Graduate Research Fellowship and D.G.K. acknowledges a Total Graduate Student Fellowship.

References

1. Bruce PG, Freunberger SA, Hardwick LJ et al (2012) Li-O_2 and Li-S batteries with high energy storage. Nat Mater 11:19–29
2. Lu Y-C, Gallant BM, Kwabi DG et al (2013) Lithium-oxygen batteries: bridging mechanistic understanding and battery performance. Energy Environ Sci 6:750–768
3. Freunberger SA, Chen Y, Drewett NE et al (2011) The lithium–oxygen battery with ether-based electrolytes. Angew Chem Int Ed 50:8609–8613
4. Gallant BM, Mitchell RR, Kwabi DG et al (2012) Chemical and morphological changes of Li-O_2 battery electrodes upon cycling. J Phys Chem C 116:20800–20805
5. McCloskey BD, Bethune DS, Shelby RM et al (2011) Solvents' critical role in nonaqueous lithium-oxygen battery electrochemistry. J Phys Chem Lett 2:1161–1166
6. McCloskey BD, Speidel A, Scheffler R et al (2012) Twin problems of interfacial carbonate formation in nonaqueous Li-O_2 batteries. J Phys Chem Lett 3:997–1001
7. Thotiyl MMO, Freunberger SA, Peng Z et al (2013) The carbon electrode in nonaqueous Li-O_2 cells. J Am Chem Soc 135:494–500
8. Xu W, Hu JZ, Engelhard MH et al (2012) The stability of organic solvents and carbon electrode in nonaqueous Li-O_2 batteries. J Power Sources 215:240–247
9. Lu Y-C, Shao-Horn Y (2012) Probing the reaction kinetics of the charge reactions of nonaqueous Li-O_2 batteries. J Phys Chem Lett 4:93–99
10. Peng Z, Freunberger SA, Chen Y et al (2012) A reversible and higher-rate Li-O_2 battery. Science 337:563–566
11. Lu Y-C, Kwabi DG, Yao KPC et al (2011) The discharge rate capability of rechargeable Li-O_2 batteries. Energy Environ Sci 4:2999–3007
12. Mitchell RR, Gallant BM, Thompson CV et al (2011) All-carbon-nanofiber electrodes for high-energy rechargeable Li-O_2 batteries. Energy Environ Sci 4:2952–2958
13. Freunberger SA, Chen YH, Peng ZQ et al (2011) Reactions in the rechargeable lithium-O_2 battery with alkyl carbonate electrolytes. J Am Chem Soc 133:8040–8047
14. Black R, Lee JH, Adams B et al (2013) The role of catalysts and peroxide oxidation in lithium-oxygen batteries. Angew Chem Int Ed 52:392–396
15. Black R, Oh SH, Lee JH et al (2012) Screening for superoxide reactivity in Li-O_2 batteries: effect on Li_2O_2/LiOH crystallization. J Am Chem Soc 134:2902–2905
16. Mitchell RR, Gallant BM, Shao-Horn Y et al (2013) Mechanisms of morphological evolution of Li_2O_2 particles during electrochemical growth. J Phys Chem Lett 4:1060–1064
17. Oh SH, Nazar LF (2012) Oxide catalysts for rechargeable high-capacity Li-O_2 batteries. Adv Energy Mater 2:903–910
18. Wang Z-L, Xu D, Xu J-J et al (2012) Graphene oxide gel-derived, free-standing, hierarchically porous carbon for high-capacity and high-rate rechargeable Li-O_2 batteries. Adv Funct Mater 22:3699–3705
19. Chen H, Armand M, Demailly G et al (2008) From biomass to a renewable $Li_xC_6O_6$ organic electrode for sustainable Li-ion batteries. ChemSusChem 4:198–198
20. Kang KS, Meng YS, Breger J et al (2006) Electrodes with high power and high capacity for rechargeable lithium batteries. Science 311:977–980

21. Girishkumar G, McCloskey B, Luntz AC et al (2010) Lithium–air battery: promise and challenges. J Phys Chem Lett 1:2193–2203
22. Lee JS, Kim ST, Cao R et al (2011) Metal-air batteries with high energy density: Li-air versus Zn-air. Adv Energy Mater 1:34–50
23. Lu Y-C, Gasteiger HA, Shao-Horn Y (2011) Catalytic activity trends of oxygen reduction reaction for nonaqueous Li-air batteries. J Am Chem Soc 133:19048–19051
24. Jung H-G, Hassoun J, Park J-B et al (2012) An improved high-performance lithium-air battery. Nat Chem 4:579–585
25. McCloskey BD, Scheffler R, Speidel A et al (2011) On the efficacy of electrocatalysis in nonaqueous Li-O_2 batteries. J Am Chem Soc 133:18038–18041
26. Gogotsi Y, Simon P (2011) True performance metrics in electrochemical energy storage. Science 334:917–918
27. Fehrenbacher K (2011) 10 companies to watch out for of ARPA-E. http://gigaom.com/cleantech/10-companies-to-watch-for-outta-arpa-e. Accessed 10 Sept 2013
28. Polyplus (2009) Advanced lithium battery technology. http://www.polyplus.com/liair.html. Accessed 10 Sept 2013
29. Gallant BM, Kwabi DG, Mitchell RR et al (2013) Influence of Li_2O_2 morphology on oxygen reduction and evolution kinetics in Li-O_2 batteries. Energy Environ Sci 6:2518–2528
30. Viswanathan V, Norskov JK, Speidel A et al (2013) Li-O_2 kinetic overpotentials: Tafel plots from experiment and first-principles theory. J Phys Chem Lett 4:556–560
31. Xu Y, Shelton WA (2011) Oxygen reduction by lithium on model carbon and oxidized carbon structures. J Electrochem Soc 158:A1177–A1184
32. Lu Y-C, Gasteiger HA, Parent MC et al (2010) The influence of catalysts on discharge and charge voltages of rechargeable Li-oxygen batteries. Electrochem Solid-State Lett 13:A69–A72
33. Fan WG, Cui ZH, Guo XX (2013) Tracking formation and decomposition of abacus-ball-shaped lithium peroxides in Li-O_2 cells. J Phys Chem C 117:2623–2627
34. Lee J-H, Black R, Popov G et al (2012) The role of vacancies and defects in $Na_{0.44}MnO_2$ nanowire catalysts for lithium/oxygen batteries. Energy Environ Sci 5:9558–9565
35. Xu D, Wang Z-l, Xu J-j et al (2012) Novel DMSO-based electrolyte for high performance rechargeable Li-O_2 batteries. Chem Commun 48:6948–6950
36. Cota LG, de la Mora P (2005) On the structure of lithium peroxide, Li_2O_2. Acta Crystallogr Sect B Struct Sci 61:133–136
37. Wang ZL (2004) Zinc oxide nanostructures: growth, properties and applications. J Phys Condens Matter 16:R829–R858
38. Goniakowski J, Finocchi F, Noguera C (2008) Polarity of oxide surfaces and nanostructures. Rep Prog Phys 71:016501–016556
39. Hanneman RE, Gatos HC, Finn MC (1962) Elastic strain energy associated with a surfaces of III-V compounds. J Phys Chem Solids 23:1553–1556
40. Radin MD, Tian F, Siegel DJ (2012) Electronic structure of Li_2O_2 0001 surfaces. J Mater Sci 47:7564–7570
41. Mo Y, Ong SP, Ceder G (2011) First-principles study of the oxygen evolution reaction of lithium peroxide in the lithium-air battery. Phys Rev B 84:205446
42. Thachepan S, Li M, Davis SA et al (2006) Additive-mediated crystallization of complex calcium carbonate superstructures in reverse microemulsions. Chem Mater 18:3557–3561
43. Wang TP, Antonietti M, Colfen H (2006) Calcite mesocrystals: "morphing" crystals by a polyelectrolyte. Chem Eur J 12:5722–5730
44. Geng X, Liu L, Jiang J et al (2010) Crystallization of $CaCO_3$ mesocrystals and complex aggregates in a mixed solvent media using polystyrene sulfonate as a crystal growth modifier. Cryst Growth Des 10:3448–3453
45. Liu Z, Wen XD, Wu XL et al (2009) Intrinsic dipole-field-driven mesoscale crystallization of core-shell ZnO mesocrystal microspheres. J Am Chem Soc 131:9405–9412

46. Cao AM, Hu JS, Liang HP et al (2006) Hierarchically structured cobalt oxide (Co_3O_4): the morphology control and its potential in sensors. J Phys Chem B 110:15858–15863
47. Come J, Taberna PL, Hamelet S et al (2011) Electrochemical kinetic study of $LiFePO_4$ using cavity microelectrode. J Electrochem Soc 158:A1090
48. Peng Z, Freunberger SA, Hardwick LJ et al (2011) Oxygen reactions in a non-aqueous Li(+) electrolyte. Angew Chem Int Ed 50:6351–6355
49. Laoire CO, Mukerjee S, Abraham KM et al (2010) Influence of nonaqueous solvents on the electrochemistry of oxygen in the rechargeable lithium-air battery. J Phys Chem C 114:9178–9186
50. Vanelp J, Wieland JL, Eskes H et al (1991) Electronic-structure of CoO, Li-doped CoO, and $LiCoO_2$. Phys Rev B 44:6090–6103
51. Yang SL, Wang DN, Liang GX et al (2012) Soft X-ray XANES studies of various phases related to $LiFePO_4$ based cathode materials. Energy Environ Sci 5:7007–7016
52. Minasian SG, Keith JM, Batista ER et al (2013) Covalency in metal-oxygen multiple bonds evaluated using oxygen k-edge spectroscopy and electronic structure theory. J Am Chem Soc 135:1864–1871
53. Degroot FMF, Faber J, Michiels JJM et al (1993) Oxygen 1s x-ray-absorption of tetravalent titanium-oxides—a comparison with single-particle calculations. Phys Rev B 48:2074–2080
54. Li YL, Wang JJ, Li XF et al (2012) Discharge product morphology and increased charge performance of lithium-oxygen batteries with graphene nanosheet electrodes: the effect of sulphur doping. J Mater Chem 22:20170–20174
55. Qiao RM, Chuang YD, Yan SS et al (2012) Soft X-ray irradiation effects of Li_2O_2, Li_2CO_3 and Li_2O revealed by absorption spectroscopy. PLoS One 7:e49182
56. Ruckman MW, Chen J, Qiu SL et al (1991) Interpreting the near edges of O_2 and O_2^- in alkali-metal superoxides. Phys Rev Lett 67:2533–2536
57. Hummelshoj J, Luntz AC, Norskov JK (2013) Theoretical evidence for low kinetic overpotentials in Li-O_2 electrochemistry. J Chem Phys 138:034703
58. Radin MD, Rodriguez JF, Tian F et al (2012) Lithium peroxide surfaces are metallic, while lithium oxide surfaces are not. J Am Chem Soc 134:1093–1103
59. Ong SP, Mo Y, Ceder G (2012) Low hole polaron migration barrier in lithium peroxide. Phys Rev B 85:081105
60. Yao KPC, Kwabi DG, Quinlan RA et al (2013) Thermal stability of Li_2O_2 and Li_2O for Li-air batteries: in situ XRD and XPS studies. J Electrochem Soc 160:A824–A831
61. Younesi R, Hahlin M, Treskow M et al (2012) Ether based electrolyte, $LiBCN_4$ salt and binder degradation in the Li-O_2 battery studied by hard X-ray photoelectron spectroscopy (HAXPES). J Phys Chem C 116:18597–18604
62. Oh SH, Yim T, Pomerantseva E et al (2011) Decomposition reaction of lithium bis(oxalato) borate in the rechargeable lithium-oxygen cell. Electrochem Solid-State Lett 14:A185–A188
63. Ong CW, Huang H, Zheng B et al (2004) X-ray photoemission spectroscopy of nonmetallic materials: electronic structures of boron and B_xO_y. J Appl Phys 95:3527–3534
64. Younesi R, Hahlin M, Bjorefors F et al (2013) Li-O_2 battery degradation by lithium peroxide (Li_2O_2): a model study. Chem Mater 25:77–84
65. Veith GM, Nanda J, Delmau LH et al (2012) Influence of lithium salts on the discharge chemistry of Li-air cells. J Phys Chem Lett 3:1242–1247
66. Lu YC, Crumlin EJ, Veith GM et al (2012) In situ ambient pressure X-ray photoelectron spectroscopy studies of lithium-oxygen redox reactions. Sci Rep 2:715
67. Harding JR, Lu Y-C, Tsukada Y et al (2012) Evidence of catalyzed oxidation of Li_2O_2 for rechargeable Li-air battery applications. Phys Chem Chem Phys 14:10540–10546
68. Viswanathan V, Thygesen KS, Hummelshoj JS et al (2011) Electrical conductivity in Li_2O_2 and its role in determining capacity limitations in non-aqueous Li-O_2 batteries. J Chem Phys 135:214704
69. Delacourt C, Ati M, Tarascon JM (2011) Measurement of lithium diffusion coefficient in Li_yFeSO_4F. J Electrochem Soc 158:A741–A749

70. Meethong N, Huang HYS, Speakman SA et al (2007) Strain accommodation during phase transformations in olivine-based cathodes as a materials selection criterion for high-power rechargeable batteries. Adv Funct Mater 17:1115–1123
71. Wen CJ, Ho C, Boukamp BA et al (1981) Use of electrochemical methods to determine chemical-diffusion coefficients in alloys: application to 'LiAl'. Int Met Rev 5:253–268
72. Xia H, Lu L, Ceder G (2006) Li diffusion in $LiCoO_2$ thin films prepared by pulsed laser deposition. J Power Sources 159:1422–1427
73. Zhu YJ, Wang CS (2010) Galvanostatic intermittent titration technique for phase-transformation electrodes. J Phys Chem C 114:2830–2841
74. Jang YI, Neudecker BJ, Dudney NJ (2001) Lithium diffusion in Li_xCoO_2 ($0.45 < x < 0.7$) intercalation cathodes. Electrochem Solid-State Lett 4:A74–A77
75. Levi MD, Aurbach D (1997) Diffusion coefficients of lithium ions during intercalation into graphite derived from the simultaneous measurements and modeling of electrochemical impedance and potentiostatic intermittent titration characteristics of thin graphite electrodes. J Phys Chem B 101:4641–4647
76. Yu P, Popov BN, Ritter JA et al (1999) Determination of the lithium ion diffusion coefficient in graphite. J Electrochem Soc 146:8–14
77. Levi MD, Aurbach D (2007) The application of electroanalytical methods to the analysis of phase transitions during intercalation of ions into electrodes. J Solid State Electrochem 11:1031–1042
78. Scharifker B, Hills G (1983) Theoretical and experimental studies of multiple nucleation. Electrochim Acta 28:879–889
79. Zhong L, Mitchell RR, Liu Y et al (2013) In situ transmission electron microscopy observations of electrochemical oxidation of Li_2O_2. Nano Lett 13:2209–2214
80. Perng Y-C, Cho J, Membreno D et al (2011) In: 11th international conference on atomic layer deposition, Cambridge, MA, 26–29 June 2011

Chapter 5
Atomistic and First Principles: Computational Studies of LiO_2 Batteries

Kah Chun Lau, Larry A. Curtiss, Maria K.Y. Chan, and Jeffrey P. Greeley

5.1 Introduction

Lithium air batteries represent a potentially revolutionary technology for energy storage, with theoretical energy densities that far exceed those of state-of-the-art lithium-ion battery [1]. A host of scientific and engineering barriers, however, have hitherto prevented the development of devices that both approach the indicated theoretical capacities and can be reversibly cycled for sufficient periods to permit commercial development. An explosion of research activity in this field has recently begun to identify some of the key problems that underlie the suboptimal performance of lithium air batteries [1–7]. Among other issues, the large overpotentials which limit energy conversion efficiencies, the low current density which may be limited by the conductivity of the lithium (per)oxide discharge products, the stability of the electrolyte, and partial oxidation of the metallic lithium anode have been pinpointed as major performance weaknesses of these systems. However, many fundamental questions of first-order importance for lithium air battery design and optimization remain answered. There is no consensus, for example, on how to remediate the weaknesses mentioned above, and other issues, such as understanding the relationship between catalysts on the cathode and battery

K.C. Lau
Material Science Division, Argonne National Laboratory, Lemont, IL, USA

L.A. Curtiss
Material Science Division, Argonne National Laboratory, Lemont, IL, USA

Center for Nanoscale Materials, Argonne National Laboratory, Lemont, IL, USA

M.K.Y. Chan
Center for Nanoscale Materials, Argonne National Laboratory, Lemont, IL, USA

J.P. Greeley (✉)
School of Chemical Engineering, Purdue University, West Lafayette, IN, USA
e-mail: jgreeley@purdue.edu

N. Imanishi et al. (eds.), *The Lithium Air Battery: Fundamentals*,
DOI 10.1007/978-1-4899-8062-5_5,

performance, are not even sufficiently well understood so as to determine whether or not they constitute an important part of the lithium air system [8].

To develop the fundamental understanding that is needed to address the formidable challenges described above, a range of experimental techniques, ranging from classical electrochemical voltammetry to advanced synthesis and characterization strategies, have been deployed to study the chemistry and physics of lithium air batteries. With these approaches, a large number of investigations have been carried out, for example, on the growth of lithium oxides and peroxides on metal surfaces [9–11], characterization of lithium (per)oxide formation in well-defined pores of MnO_2s [12], and detailed characterization of the magnetic structures of lithium peroxides nucleated on various carbon-based surfaces [13–15]. Such studies are beginning to tease out some of the fundamental, molecular-level mechanisms that underlie discharging and charging in lithium air systems.

In spite of these advances, many details of the molecular-level processes underlying lithium air chemistry and physics remain poorly understood. To make progress in these areas, researchers have turned to theoretical and computational analyses of diverse facets of these complex systems, ranging from the catalytic properties of metal and metal oxide surfaces to the stability of electrolytes under given concentration and potential regimes. As described further below, electronic structure-based calculations have been widely employed, with different flavors of density functional theory (DFT), appropriate for bulk-phase electrolyte systems, electrode–electrolyte interfaces, or bulk electrode materials, being among the most popular. High-level quantum chemical calculations have also been employed to provide data of even higher accuracy, while classical potential molecular dynamics and mesoscale phase field models are also being developed to provide information about dynamical effects and larger length scale effects.

In this chapter, we summarize some recent developments in computational lithium air research, with a specific focus on the properties of bulk and surface lithium peroxides and the growth of lithium-oxide subunits in transition metal oxide lattices. Due to space limitations and discussions in other chapters of this volume, we discuss only briefly studies on electrolyte chemistry and noble metal catalysts. We consider treatments of electronic, energetic, and dynamical properties of the above systems, and we discuss how theory is providing a compelling molecular-scale view of their properties that complements and extends insights available from experiments.

5.2 Lithium Peroxide Properties

5.2.1 Overpotentials

For a working lithium air battery to be practical, the discharge–charge efficiency in discharge/charge voltage profiles [4, 16–20] of the cell is critical. The discharge–charge efficiency depends on the basic electrochemical reactions and the overpotentials, η, that can be attributed to the electrochemical reactions at the surface and interface, internal resistance within the lithium air cell, reactant mass transport, and charge transport. In addition, current densities of only ~0.1–1 mA/cm^2 are currently achievable in lithium air batteries [1, 11, 16, 21, 22], far below that required for electric vehicle applications [4]. Similarly for a practical lithium air cell, the polarization has to be reduced to achieve high density and efficiency. The efficiency of discharge/charge cycles in lithium air cells requires Li_2O_2 as a discharge product to have sufficiently high electronic conductivity in order to reduce the overall cell resistance. In order to solve these problems, a basic understanding of the fundamental mechanisms of the oxygen reduction reaction (ORR) during discharge that dictates the Li_2O_2 growth and of the OER during charge that determines the dissolution of Li_2O_2 at the cathode is necessary. Similarly, a basic understanding of the electronic conductivity of bulk Li_2O_2 and its surfaces at interfaces is needed to optimize future lithium air cell design.

For the electrochemical growth and dissolution of a discharge product (e.g., Li_2O_2) in a lithium air battery, the mechanisms involved can be very complicated due to the complex solid/liquid (electrolyte)/gas three-phase boundary interfaces at the cathode where the oxygen reduction/evolution process occurs. However, simple yet reasonable theoretical models can be derived based on DFT calculations to describe the electrochemical growth and dissolution of Li_2O_2 surfaces, assuming no limitations on the charge transfer processes and that kinetic barriers are negligible [18, 23, 24]. Using this model, discharge/charge overpotentials can be determined from the relevant chemical and electrochemical reaction steps.

In the models developed by Hummelshøj et al. [18, 24], the fundamental electrochemistry of Li_2O_2 surfaces is determined by the adsorption and desorption of Li^+ and an electron [24] on a Li_2O_2 facet:

$$\text{(anode)} \quad 2\text{Li} \leftrightarrow 2(\text{Li}^+ + e^-) \tag{5.1}$$

$$\text{(cathode)} \quad \text{Li}^+ + e^- + \text{O}_2 + * \leftrightarrow \text{LiO}_2{}^* \tag{5.2}$$

$$\text{(cathode)} \quad \text{Li}^+ + e^- + \text{LiO}_2{}^* \leftrightarrow \text{Li}_2\text{O}_2{}^* \tag{5.3}$$

where * refers to an adsorbed species on a Li_2O_2 surface. According to Hummelshøj et al. [24], the overall reaction, i.e., $2Li + O_2 \leftrightarrow Li_2O_2$, occurring at the cathode can be modeled based on the following approximations (Fig. 5.1), assuming the dynamics, mobility, and diffusivity of the ad-species are minimal in the model:

1. O_2 is not dissociated with the assumption that the consumption of electrolyte and cathode is minimal during reaction.
2. Growth and dissolution of Li_2O_2 are dominant only at Li_2O_2 surfaces after nucleation is initiated.
3. Li^+ ions in solution are in equilibrium with Li–metal, and O_2 molecules in solution are in equilibrium with both gas-phase oxygen and O_2 adsorption at the cathode surface: $O_2(g) \leftrightarrow O_2\ (liq) \leftrightarrow O_2{}^*$ with $\Delta G = 0$.
4. Chemical and electrochemical interactions with the electrolyte at cathode interfaces are ignored.

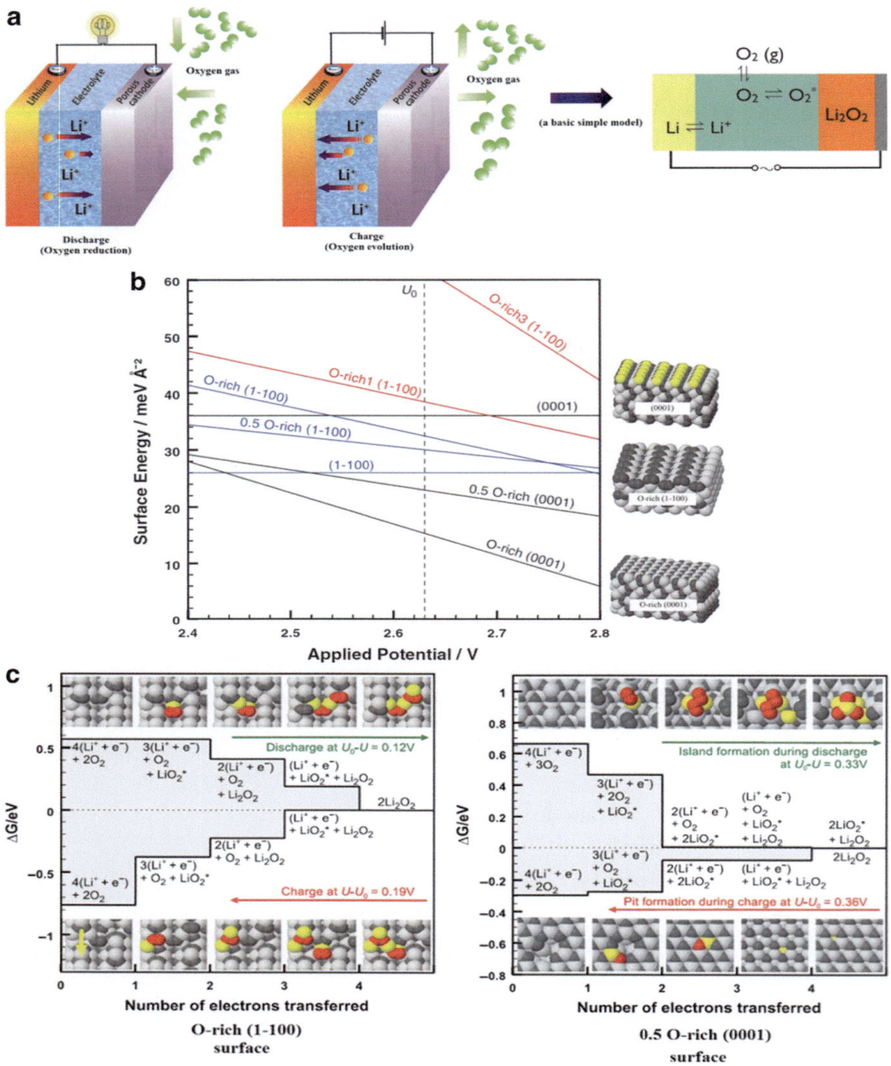

Fig. 5.1 (**a**) A schematic model of the basic theoretical model of ORR and OER process in lithium air cell following the approximations as outline by Hummelshøj et al. [24], (**b**) potential plot of the few low-energy Li_2O_2 surfaces reported, and (**c**) the free energy thermodynamic steps of discharge and charge of O-rich Li_2O_2 (1-100) and 0.5 O-rich (0001) surfaces reported [24]. Reprinted with permission from [24]. Copyright 2013, AIP Publishing LLC

Using these assumptions, the basic electrochemistry at both (1-100) and (0001) facets of Li_2O_2 was studied. From this model, the lowest overpotential is the one where all the individual adsorption energies are found to be as closest to eU_0 where U_0 is the equilibrium thermodynamic potential of Li_2O_2 bulk crystal. Accordingly, the free energy diagram is potential dependent and the discharge overpotential η_{dis} is defined to be $eU_0 - \min[G(Li + nO_2)]$, whereas for the charge overpotential η_{chg}, it is defined to be $|eU_0 - \max[G(Li + nO_2)]|$ [24].

Based on DFT non-spin-polarized calculations [24], the potential (U)-dependent stability of (1-100) and (0001) facets of Li_2O_2, nonstoichiometric surfaces (i.e., 0.5 O-rich and O-rich surfaces), including their structural defects (i.e., steps, kinks), was determined (Fig. 5.1). Regardless of the structural defects (e.g., steps, kinks), the thermodynamically most favorable Li_2O_2 facets are all found to be O-rich terminations from a wide variety of surface structures. In particular, the η_{dis} and η_{chg} are predicted to be quite small at kink and step sites, i.e., <0.2 V from an O-rich (1-100) surface (Fig. 5.1), and nearly zero at an O-rich (0001) facet. This is smaller than the previously reported $\eta_{dis} = 0.44$ V and $\eta_{chg} = 0.61$ V based on the stable reconstructed stoichiometric Li_2O_2 (1-100) surface [24]. For charge transfer taking place on the terrace sites, presumably dominant at crystal surfaces, all the reaction paths are reported to have higher overpotentials than the kinks and steps of the same facet. For the (1-100) terrace, $\eta_{dis} = 0.68$ V and $\eta_{chg} = 0.20$ V is the lowest overpotential process occurring on the stoichiometric facets, whereas for the more stable O-rich facets (e.g., O-rich (0001)) around U_0 (Fig. 5.1), smaller values of η_{dis} and η_{chg} (~0.3 V) are predicted [24].

In addition, several other authors have considered the reaction mechanism and other possible factors that may affect the overpotentials in a lithium air battery, including nucleation and stability of small clusters of Li_2O_2 on a metal substrate [25, 26], growth of clusters and nanoparticles [14], the stability of nonstoichiometric metallic Li_2O_2 surfaces [27], and the decomposition processes that dictate the OER process on several low index thermodynamically favorable Li_2O_2 surfaces [23]. A systematic study of the OER process on Li_2O_2 crystal surfaces was recently carried out by the Ceder group [23]. According to Mo et al. [23], the fundamental OER process of Li_2O_2 is dictated by the thermodynamically most stable Li_2O_2 crystal surfaces. By ignoring the effects of electrolyte media and electrolyte/cathode interfaces, the basic dissolution processes at Li_2O_2 surfaces are essentially assumed to be equivalent to the decomposition of a fundamental Li_2O_2 unit from its surfaces, which means the reaction path and desorption barrier will be affected by the surface structure. Based on DFT results, Mo et al. [23] determined the lowest energy surfaces and the respective Wulff shape of Li_2O_2 crystals. From this the kinetic energy barriers for oxygen evolution on the thermodynamic dominant surfaces are determined.

Mo et al. [23] studied a few low index Li_2O_2 surfaces including (0001), (11-20), (1-100), (11-21), (1-101) stoichiometric, and nonstoichiometric surfaces. DFT calculations showed that the kinetic rate of OER is highly dependent on surface orientation of Li_2O_2, a result corroborated by Hummelshøj et al. [24]. From the results, the (0001) and (11-20) surfaces are predicted to dominate the Wulff shape of Li_2O_2 crystal at both the most oxidizing (i.e., the O_2 gas limit) and the most reducing condition (i.e., the Li_2O limit) [23]. In all the surfaces Mo et al. [23] studied, the decomposition pathway is found to be likely initiated with Li removal and subsequent formation of a superoxide-like stoichiometry or structure (e.g., Li_2O_4) at the surface. Interestingly, they found that oxygen evolution has the highest energy barrier among the decomposition paths studied. The kinetics of OER is slow on the thermodynamically most abundant surfaces (i.e., (0001) and (11-20)) from the Wulff construction crystal of Li_2O_2, but is faster on the other thermodynamically higher energy surfaces (e.g., (1-100), (11-21), (1-101)). The energy barriers for Li desorption are predicted to be ~0.13–0.32 eV among these low index surfaces.

Mo et al. calculated the activation energy for the desorption of O_2 from Li_2O_2 surfaces to be ~0.17–1.57 eV. From all the surfaces, the overpotentials are predicted to be ~0.20–0.61 V (with respect to $U_0 = 2.82$ V) with the highest found to be from the (11-20) surface (i.e., 0.61 V), followed by the (0001) surface (i.e., 0.45 V), which are both predicted to be lowest energy surfaces of Li_2O_2. In this case, the rate of O_2 desorption, which carries the charge (i.e., corresponds to two electrons with respect to two Li ions per O_2), is used to estimate the surface current density by Mo et al. [23] with the assumption that the intrinsic surface electronic conductivity can be ignored. Using the Arrhenius relation, the current density contributed by these surfaces is found to be vastly different. For the (1-100), (11-21), (1-101) surfaces, which contribute a small fraction of the surface in the Wulff shape of Li_2O_2, the current density is estimated to be within a range of 10^7–10^{10} μA/cm^2. Whereas for the dominant (11-20) and (0001) surfaces on a Wulff crystal of Li_2O_2 in both oxidizing and reducing condition, the surface current density is negligible (~10^{-14}–10^2 μA/cm^2). Given that a current density of more than 100 mA/cm^2 in discharge is required for electric vehicle applications [4], this suggests that the low charging rate and high overpotential of the lithium air batteries may be caused by the poor kinetics for the OER possibly due to the thermodynamically dominant surfaces of Li_2O_2 crystals. Therefore these theoretical results imply that the intrinsic electronic conductivity of Li_2O_2 is critical.

5.2.2 Electronic Conductivity

The electronic conductivity of Li_2O_2 has also recently been investigated using DFT. Bulk Li_2O_2 crystal is an insulator with an electronic band gap predicted to be approximately 4.9 eV from theory [18]. Therefore, the formation of Li_2O_2 bulk crystal as a discharge product is undesirable from the perspective of electronic conductivity and most likely will cause problems, including electrical passivation, pore clogging, etc. [28]. This suggests that any electronic conductivity through insulating layers of Li_2O_2 would have to be assisted by the presence of dopants, defects, and polarons or by surface conductivity.

According to Hummelshøj et al. [18], based on DFT calculations, the poor electrical conductivity of Li_2O_2 bulk crystal can be improved through the introduction of Li vacancies in the crystal lattice. They find [18] that the Li_2O_2 exhibits metallic behavior in the limit of very high concentrations (~ 6 at.%) of Li vacancies. Consequently, if surfaces and other defects such as grain boundaries exhibit sufficient lithium-poor stoichiometries, then metallic features and facile electron conductivity could result. From the calculations of Hummelshøj et al. [18], the bulk vacancy formation energy is predicted to be 2.85 eV, with Li-vacancy diffusion activation energy found to be approximately 0.35 eV along the (100) crystalline plane. From Hummelshøj et al., the presence of Li vacancies eventually produces holes in the valence band, thus pointing to a band-like bulk transport mechanism in p-doped Li_2O_2 which would then be metallic.

An alternative model of bulk electrical conductivity in Li_2O_2 has also recently been proposed based on polaronic conduction [23, 29, 30], i.e., via hole-polaron and electron-polaron hopping mechanisms. A polaron is a quasiparticle consisting of an electronic charge coupled to lattice distortions and accompanying polarizations [31–33]. For a Li_2O_2 crystal, the existence of a polaron is related to the presence of highly localized electrons in the peroxide anions (O_2^{2-}) [23, 29, 30, 34]. The strong excitonic and vibronic effects of the optical properties of Li_2O_2 crystal and the unique molecular orbitals of the peroxide anions (Fig. 5.2) can probably accommodate the localization of polaron that has a strong coupling with the outermost valence electrons. As shown in Fig. 5.2, the structure of the Li_2O_2 crystal consists of an array of O_2^{2-} molecular anions connected to a matrix of Li^+ ions. Consequently, this unique atomic arrangement of O_2^{2-} leads to an ordering sequence of the molecular orbitals of σ_p, π_p, π_p^*, and σ_p^* as shown in Fig. 5.2. For Li_2O_2 crystal, the fully occupied orbitals, σ_p, π_p, and π_p^*, comprise the top valence bands of bulk Li_2O_2, whereas the empty unoccupied antibonding σ_p^* states, as well as the Li s-states, constitute the bottom conduction bands, with a band gap of ~4.9 eV in solid according to a most recent PBE + U calculation [30].

As pointed out by Kang et al. [29], the Li s-states that lies above the σ_p^* oxygen $2p$-derived conduction states can lead to a unique electron–lattice coupled interaction in Li_2O_2 crystal lattice that forms a small polaron. Based on DFT-HSE (Heyd–Scuseria–Ernzerhof) calculations, Kang et al. [29] found that an excess electron in Li_2O_2 causes elongation of one of the peroxide bonds ($d_{O\text{–}O} \sim 1.5$ Å), which eventually leads to cleavage of the O–O bond. The excess electron is localized at this O–O site (Fig. 5.2) and forms an electron polaron. This leads to a system with a magnetic moment of $\mu = 1\ \mu_B$ and a localized spin-up state inside the π_p, π_p^* valence band gap (Fig. 5.2). If two extra electrons are added to the system, the polarons favor a uniform distribution in Li_2O_2, rather than a localized bipolaron, due to strong Coulomb repulsion. In this case, the electron conduction is associated with the polarons hopping, in conjunction with breaking and reformation of O_2 bonds, through a thermally activated process. The hopping process can be described as an adiabatic electron transfer process [35], and it has a predicted activation energy barrier (E_a) of ~0.54 and 0.66 eV for interlayer (along (001)) and intra-layer (along (100)) hopping, respectively. With these values, the electron mobility is predicted to be extremely low ($\mu = 10^{-10} - 10^{-9}$ cm^2/V s) at room temperature and suggests that this leads to substantial overpotential (>0.5 V) at high current density ($J_{Li_2O_2} > 0.01\,\mathrm{mA/cm^2}$), which limits the power density in lithium air battery.

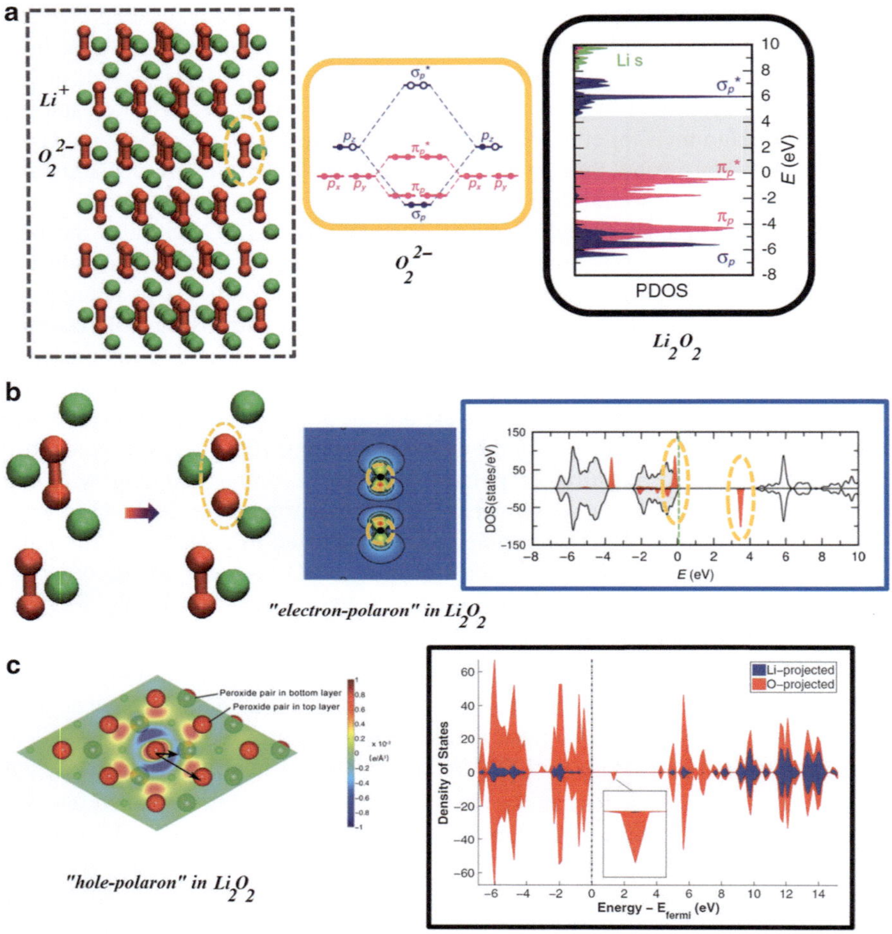

Fig. 5.2 (**a**) The array of Li^+ and O_2^{2-} ions in Li_2O_2 crystal and the electronic density of states (DOS) obtained from DFT-HSE calculations, together with the schematic molecular orbitals of an isolated O_2^{2-} peroxide ion. (**b**) The electron polaron attributed to a one-electron excess in the Li_2O_2 crystal and its localized electronic states reported by Kang et al. [29]. (**c**) The localized electronic hole-polaron state attributed to one-electron removal from Li_2O_2 crystal reported by Ong et al. [35]. Reprinted with permission from [29, 35]. Copyright 2012 by the American Physical Society

For the hole polarons, the migration barrier is predicted to be smaller than that for electron polarons and may be sufficiently small to give rise to a path for electrical conduction [30]. Based on DFT-HSE calculations by Ong et al. [35], a π^* hole polaron is created by removing an electron from a crystal cell, with overall charge neutrality preserved. In this case, the polaronic peroxide species is found to be an O_2^- species as opposed to a peroxide anion. From the electronic density of states (DOS) analysis, the presence of a hole polaron can be represented by an oxygen hole state (Fig. 5.2) just above the Fermi level and hence insufficient to allow the band-like transport in metallic state proposed earlier [18]. Through the hole-polaron hopping mechanism, the electronic conductivity in a Li_2O_2 crystal is found to be highly anisotropic. The polaron migration barrier for intra-layer hopping (e.g., (1–100)) is ~68 meV, which is smaller than interlayer polaron hopping barrier (i.e., ~152 meV along (0001)) for free hole polaron. When a Li vacancy (V_{Li^+}) is present, the bound hole-polaron migration is coupled to polaron -V_{Li^+} migration with a binding energy of around 468 meV, slightly larger than Chen et al. [9] predicted (~360 meV).

The polaronic state is very likely to be susceptible to thermal fluctuations and crystalline lattice expansion from temperature variations as the metastable non-polaronic state and polaronic state are only separated by an extremely small perturbation barrier (~20 meV) [23]. In addition, if nonstoichiometric and defective Li_2O_2 is formed during ORR in a lithium air battery, it could mean that the limited electronic conductivity due to the polaronic conduction might not be a limiting factor as originally thought to be. Work by Radin et al. and Yang et al. [13, 27, 36] indicates that the metallic state of stoichiometric and nonstoichiometric Li_2O_2 surfaces could possibly facilitate charge transport if finite-size nanoparticles of Li_2O_2 are formed with a substantial surface to volume ratio. According to Radin et al. [27, 36], oxygen-rich facets are most likely in Li_2O_2 crystallites compared to the Li_2O crystallite that prefer stoichiometric facets. Interestingly, surface state analysis reveals that the thermodynamic stable surfaces of Li_2O_2 are most likely to be metallic, despite the fact that Li_2O_2 bulk crystal is known to be an insulator (Fig. 5.3). In contrast, the stable surfaces of Li_2O are insulating. This surface property may explain the reversibility of Li_2O_2 discharge product and the irreversibility of the system if Li_2O is formed in discharge product.

According to Radin et al. [36], the electron transport properties attributed to "metallic-like" Li_2O_2 surfaces can be substantial (~$10^5\ \Omega^{-1}$/m), based on surface band structure and the electronic DOS and assuming the surface states to be isotropic. Therefore, the presence of conductive surface pathways in Li_2O_2 could possibly offset the limited electron transport through the polaronic transport in bulk. Although no experimental measurement of the surface conductivity of Li_2O_2 has yet reported, the presence of the oxygen-rich Li_2O_2 in the discharge product of lithium air cell has recently been reported [13]. According to a recent study reported by Yang et al. [13], the Li_2O_2 discharge product based on a large surface area activated carbon electrode, and TEGDME-$LiCF_3SO_3$ electrolyte is predicted to consist of oxygen-rich Li_2O_2 and stoichiometric Li_2O_2 that yields a two-plateau (~3.2–3.5 and ~4.2–4.3 V vs. Li^+/Li) galvanostatic charge profile. As reported [13], the low plateau (i.e., ~3.2–3.5 V) corresponds to a form of oxygen-rich Li_2O_2 with superoxide-like properties that can be characterized by a low-temperature magnetic phase transition and a high O–O stretching frequency (~1,125 cm^{-1}) from Raman spectroscopy in experiment (Fig. 5.3). Interestingly, the magnetic phase transition and the high O–O stretching frequency both disappear when the cell charged to

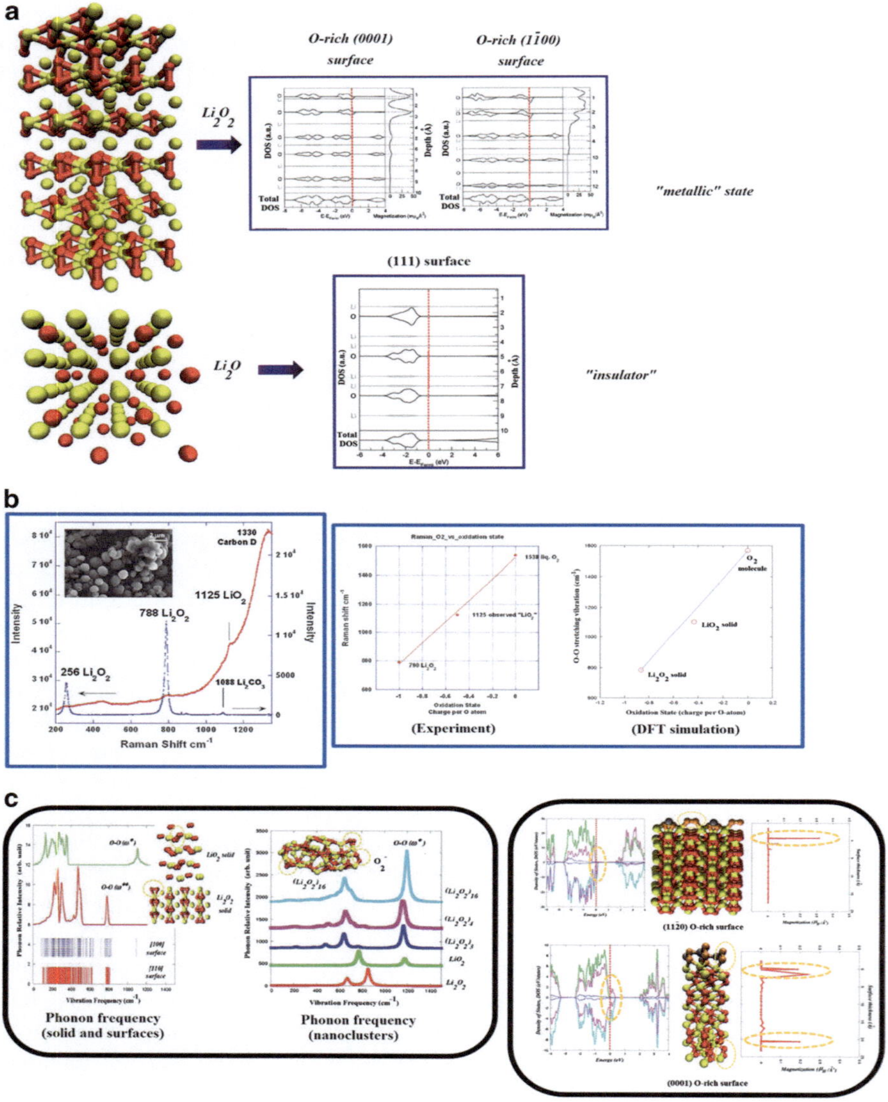

Fig. 5.3 (**a**) The distinct electronic features of Li_2O_2 and Li_2O surfaces reported by Radin et al. [27]. (**b**, **c**) The experimental Raman spectra, DFT phonon vibration frequency of peroxide and superoxide species, and unsaturated spins that correspond to the observed magnetism in oxygen-rich Li_2O_2 discharge product in lithium air cell reported by Yang et al. [13]

above 3.7 V, and these observation can be explained based on DFT calculations. From Yang et al. [13], DFT calculations indicate that a surface superoxide structure on lithium peroxide clusters [14] and some lithium peroxide surfaces have an unpaired electron and favor magnetic states, consistent with previously reported metallic Li_2O_2 surfaces [27] with finite magnetic moments. The observed magnetic moments from the experiments can be attributed to magnetic states resulting from unsaturated spins populated on the surface oxygen species (i.e., O_2^- superoxide species) that are predicted to be thermodynamically stable in an O-rich environment at the bulk limit [37]. In the meantime, the experimentally observed high O–O stretching frequency is also found to match very well with the DFT predicted superoxide species O–O stretching frequency predicted to be in metallic Li_2O_2 surfaces, Li_2O_2 clusters, and LiO_2 crystalline bulk (Fig. 5.3) [13, 14, 37]. These results provide the first evidence that the form of the lithium peroxide discharge product is important to obtaining low charge overpotential and thus improving the round-trip efficiency between discharge and charge. However, direct evidence of metal-like electrical conductivity from Li_2O_2 surfaces requires further verification from experimental studies in the future.

5.3 Electrocatalysts

The role of catalysts in lithium air electrochemistry is still under intense debate [4, 7]. In a lithium air cell, studies have shown that different types of carbon [38, 39], noble metals [11, 19, 21], and transition metal oxides [16, 17, 40–42] may act as electrocatalysts to facilitate the discharge and/or charge process. DFT calculations can provide molecular-level insights into the thermodynamics of catalysts or electrocatalysts that either facilitate OER/ORR or accommodate Li_xO_y species in their structures.

5.3.1 Carbon

The ORR on carbon was studied by Xu et al. [26]. They carried out a systematic study on a simple model of graphitic carbon. From DFT studies, Xu et al. found that the chemical inertness of the graphitic basal plane does not stabilize the important intermediate (i.e., LiO_2) and therefore limits the reversible potential of O_2 reduction to 1.1 V. For the defect sites of carbon, the armchair edge and di-vacancy in basal plane are found to be highly reactive and can be significantly oxidize at ambient conditions to form various CO_x groups. The oxidized carbon species (CO_x) subsequently can be reduced by Li via redox mechanisms at 1.2–1.4 V according to DFT results. According to Xu et al. [26], these CO_x groups can also serve as the active sites for catalyzing O_2 reduction and will occur at ~1.8–2.3 V by stabilizing the LiO_2 intermediate. However it is noteworthy to point out that the effects of electrolytes are absent in this study.

5.3.2 Noble Metals

As recently reported by Dathar et al. [10], the intrinsic activity of Au, Ag, Pt, Pd, Ir, and Ru for the ORR by Li forms a volcano-like trend with respect to the adsorption energy analogous to H–ORR in hydrogen fuel cell [43]. By ignoring the effect of impurities, atomic defects, and electrolyte interfaces, Pt and Pd are predicted to be the most active metal with respect to O_2 adsorption energy. The Au and Ag metal surfaces are found to be chemically inert relative to other metals with a relatively small sticking probability of O_2. Therefore the O_2 reduction involves an associative mechanism that favors the disproportionation of LiO_2 to form Li_2O_2. Whether Li_2O_2 forms electrochemically or thermodynamically remains an open question. For Pt, Pd, Ir, and Ru, the dissociative mechanism involves the reduction of atomic O-atom [10] as these metals are expected to be already covered by O atoms when exposed in an air electrode. Thus, according to Dathar et al. [10], the most probable electron transfer step is to form $(Li_xO)_n$ surface species that eventually initiate further nucleation for the formation of bulk Li_2O. From the trend Dathar et al. found [10], strong oxygen–metal interactions (e.g., Ru, Ir) will most probably diminish the capacity of oxygen to bond with lithium, whereas too weak oxygen–metal interactions (e.g., Au, Ag) hinder O_2 activation and therefore also hinder the formation of oxygen–lithium bonding. Thus, the intermediate oxygen–metal interaction strengths of Pt and Pd appear to make them optimal candidates as the choice of metal electrocatalyst in lithium air electrochemistry, despite the fact that these metals general tend to be oxidized when exposed to air.

5.3.3 Transition Metal Oxides

While noble metals can be effective catalysts in lithium air batteries [19], the amount of catalyst per watt needed for a practical lithium air cell is at least 100-fold greater than that in fuel cells [7], making it imperative that more economical alternatives be identified. Since transition metal oxides can be cost effective and have been widely used as catalysts for the hydrogen analog to the $Li + O_2 \leftrightarrow Li_2O_2$ reaction, they have been investigated for use in lithium air cells. Most of the experimental work on using transition metal oxides as electrocatalysts in lithium air cells has focused on MnO_2 [16, 17, 42, 44, 45], although others such as Fe oxides [40, 46], Co and Cu oxides [40, 47], and perovskites [48] have also been studied. Of the many polymorphs of MnO_2 [49], αMnO_2 in particular, in the Hollandite structure containing 5-Å wide 2×2 tunnels (Fig. 5.4a), has been shown to lower the charge overpotential, facilitate cycle ability, and increase capacity in a lithium air cell [17, 42, 45]. Computational work on transition metal oxide electrocatalysts for lithium air batteries has been scarce and thus far also concentrated on MnO_2 [42, 50].

An interesting feature of the transition metal oxide electrocatalysts is that many of these compounds are also lithium intercalation compounds, often at a similar

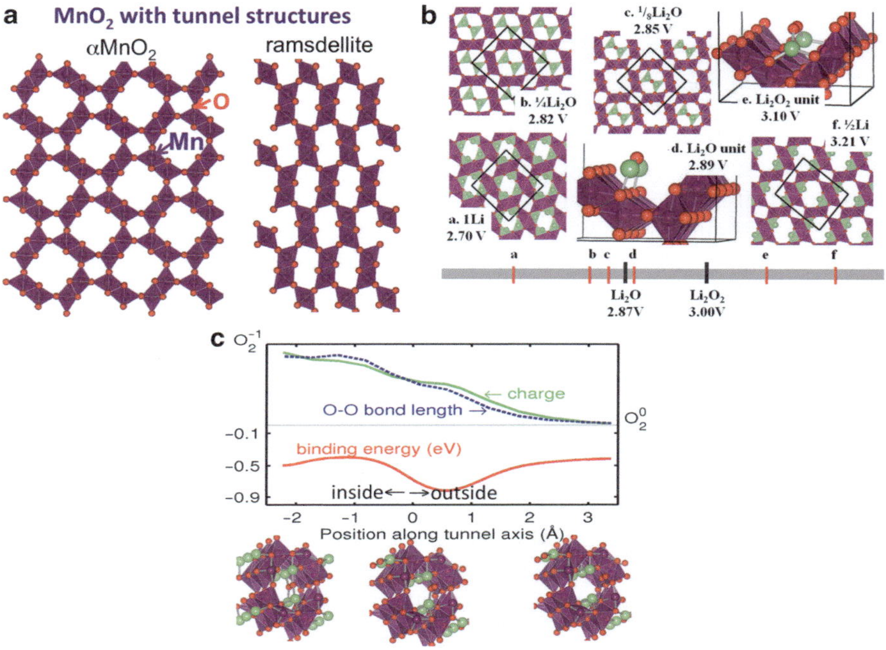

Fig. 5.4 (**a**) Structures of α- and R-MnO_2 showing 2 × 2 and 2 × 1 tunnels. (**b**) DFT + U predicted equilibrium voltages for Li and Li_xO_y insertion into αMnO_2, compared to those of Li_2O and Li_2O_2 formation. (**c**) Charge (*green*) on the O_2 molecule and O–O distance (*blue*), as well as binding energy (*red*) between the O_2 molecule and αMnO_2 surface, as a function of the position of an O_2 molecule inside/outside of the tunnel. The charge and O–O distance are scaled to vary between that of a neutral O_2 molecule and a superoxide anion

voltage to the Li–O_2 reactions, raising the prospect of simultaneous lithium intercalation and lithium–oxygen reaction. In these cases, the transition metal oxides in lithium air cells are acting not as a "catalysts" in the traditional sense of facilitating a chemical reaction while remaining unchanged, but undergo chemical and structural changes themselves as well.

The concept of a "dual-purpose" transition metal oxide acting as a lithium intercalation compound as well as facilitating Li–O_2 reactions was first suggested by Trahey et al. [42]. Using DFT + U, they studied the insertion of Li and Li_xO_y into the 2 × 2 and 2 × 1 tunnels of αMnO_2 and ramsdellite (R)-MnO_2, respectively. They found that the Li intercalation into α- and R-MnO_2 to form Li_xMnO_2 occurs at 2.6–3.5 V, with voltages above 3 V for $x \leq 0.5$ and below 3 for $x > 0.5$. Since the equilibrium voltage of Li_2O_2 and Li_2O formation is approximately 3 V, Li intercalation α- and R-MnO_2 thus straddles the Li–O_2 reactions. During first discharge, the facile diffusion of Li in the tunnels means that $Li_{0.5}MnO_2$ readily forms prior to any Li–O_2 reactions. The electrocatalyst that is present during the Li–O_2 discharge is therefore a partially reduced Mn oxide. Analysis of the magnetic moments shows that the Mn ions are of mixed 3+/4+ valence, which is believed to catalyze oxygen reduction. Moreover, DFT + U calculations show that the adsorption of O_2 on the surface of αMnO_2, both on exterior and internal surfaces, is enhanced by Li intercalation.

Insertion of Li_xO_y into the tunnels, on the other hand, has equilibrium voltages of 2.8–3.2 V, the lower limit of which is similar to the first discharge voltage observed experimentally [42]. Therefore, it is possible that Li_xO_y species are incorporated into the tunnels during the lithium air discharge process. Charge voltages vary [17, 42, 45], but are typically circa 3.5 V. Therefore, any intercalated lithium ions would likely be removed before or during the Li–O_2 charge process. The proximity of the voltages of Li_xO_y insertion into the tunnels of MnO_2 to the voltages of Li_2O or Li_2O_2 formation (Fig. 5.4b) has led Trahey et al. to propose that MnO_2 can act as a "Li_xO_y-storage material," reversibly storing lithium-oxide species during the Li–O_2 reactions. These results were later corroborated by Ling and Mizuno [50].

The equilibrium voltages of Li_xO_y insertion into the MnO_2 tunnels demonstrate only the thermodynamic favorability of these reactions, but kinetic factors will determine the likelihood of the reactions actually occurring. The kinetics of the uptake and release of O_2 from the tunnel entrances is particularly important, since it would represent the first step of ORR and last step of OER. DFT calculations of O_2 entering through the tunnel entrance of partially lithiated αMnO_2 (Fig. 5.4c) show that the barriers for O_2 entering and exiting the tunnel are low, approximately 0.5 and 0.2 eV, respectively. Moreover, as shown in Fig. 5.4c, upon entering the tunnel, the O_2 molecule is reduced to a superoxide species both in terms of O–O bond length and in terms of the Bader charges. And conversely, an O_2^- inside the tunnel can exit the tunnel easily into a neutral O_2 molecule. This shows that the tunnel entrances of partially lithiated αMnO_2 can be effective conduits for the ORR and OER.

5.4 Electrolytes

As reported in several seminal papers from the literature [4, 51], one of the key problems in lithium air batteries is the stability of the electrolyte and the electrochemical selectivity at the interface. The electrolyte stability is of the main concern during both charge and discharge at both the cathode and anode. Most experimental and computational studies have focused on the stability of solvents during charge and discharge at the cathode. Although there is strong evidence that carbonates decompose at the cathode to form solid products such as Li_2CO_3, it is still not yet completely clear how the reduced O_2 species are directly related to the decomposition mechanism. Due to the failure of carbonate solvents, recent experimental and computational work has focused on the stability of ether-based and DMSO-based electrolytes [2, 52].

In contrast to the extensive theoretical studies of Li_2O_2 solids and surfaces, the theoretical modeling of electrolytes and their interfaces in lithium air systems is a relatively unexplored area. In parallel with the experiments, most computational studies so far have focused on the stability of solvents during charge and discharge at both electrodes. DFT studies of the stability of carbonate-based solvents (e.g., propylene carbonate (PC)) against reduced oxygen species (e.g., O_2^-) have been reported by Bryantsev et al. [53], Zhang et al. [54], and Laino et al. [55]. All of these studies have found that the carbonates are unstable with respect to reduced oxygen species such as O_2^-. Relative to the carbonate-based solvents, the ether-based electrolytes have been found to have increased stability for oxygen reduction products during discharge as most studies do not see evidence for solid decomposition products during the initial cell cycling. The relatively high stability of ether-based electrolytes against the reduced O_2 species compared to PC is supported by a few theoretical studies [54, 56, 57] with higher reaction barriers during the solvent decomposition reactions. However, the compatibility of the Li-salt in electrolytes is also found to be important as recently reported by Du et al. [58] in a combined theoretical and experimental study. In addition, decomposition reactions may occur at the anode involving an oxygen cross-over effect suggested by a recent theoretical study [59]. Whether the solvent decomposition reaction proceeds via O_2^{2-}, LiO_2, or other reactive species, it is now clear that a solvent that is resistant to attack by reduced O_2 species must be discovered in order to achieve a practical lithium air cell with a long cycle life. In addition, a recent study [15] has shown that ether molecules will decompose on some sites on a lithium peroxide surface.

5.5 Conclusions

In this chapter, we have discussed the state-of-the-art of first-principles computational work on lithium air batteries. Through DFT calculations of bulk, surfaces, kinks, steps, nanoparticles, defects, and charged lattice distortions (polarons), a preliminary understanding of the mechanisms of Li_2O_2 growth and dissolution has emerged. It is now known that oxygen-rich surfaces of Li_2O_2 are instrumental in controlling the OER overpotential and that conductivity and hence current densities are limited by the particle morphology and defects. Similarly, through DFT calculations, some understanding of the roles of the electrocatalysts in facilitating $Li–O_2$ reactions has been obtained. For carbon, oxidized defect sites are key for catalyzing the $Li–O_2$ reaction, whereas for noble metals, Pt and Pd are favored. Transition metal oxides such as MnO_2 can act as dual-purpose lithium and lithium-oxide incorporating structures, facilitating the reduction and oxidation of O_2/O_2^- when lithiated. Finally, computational work on electrolytes highlights the need for electrolytes that are stable in the presence of reactive peroxide and superoxide species.

This brief overview of selected computational studies highlights the contributions that theoretical analyses can make to the understanding and development of lithium air batteries. Continued advances in theoretical treatments at a variety of length and time scales, with applications ranging from the fundamental properties of lithium peroxides to the identification of optimal electrolytes for lithium air batteries, will undoubtedly enhance these contributions and ultimately facilitate the identification of novel lithium-based battery chemistries.

Acknowledgment This work was supported by the Center for Electrical Energy Storage (CEES): Tailored Interfaces, an Energy Frontier Research Center at Argonne National Laboratory, Northwestern University, and University of Illinois at Urbana Champaign, funded by the U.S. Department of Energy, Office of Science, Office of Basic Energy Sciences. Use of the Center for Nanoscale Materials was supported by the U. S. Department of Energy, Office of Science, Office of Basic Energy Sciences, under contract No. DE-AC02-06CH11357. The authors also acknowledge grants of computer time from EMSL, a national scientific user facility located at Pacific Northwest National Laboratory, and the Fusion cluster in the Laboratory Computing Resource Center at Argonne National Laboratory.

References

1. Abraham KM, Jiang Z (1996) A polymer electrolyte-based rechargeable lithium/oxygen battery. J Electrochem Soc 143(1):1–5
2. Bruce PG, Freunberger SA, Hardwick LJ, Tarascon J-M (2012) Li-O_2 and Li-S batteries with high energy storage. Nat Mater 11(02):172–172
3. Christensen J, Albertus P, Sanchez-Carrera RS, Lohmann T, Kozinsky B, Liedtke R, Ahmed J, Kojic A (2012) A critical review of Li/air batteries. J Electrochem Soc 159(2):R1–R30. doi:10.1149/2.086202jes
4. Girishkumar G, McCloskey B, Luntz AC, Swanson S, Wilcke W (2010) Lithium–air battery: promise and challenges. J Phys Chem Lett 1(14):2193–2203. doi:10.1021/jz1005384
5. Kraytsberg A, Ein-Eli Y (2011) Review on Li-air batteries—opportunities, limitations and perspective. J Power Sources 196(3):886–893. doi:10.1016/j.jpowsour.2010.09.031
6. Shao YY, Ding F, Xiao J, Zhang J, Xu W, Park S, Zhang JG, Wang Y, Liu J (2013) Making Li-air batteries rechargeable: material challenges. Adv Funct Mater 23(8):987–1004. doi:10.1002/adfm.201200688
7. Shao Y, Park S, Xiao J, Zhang J-G, Wang Y, Liu J (2012) Electrocatalysts for nonaqueous lithium–air batteries: status, challenges, and perspective. ACS Catal 2(5):844–857. doi:10.1021/cs300036v
8. Cao R, Lee JS, Liu ML, Cho J (2012) Recent progress in non-precious catalysts for metal-air batteries. Adv Energy Mater 2(7):816–829. doi:10.1002/aenm.201200013
9. Chen JZ, Hummelshoj JS, Thygesen KS, Myrdal JSG, Norskov JK, Vegge T (2011) The role of transition metal interfaces on the electronic transport in lithium-air batteries. Catal Today 165(1):2–9. doi:10.1016/j.cattod.2010.12.022
10. Dathar GKP, Shelton WA, Xu Y (2012) Trends in the catalytic activity of transition metals for the oxygen reduction reaction by lithium. J Phys Chem Lett 3(7):891–895. doi:10.1021/jz300142y
11. Lu Y-C, Gasteiger HA, Crumlin E, McGuire R, Shao-Horn Y (2010) Electrocatalytic activity studies of select metal surfaces and implications in Li-air batteries. J Electrochem Soc 157(9):A1016–A1025
12. Zeng JQ, Nair JR, Francia C, Bodoardo S, Penazzi N (2013) Li-O_2 cells based on hierarchically structured porous alpha-MnO2 catalyst and an imidazolium based ionic liquid electrolyte. Int J Electrochem Sci 8(3):3912–3927
13. Yang J, Zhai D, Wang H-H, Lau KC, Schlueter JA, Du P, Myers DJ, Sun Y-K, Curtiss LA, Amine K (2013) Evidence for lithium superoxide-like species in the discharge product of a Li-O2 battery. Phys Chem Chem Phys 15(11):3764–3771

14. Lau KC, Assary RS, Redfern P, Greeley J, Curtiss LA (2012) Electronic structure of lithium peroxide clusters and relevance to lithium–air batteries. J Phys Chem C 116(45):23890–23896. doi:10.1021/jp306024f
15. Assary RS, Lau KC, Amine K, Sun Y-K, Curtiss LA (2013) Interactions of dimethoxy ethane with Li_2O_2 clusters and likely decomposition mechanisms for Li-O_2 batteries. J Phys Chem C. doi:10.1021/jp400229n
16. Ogasawara T, Débart A, Holzapfel M, Novák P, Bruce PG (2006) Rechargeable Li_2O_2 electrode for lithium batteries. J Am Chem Soc 128(4):1390
17. Débart A, Paterson AJ, Bao J, Bruce PG (2008) α-MnO_2 nanowires: a catalyst for the O_2 electrode in rechargeable lithium batteries. Angew Chem Int Ed 47(24):4521–4524. doi:10.1002/anie.200705648
18. Hummelshøj JS, Blomqvist J, Datta S, Vegge T, Rossmeisl J, Thygesen KS, Luntz AC, Jacobsen KW, Norskov JK (2010) Communications: elementary oxygen electrode reactions in the aprotic Li-air battery. J Chem Phys 132(7):071101–071104
19. Lu Y-C, Xu Z, Gasteiger HA, Chen S, Hamad-Schifferli K, Shao-Horn Y (2010) Platinum–gold nanoparticles: a highly active bifunctional electrocatalyst for rechargeable lithium–air batteries. J Am Chem Soc 132(35):12170–12171. doi:10.1021/ja1036572
20. Hassoun J, Croce F, Armand M, Scrosati B (2011) Investigation of the O_2 electrochemistry in a polymer electrolyte solid-state cell. Angew Chem Int Ed 50(13):2999–3002. doi:10.1002/anie.201006264
21. Lu Y-C, Gasteiger HA, Parent MC, Chiloyan V, Shao-Horn Y (2010) The influence of catalysts on discharge and charge voltages of rechargeable Li–oxygen batteries. Electrochem Solid-State Lett 13(6):A69–A72
22. Lu Y-C, Kwabi DG, Yao KPC, Harding JR, Zhou J, Zuin L, Shao-Horn Y (2011) The discharge rate capability of rechargeable Li-O_2 batteries. Energy Environ Sci 4(8):2999–3007
23. Mo Y, Ong SP, Ceder G (2011) First-principles study of the oxygen evolution reaction of lithium peroxide in the lithium-air battery. Phys Rev B 84(20):205446
24. Hummelshøj JS, Luntz AC, Norskov JK (2013) Theoretical evidence for low kinetic overpotentials in Li-O_2 electrochemistry. J Chem Phys 138(3):034703–034712
25. Xu Y, Shelton WA (2010) O_2 reduction by lithium on Au(111) and Pt(111). J Chem Phys 133 (2):024703–024709
26. Xu Y, Shelton WA (2011) Oxygen reduction by lithium on model carbon and oxidized carbon structures. J Electrochem Soc 158(10):A1177–A1184
27. Radin MD, Rodriguez JF, Tian F, Siegel DJ (2011) Lithium peroxide surfaces are metallic, while lithium oxide surfaces are not. J Am Chem Soc 134(2):1093–1103. doi:10.1021/ja208944x
28. Albertus P, Girishkumar G, McCloskey B, Sánchez-Carrera RS, Kozinsky B, Christensen J, Luntz AC (2011) Identifying capacity limitations in the Li/oxygen battery using experiments and modeling. J Electrochem Soc 158(3):A343–A351
29. Kang J, Jung YS, Wei S-H, Dillon AC (2012) Implications of the formation of small polarons in Li_2O_2 for Li-air batteries. Phys Rev B 85(3):035210
30. Garcia-Lastra JM, Myrdal JSG, Christensen R, Thygesen KS, Vegge T (2013) DFT+U study of polaronic conduction in Li_2O_2 and Li_2CO_3: implications for Li–air batteries. J Phys Chem C 117(11):5568–5577. doi:10.1021/jp3107809
31. Holstein T (1959) Studies of polaron motion: Part I. The molecular-crystal model. Ann Phys 8 (3):325–342. doi:10.1016/0003-4916(59)90002-8
32. Holstein T (1959) Studies of polaron motion: Part II. The "small" polaron. Ann Phys 8 (3):343–389. doi:10.1016/0003-4916(59)90003-X
33. Emin D, Holstein T (1969) Studies of small-polaron motion IV. Adiabatic theory of the Hall effect. Ann Phys 53(3):439–520. doi:10.1016/0003-4916(69)90034-7
34. Garcia-Lastra JM, Bass JD, Thygesen KS (2011) Communication: strong excitonic and vibronic effects determine the optical properties of Li_2O_2. J Chem Phys 135 (12):121101–121104

35. Ong SP, Mo Y, Ceder G (2012) Low hole polaron migration barrier in lithium peroxide. Phys Rev B 85(8):081105
36. Radin MD, Rodriguez JF, Siegel DJ (2011) Lithium peroxide surfaces and point defects: relevance for Li-air batteries. In: Uddin MN (ed) Battery congress, p 6
37. Lau KC, Curtiss LA, Greeley J (2011) Density functional investigation of the thermodynamic stability of lithium oxide bulk crystalline structures as a function of oxygen pressure. J Phys Chem C 115(47):23625–23633. doi:10.1021/jp206796h
38. Ottakam Thotiyl MM, Freunberger SA, Peng Z, Bruce PG (2012) The carbon electrode in nonaqueous Li–O2 cells. J Am Chem Soc 135(1):494–500. doi:10.1021/ja310258x
39. Ren X, Zhang SS, Tran DT, Read J (2011) Oxygen reduction reaction catalyst on lithium/air battery discharge performance. J Mater Chem 21(27):10118–10125
40. Débart A, Bao J, Armstrong G, Bruce PG (2007) An O_2 cathode for rechargeable lithium batteries: the effect of a catalyst. J Power Sources 174(2):1177–1182. doi:10.1016/j.jpowsour.2007.06.180
41. Thapa AK, Ishihara T (2011) Mesoporous α-MnO2/Pd catalyst air electrode for rechargeable lithium–air battery. J Power Sources 196(16):7016–7020. doi:10.1016/j.jpowsour.2010.09.112
42. Trahey L, Karan NK, Chan MKY, Lu J, Ren Y, Greeley J, Balasubramanian M, Burrell AK, Curtiss LA, Thackeray MM (2013) Synthesis, characterization, and structural modeling of high-capacity, dual functioning MnO2 electrode/electrocatalysts for Li-O_2 cells. Adv Energy Mater 3(1):75–84. doi:10.1002/aenm.201200037
43. Nørskov JK, Rossmeisl J, Logadottir A, Lindqvist L, Kitchin JR, Bligaard T, Jónsson H (2004) Origin of the overpotential for oxygen reduction at a fuel-cell cathode. J Phys Chem B 108 (46):17886–17892. doi:10.1021/jp047349j
44. Read J (2002) Characterization of the lithium/oxygen organic electrolyte battery. J Electrochem Soc 149(9):A1190–A1195
45. Giordani V, Freunberger SA, Bruce PG, Tarascon JM, Larcher D (2010) H2O2 decomposition reaction as selecting tool for catalysts in Li–O2 cells. Electrochem Solid-State Lett 13(12): A180–A183
46. Trahey L, Johnson CS, Vaughey JT, Kang SH, Hardwick LJ, Freunberger SA, Bruce PG, Thackeray MM (2011) Activated lithium-metal-oxides as catalytic electrodes for Li–O_2 cells. Electrochem Solid-State Lett 14(5):A64–A66
47. Lee SW, Carlton C, Risch M, Surendranath Y, Chen S, Furutsuki S, Yamada A, Nocera DG, Shao-Horn Y (2012) The nature of lithium battery materials under oxygen evolution reaction conditions. J Am Chem Soc 134(41):16959–16962. doi:10.1021/ja307814j
48. Suntivich J, May KJ, Gasteiger HA, Goodenough JB, Shao-Horn Y (2011) A perovskite oxide optimized for oxygen evolution catalysis from molecular orbital principles. Science 334 (6061):1383–1385
49. Julien C, Massot M (2002) Lattice dynamics of manganese oxides and their intercalated compounds. In: Julien C, Pereira-Ramos JP, Momchilov A (eds) New trends in intercalation compounds for energy storage, vol 61, NATO science series. Springer, Dordrecht, pp 235–252. doi:10.1007/978-94-010-0389-6_16
50. Ling C, Mizuno F (2012) Capture lithium in αMnO_2: insights from first principles. Chem Mater 24(20):3943–3951. doi:10.1021/cm302347j
51. Mizuno F, Nakanishi S, Kotani Y, Yokoishi S, Iba H (2010) Rechargeable Li-air batteries with carbonate-based liquid electrolytes. Electrochemistry 78(5):403–405
52. Peng Z, Freunberger SA, Chen Y, Bruce PG (2012) A reversible and higher-rate Li-O_2 battery. Science 337(6094):563–566
53. Bryantsev VS, Blanco M (2011) Computational study of the mechanisms of superoxide-induced decomposition of organic carbonate-based electrolytes. J Phys Chem Lett 2(5):379–383. doi:10.1021/jz1016526
54. Zhang Z, Lu J, Assary RS, Du P, Wang H-H, Sun Y-K, Qin Y, Lau KC, Greeley J, Redfern PC, Iddir H, Curtiss LA, Amine K (2011) Increased stability toward oxygen reduction products for

lithium-air batteries with oligoether-functionalized silane electrolytes. J Phys Chem C 115 (51):25535–25542. doi:10.1021/jp2087412
55. Laino T, Curioni A (2012) A new piece in the puzzle of lithium/air batteries: computational study on the chemical stability of propylene carbonate in the presence of lithium peroxide. Chem Eur J 18(12):3510–3520. doi:10.1002/chem.201103057
56. Assary RS, Curtiss LA, Redfern PC, Zhang Z, Amine K (2011) Computational studies of polysiloxanes: oxidation potentials and decomposition reactions. J Phys Chem C 115 (24):12216–12223. doi:10.1021/jp2019796
57. Bryantsev VS, Faglioni F (2012) Predicting autoxidation stability of ether- and amide-based electrolyte solvents for Li–air batteries. J Phys Chem A 116(26):7128–7138. doi:10.1021/jp301537w
58. Du P, Lu J, Lau KC, Luo X, Bareno J, Zhang X, Ren Y, Zhang Z, Curtiss LA, Sun Y-K, Amine K (2013) Compatibility of lithium salts with solvent of the non-aqueous electrolyte in Li-O2 batteries. Phys Chem Chem Phys 15(15):5572–5581
59. Assary RS, Lu J, Du P, Luo X, Zhang X, Ren Y, Curtiss LA, Amine K (2013) The effect of oxygen crossover on the anode of a Li–O_2 battery using an ether-based solvent: insights from experimental and computational studies. ChemSusChem 6(1):51–55. doi:10.1002/cssc.201200810

Chapter 6
Lithium Air Batteries Based on Protected Lithium Electrodes

Steven J. Visco, Vitaliy Nimon, Alexei Petrov, Kirill Pridatko, Nikolay Goncharenko, Eugene Nimon, Lutgard De Jonghe, Mary Hendrickson, and Edward Plichta

6.1 Introduction

As the world moves towards electrification of transportation and increased use of renewables for grid power, the demand for advanced secondary batteries is prompting a global race for dominance in these key energy technologies. In order to maintain a competitive position, battery developers will need to innovate both in the development of step-change energy storage technologies and in the cost-competitive manufacturing of those products. At present, the key barrier to wide-spread adoption of battery systems for energy storage in electric vehicles and for the electric grid is cost. The path to significant battery cost reduction and global competitiveness will rely on either the incremental engineering improvements to commercial batteries or the introduction of a step-change battery technology. In this vein, we have invented and patented a new generation of water-stable protected lithium metal electrodes that are enabling for lithium air, lithium–water, and lithium sulfur batteries [1–4]. Protected lithium electrodes (PLEs) utilize water-stable, high Li^+ conductivity solid electrolytes such as $Li_{1.3}Al_{0.3}Ti_{1.7}(PO_4)_3$ (LATP) to chemically isolate a lithium core from the external environment. In this way lithium electrodes can be used in combination with aqueous or aggressive nonaqueous electrolytes, a practical necessity for lithium air batteries. In this chapter we discuss the state of the art in lithium air batteries and the road map to commercial systems.

S.J. Visco (✉) • V. Nimon • A. Petrov • K. Pridatko • N. Goncharenko
E. Nimon • L. De Jonghe
PolyPlus Battery Company, 2431 5th Street, Berkeley, CA 94710, USA
e-mail: SVisco@polyplus.com

M. Hendrickson • E. Plichta
US Army CERDEC, APG, 5100 Magazine Road, Aberdeen Proving Ground,
Aberdeen MD 21005, USA

N. Imanishi et al. (eds.), *The Lithium Air Battery: Fundamentals*,
DOI 10.1007/978-1-4899-8062-5_6, © Springer Science+Business Media New York 2014

Table 6.1 Specific energy for various metal-air couples

Metrics	Li	Mg	Al	Ca	Zn	Fe
Specific capacity (Ah/g)	3.86	2.21	2.98	1.34	0.82	0.96
Anode potential (V) vs. SHE	−3.05	−2.69	−2.30	−2.84	−1.25	−0.88
Typical discharge voltage (V)	3.0	1.4	1.3	2.0	1.2	1.0
Specific energy of anode (Wh/kg)	11,600	3,100	3,900	2,700	980	960
Specific energy of anode + H_2O (Wh/kg)	5,040	1,770	1,940	1,850	770	730
End of discharge specific energy (Wh/kg)	3,360	1,290	1,340	1,450	660	600

6.2 The Promise and Problems with Metal-Air Batteries

Galvanic couples based on the use of ambient oxygen offer significant gravimetric and volumetric advantages relative to conventional couples. Since the positive electrode (O_2) contributes no weight to the battery (prior to cell discharge), the theoretical energy density for metal-air couples, $M + xO_2 = MO_{2x}$, as shown in Table 6.1, can be exceptionally high. Metal-air battery chemistry development has been based almost exclusively on the use of water-based electrolytes; certainly, all commercial zinc air batteries use aqueous electrolyte. However, since the metals listed in Table 6.1 corrode rapidly at low pH, aqueous electrolytes for metal-air cells use concentrated KOH to limit self-discharge. Unfortunately, once conventional aqueous metal-air cells are activated and oxygen is allowed to enter the cathode, O_2 reacts directly with the zinc leading to relatively high self-discharge rates which limits the commercial markets available to Zn-air batteries. Another major problem with metal-air batteries using alkaline electrolytes is the tendency to absorb CO_2 from ambient air which reacts with the KOH electrolyte to form practically insoluble K_2CO_3, a process known as carbonation, which degrades cell performance. Battery scientists have been aware of these problems for decades, but neither issue has been solved to date. The carbonation problem could be addressed by scrubbing CO_2 from the ambient air stream, but this is impractical for most battery applications.

When one considers replacement of the zinc anode with a lithium anode, the problem of self-discharge appears to be so untenable that this galvanic couple would seem to be unworthy of serious consideration. Nonetheless, the extremely high energy density for the lithium air couple has attracted the interest of battery developers for decades and this interest has peaked over the past few years. There are several approaches to the development of rechargeable (and non-rechargeable) lithium air batteries, and these are outlined below. Notably, due to the invention of the water-stable PLE, it now appears that the lithium air couple may be the only metal-air couple where self-discharge can be completely eliminated.

6.3 The Lithium Air Battery: Aqueous or Nonaqueous?

In conventional aqueous metal-air cells, the products of the cell reaction are contained in the anode; in zinc air cells the zinc suspension is converted to ZnO as the cell reaction proceeds whereas the structure and composition of the air

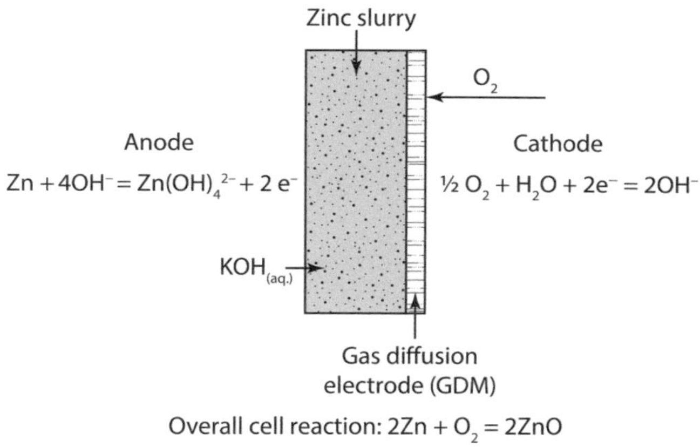

Fig. 6.1 Schematic of Zn-air cell

Table 6.2 Lithium air cell types

Electrolyte	Anode	Type	OCV	Energy density (end of discharge)
Nonaqueous $Li + O_2 = Li_2O_2$	Unprotected (bare Li)	Type A1	2.96 V	3,450 Wh/kg
	Protected (PLE)	Type A2		8,000 Wh/L
Aqueous $4Li + O_2 + 2H_2O = 4LiOH$ (simplified)	Unprotected (bare Li)	Not possible	3.0 V	3,360 Wh/kg
	Protected (PLE)	Type B		6,100 Wh/L

electrode is essentially unchanged as the cell is discharged (Fig. 6.1). This is fundamentally different in lithium air technology since the lithium anode cannot accommodate the product of cell reaction. In a lithium air cell the air electrode must facilitate oxygen reduction and evolution as well as accommodate a substantial amount of reaction product while still functioning as an air electrode throughout cell discharge. The engineering of such a structure is certainly nontrivial. There are also several types of lithium air technology [1, 5, 6] which can be broadly divided by the choice of electrolyte, either aqueous or nonaqueous (Table 6.2). Although the choice of nonaqueous electrolyte allows the use of bare lithium foil, it also introduces a number of complex problems including the extremely low solubility of the cell reaction product, Li_2O_2, as well as reactivity of peroxide (and/or superoxide) with the most commonly used lithium battery electrolytes. For the case of nonaqueous lithium air, most researchers use anhydrous oxygen to avoid corrosion of the bare Li electrode by ambient moisture. The use of bare unprotected lithium greatly limits the choice of electrolytes (due to reaction with lithium) and precludes the use of air due to rapid corrosion of the Li metal electrode by moisture. Some groups have reported ex situ passivation of lithium by carbonate electrolytes [7] or the use of oxidizing additives [8] like $LiNO_3$ to form a protective solid electrolyte

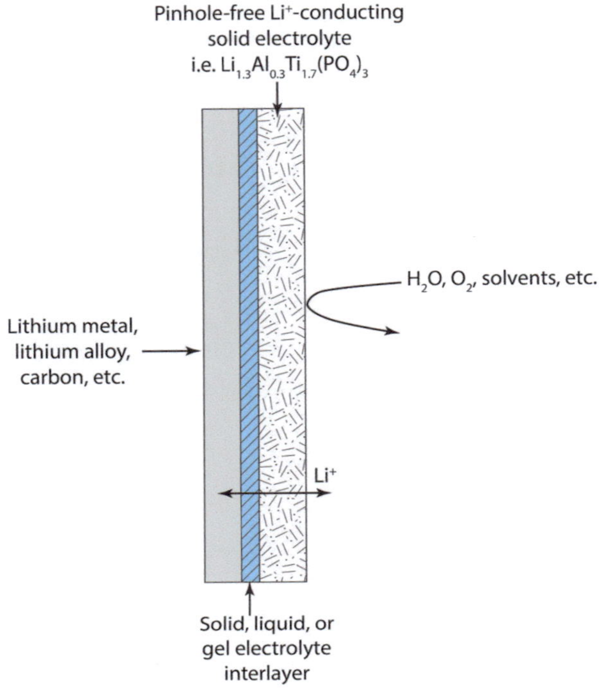

Fig. 6.2 Schematic of protected lithium electrode

interface (SEI) allowing the use of aggressive solvents like amides and sulfoxides. Although it is unclear that this approach is fully protective or that the SEI will survive extended cycling, preliminary results are interesting. PolyPlus invented and patented water-stable PLEs [2, 3]. We described PLEs and lithium air primary and rechargeable batteries based on them publicly for the first time in Nara, Japan [9–11]. The essential concept is to use a dense, pinhole-free Li^+ solid electrolyte to chemically isolate the lithium electrode (Li, Li_xC, Li_xSi, etc.) from the positive electrode (Fig. 6.2). In this way, lithium electrodes can be used in combination with aggressive solvents, including aqueous electrolyte. The ceramic electrolyte most commonly used is based on the lithium analog of NASICON (LISICON) with a composition of $Li_{1+x+y}(M, Al, Ga)_x(Ge_{1-q}Ti_q)2_{-x}Si_yP_{3-y}O_{12}$ and typically is of the composition $Li_{1.3}Al_{0.3}Ti_{1.7}(PO_4)_3$ (LATP) produced in Japan by Ohara Corporation and in the USA by Corning. Although LISICON is stable to water, it is not stable to lithium, requiring the incorporation of an interlayer which can be solid state (i.e., Li_3N), polymeric, ionic liquid, or nonaqueous liquid electrolyte. Some groups prefer to call this a "dual electrolyte" system, but it all originates from the original PLE concept and patents. This advance now appears to be critical to the practical development of both nonaqueous and aqueous lithium air and lithium sulfur batteries. The PLE also affords an elegant solution to the investigation and development of the air electrode for both aqueous and nonaqueous lithium air chemistries. Among the key differences between the nonaqueous and aqueous air electrodes is that for aqueous systems this electrode functions as a gas diffusion electrode (GDE) with hydrophobic domains creating channels for O_2 gas transport whereas in the nonaqueous case the electrode is usually flooded. This fact highlights one of the key advantages of aqueous lithium air over nonaqueous: the existence of three-phase boundaries throughout the aqueous air electrode leads to much faster oxygen transport and better electrode kinetics and therefore much higher sustained current densities. Another issue for the nonaqueous chemistry is solvent loss from the air electrode which would be unavoidable in an open system. Accordingly, any practical nonaqueous lithium air battery would almost certainly need to be based on a closed system running on anhydrous oxygen, and such constraints may override the energy density advantages that lithium air chemistry offers. We have studied both the nonaqueous and the aqueous lithium air chemistries in some detail to determine baseline performances for both approaches and to initiate commercialization of ultra-lightweight rechargeable batteries. Assuming the use of commercially available components and reasonable values for components still in development, one can project lithium air cell energy densities of about 1,000 Wh/L and 600 Wh/kg for a reversible electrode capacity of 5 mAh/cm^2, implying that lithium air cells should sustain current densities of at least 1 mA/cm^2 (C/5) in order to realize broad commercial application.

6.3.1 Nonaqueous Lithium Air

In the mid-2000s interest in lithium air batteries rose tremendously as a number of research groups claimed to be on the path to batteries with an order of magnitude improved energy density, motivated by the extremely high energy density of the Li/O_2 couple. If one does not include the weight of the oxygen consumed as the discharge reaction progresses, the specific energy for the Li/O_2 couple is 11,600 Wh/kg. Although metal-air batteries gain weight as the cell reaction progress and O_2 is converted to lithium peroxide (yielding a theoretical maximum of 3,450 Wh/kg), the average specific energy over the course of cell discharge is well over 7,000 Wh/kg, approaching that of gasoline and providing ample motivation for cells developers.

The earliest work in nonaqueous lithium air chemistry was reported by K.M. Abraham [12]. These studies were done with bare unprotected lithium, dry oxygen, and polymer electrolytes plasticized by organic carbonates. In fact most early studies of nonaqueous lithium air batteries relied heavily on the use of electrolytes based on carbonate electrolytes typically used in lithium ion batteries. However, an early paper by Aurbach and Yeager [13] pointed out that carbonate electrolytes are unstable to nucleophilic attack by superoxide anion radical ($O_2^{\bullet -}$), an intermediate in the reduction of oxygen. Consequently, early results that demonstrated extended cycling of Li–O_2 batteries and improvements due to electrocatalysts were greatly affected by solvent degradation and oxidation of those degradation products. Since that time, a number of groups have conducted

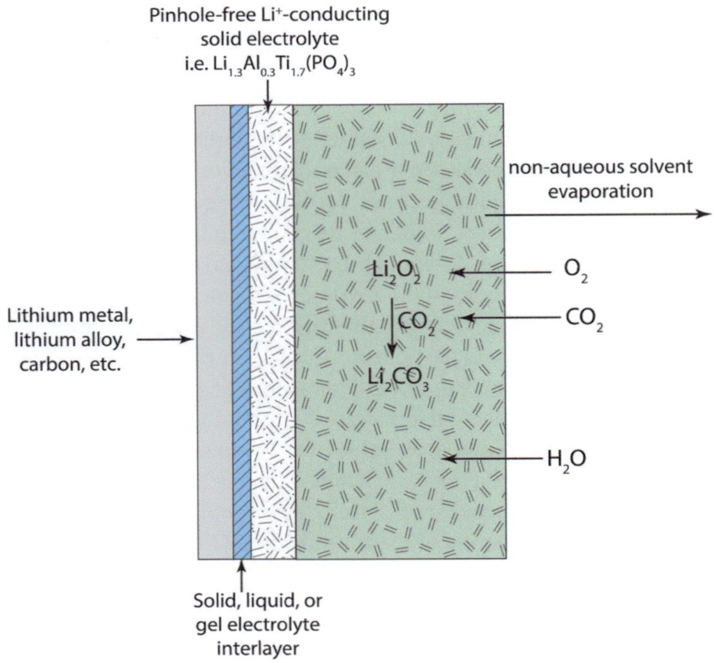

Fig. 6.3 Schematic of nonaqueous lithium air cell with PLE

careful investigations of those degradation phenomena [14, 15] followed by a search for stable nonaqueous electrolytes for Li–O_2 batteries [16].

In 2005, PolyPlus Battery Company expanded the exploration of a number of nonaqueous lithium air chemistries based on the use of PLEs in combination with solvents known to be stable to peroxide (and superoxide) but not necessarily to lithium metal including amides, sulfoxides, and nitriles [17]. We explored a variety of nonaqueous electrolytes using lithium air cells having PLEs (as shown schematically in Fig. 6.3). Conductivities of nonaqueous electrolytes based on aprotic solvents such as dimethylformamide (DMF), dimethylacetamide (DMA), dimethyl sulfoxide (DMSO), diglyme (DG), and sulfolane and three different supporting salts are presented in Fig. 6.4. Although the theoretical energy densities for lithium air chemistries are certainly large (Table 6.2), practical cell performance will not exceed that of commercial lithium ion batteries until lithium air cells reach certain threshold current densities and capacities. Assuming commercially available components and reasonable values for components that are still in development, one can project lithium air cell energy densities of about 1,000 Wh/L and 600 Wh/kg for a reversible electrode capacity of 5 mAh/cm^2 at a current density of 1 mA/cm^2 (C/5). However, nonaqueous lithium air performance is typically limited to much lower current densities, often below 0.1 mA/cm^2, and cycled capacity is typically less than 1.0–1.5 mAh/cm^2. We tested lithium air cells with glyme-based electrolytes,

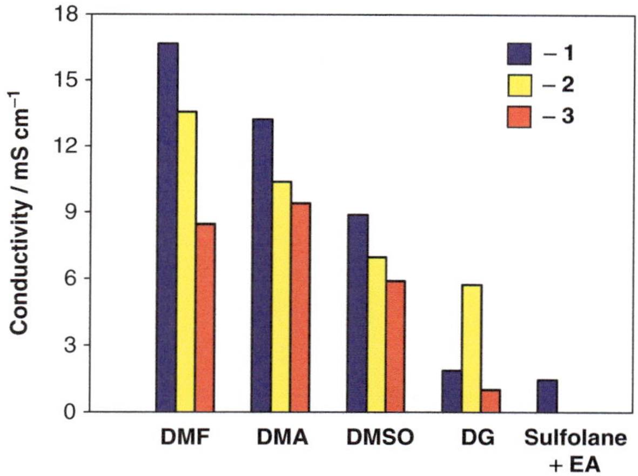

Fig. 6.4 Conductivity of nonaqueous electrolytes based on aprotic solvents tested in PolyPlus lithium air cells with PLEs: (*1*) lithium tetrafluoroborate ($LiBF_4$), (*2*) lithium bis(trifluoromethylsulfonyl)imide ($LiN(CF_3SO_2)_2$) [LiTFSI], and (*3*) lithium trifluoromethanesulfonate ($LiCF_3SO_3$). Lithium salt molarity was 0.5 M. Data courtesy of PolyPlus Battery Company

in particular with diglyme, as well as electrolytes based on sulfolane. Diglyme was chosen due to its lower volatility compared to DME (monoglyme), and sulfolane (in combination with ethyl acetate (EA) cosolvent and $LiBF_4$ salt) was chosen because of its exceptional oxidative stability. Cycling of a nonaqueous lithium air cell with diglyme-based electrolyte is shown in Fig. 6.5. The cell exhibits a reversible capacity of slightly more than 1 mAh/cm^2 at a discharge current density of 0.25 mA/cm^2 and charge current density of 0.125 mA/cm^2. Although preliminary results with sulfolane (Fig. 6.6) are also interesting, we observed increasing polarization on charge and limited capacity on cycling. In order to achieve high specific energy and reasonable power density in an engineered battery, the lithium air cells based on diglyme electrolytes would require a fourfold improvement in reversible capacity at four times the current density shown here. Certainly one of the key problems with nonaqueous Li–O_2 chemistry is the extremely low solubility of the cell discharge product, Li_2O_2. PolyPlus explored increasing the solubility of Li_2O_2 by the addition of anion receptors [18] to the nonaqueous electrolyte in an attempt to improve the performance of the air electrode. We synthesized a variety of boron esters reported in the literature but found them to be both insoluble in the electrolyte and easily hydrolyzed. We had better success with a commercially available anion receptor, tris(pentafluorophenyl)borane (TPFB). We found that Li_2O_2 is nearly completely insoluble in DMF and only in the presence of an anion receptor TPFB is there any appreciable solubility of Li_2O_2 in DMF. In particular, in the presence of 0.25 M TPFB, solubility of Li_2O_2 increased to approximately 17 mmol/L. This implies that on cell discharge the reaction product Li_2O_2 immediately precipitates in the porous air electrode structure. As can be seen from the data in Fig. 6.7, the addition of TPFB had a dramatic effect on the reversibility of air electrodes

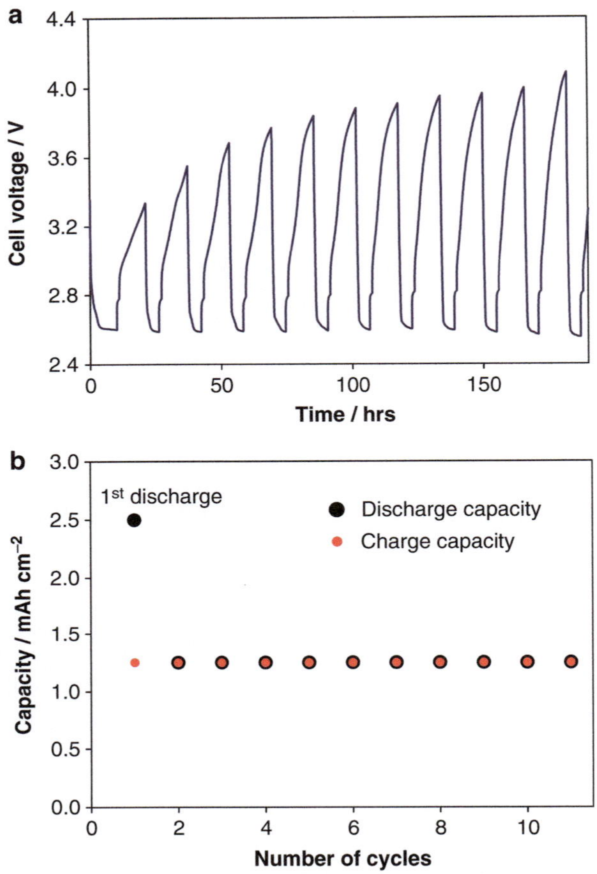

Fig. 6.5 Cycling performance of nonaqueous lithium air cells in dry oxygen atmosphere. Electrolyte: 0.5 M LiTFSI in DG. Discharge current density was 0.25 mA/cm^2. Charge current density was 0.125 mA/cm^2. The first discharge lasted 10 h. The first charge capacity was 50 % of the first discharge capacity. Capacities of the second and the following charges were equal to the capacity of the second charge or were limited by a cutoff voltage of 4.15 V. Capacities of the second and the following discharges were equal to the capacity of the second charge or were limited by a cutoff voltage of 2.2 V. (**a**) Voltage profiles. (**b**) Cycling capacity as a function of cycle number. Data courtesy of PolyPlus Battery Company

based on DMF. It should be mentioned that this approach could not be used for lithium air cells without a PLE since TPFB reacts directly with bare lithium. However, the improvement seems to be short-lived and insufficient to meet the performance targets. Recently, several other groups have reported encouraging cycling of the O_2 electrode in nonaqueous electrolytes such as sulfoxides [7, 19] and amides [8] neither of which is stable to bare metallic lithium metal. Assuming that developers are successful in identifying nonaqueous electrolytes that permit reversible cycling of O_2, it appears likely practical that cells will require a fully PLE

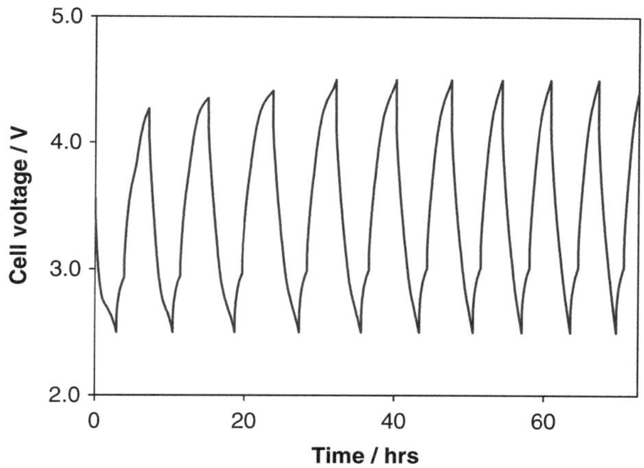

Fig. 6.6 Cycling performance of nonaqueous lithium air cells in dry oxygen atmosphere. Electrolyte: 0.5 M $LiBF_4$ in sulfolane and EA mixture (1 to 1 by volume). Discharge and charge current densities were 0.25 mA/cm^2. Discharge capacities were limited by a cutoff voltage of 2.5 V. Charge capacities were equal to the capacity of the previous discharge or were limited by a cutoff voltage of 4.5 V. Data courtesy of PolyPlus Battery Company

(an absolute for aqueous lithium air). Still, even with the incorporation of a PLE, there are unresolved technical issues for nonaqueous lithium air including (1) evaporation of nonaqueous solvent from the air electrode to the external environment, (2) uptake of water by the nonaqueous electrolyte, (3) irreversible reaction of Li_2O_2 with CO_2 leading to the formation of Li_2CO_3 and loss of electrode capacity, (4) flooding of the air electrode with nonaqueous electrolyte leading to slow oxygen transport, and (5) precipitation of insoluble Li_2O_2 in the pores of the air electrode leading to pore clogging and polarization of the air electrode. Some of the technical issues for nonaqueous lithium air could be addressed with a closed system (no solvent loss) and anhydrous oxygen (no water or CO_2 contamination), although this approach would certainly degrade the system energy density. Assuming successful identification of a nonaqueous electrolyte stable to products of O_2 reduction and the use of a closed system with dry oxygen, there still remains the issue of poor solubility of Li_2O_2, flooding of the air electrode leading to poor electrode kinetics, and clogging of the porous oxygen electrode structure as cell discharge proceeds. PolyPlus investigated the discharge behavior of nonaqueous Li–O_2 cells based on DMF electrolyte and, in collaboration with Yu. M. Volfkovich and D. A. Bograchev, performed mathematical modeling of this system [20]. We employed the method of standard contact porosimetry to characterize the porous structure of undischarged, partially discharged, and fully discharged carbon black-based air electrodes and determined that deposition of solid discharge products occurs within a specific air electrode pore fraction (small mesopores with radii of 4–10 nm). This was accomplished by determining the pore size distribution of pristine air electrodes, followed by assembly into working nonaqueous lithium air cells and then discharge of the air electrodes. Air electrodes from discharged

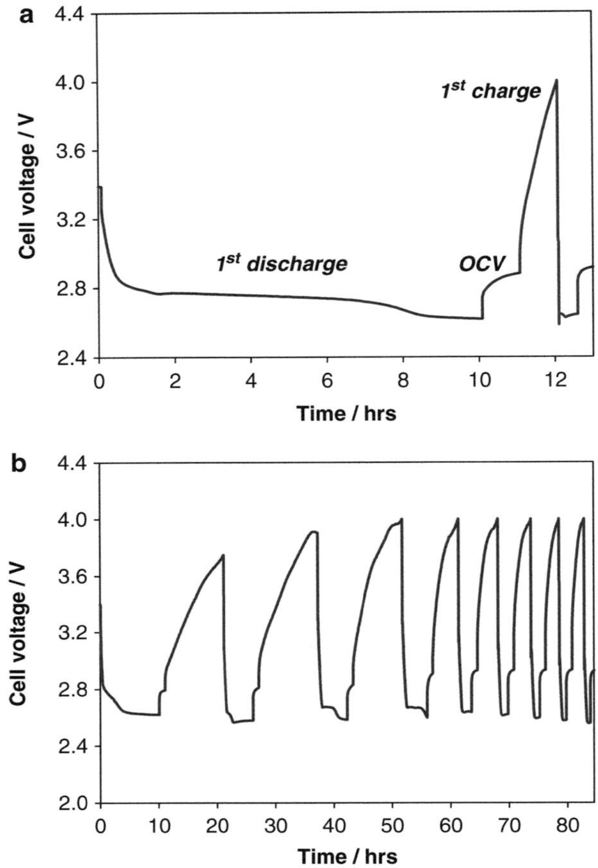

Fig. 6.7 Cycling performance of nonaqueous lithium air cells with electrolytes based on DMF in dry oxygen atmosphere. (**a**) Electrolyte: 0.5 M LiTFSI in DMF. Discharge current density was 0.25 mA/cm^2. Charge current density was 0.125 mA/cm^2. The first discharge lasted 10 h. The charge capacities were limited by a cutoff voltage of 4.0 V. Data courtesy of PolyPlus Battery Company. (**b**) Electrolyte: 0.5 M LiTFSI, 0.25 M TPFB in DMF. Discharge current density was 0.25 mA/cm^2. Charge current density was 0.125 mA/cm^2. The first discharge lasted 10 h. The first charge capacity was 50 % of the first discharge capacity. Capacities of the second and the following discharges were equal to capacity of the previous charge. Capacities of the second and the following charges were equal to the capacity of the previous discharge or were limited by a cutoff voltage of 4.0 V. Data courtesy of PolyPlus Battery Company

lithium air cells were then rinsed with DMF and the pore size distribution was determined for partially and fully discharged air electrodes. Notably, we did not find any experimental support for a continuous coating of the air electrode surface with insoluble discharge products, which would have manifested itself as a gradual decrease in the radii of all air electrode pores, but instead found gradual reduction in overall volume fraction of mesopores with radii in the 4–10 nm range as shown in Fig. 6.8. According to our experimental data, in the Li–O_2 system having a

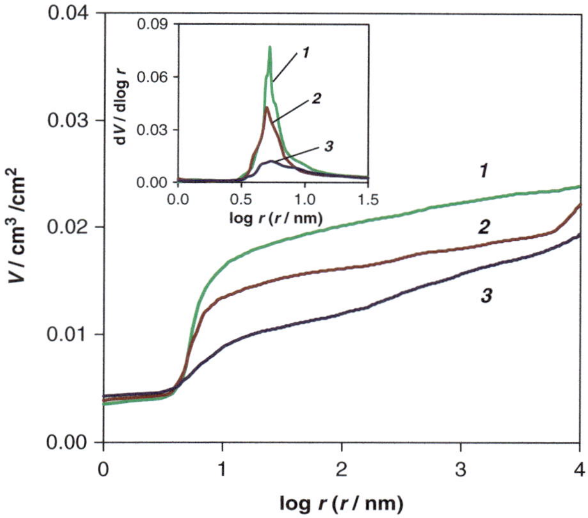

Fig. 6.8 Integral and differential (*inset*) porograms for air cathodes before discharge (*1*), after partial discharge of 48 % at 0.25 mA/cm^2 (*2*), and after full discharge at 0.25 mA/cm2 (*3*). Electrolyte: 0.5 M LiTFSI in DMF. Data courtesy of PolyPlus Battery Company

DMF-based electrolyte and Li_2O_2 as a solid discharge product, the volume fraction of the smallest air electrode pores decreases as the discharge progresses. This can be explained by a strong adhesion of the solid discharge product (Li_2O_2) to the carbon on the surface of the air electrode as described by the Kelvin equation, resulting in the progression of pore filling from the smallest to the largest pores. We further discovered that the discharge capacity of the air electrode is limited by the amount of solid discharge product that can be accommodated within the volume of small mesopores with radii of up to 10 nm. Our data suggest that the end of discharge corresponds to the liquid electrolyte volume fraction falling below the percolation threshold and thus impairing oxygen transport.

In light of the above discussion, the list of promising electrolytes for nonaqueous lithium air batteries is short at this time including amides, sulfoxides, and glymes, as described in a PolyPlus patent application in early 2006 [17]. To date, none of these electrolytes demonstrate the rate capability, cycling capacity (mAh/cm^2), or cycle life required for practical applications.

6.3.2 Aqueous Lithium Air

The first examples of aqueous lithium air batteries were reported by PolyPlus Battery Company [2, 9–11] and enabled by the invention of the PLE. One of the commonly recognized problems with aqueous metal-air chemistries is the high rate of self-discharge due to direct reaction of oxygen with the negative electrode. In stark contrast, the self-discharge rate of our aqueous lithium air cell is effectively zero, due to the presence of a solid electrolyte membrane which chemically isolates the metal electrode from the positive electrode compartment (see schematic

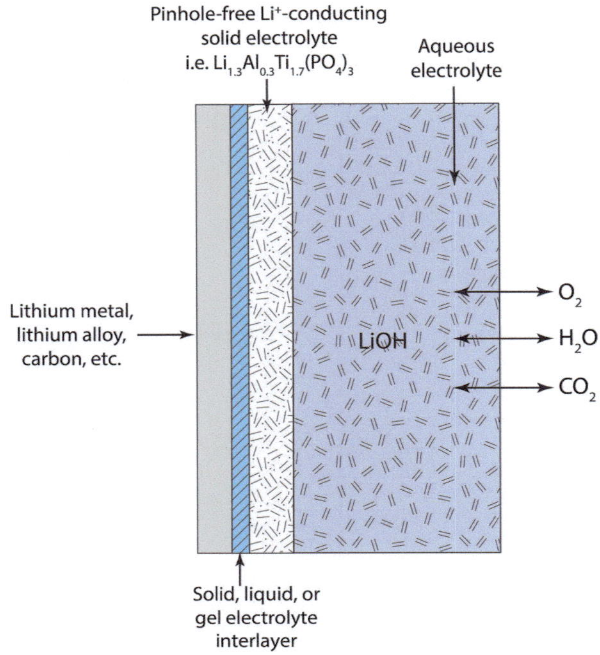

Fig. 6.9 Schematic diagram of aqueous lithium air cell having PLE

drawing in Fig. 6.9). The use of a Li^+-conductive solid electrolyte in aqueous electrolyte also begs the question of stability of the ceramic LATP as a function of pH.

6.3.2.1 Stability of the Solid Electrolyte Membrane

Determination of stability of the ceramic membrane in aqueous electrolyte is not trivial and is certainly not measured properly by immersing the membrane into a large volume of concentrated acid or base, as some groups have done. We have determined the stability of the ceramic membrane in a variety of aqueous electrolytes by in situ impedance spectroscopy using a volume of electrolyte corresponding closely to that used in a practical cell. There is no doubt that LATP membranes degrade in strong inorganic acid or base, particularly when there is a large excess of aqueous electrolyte, but this has little to do with realistic cell conditions. We have found that LATP membranes exhibit exceptional stability at low and high pH, depending critically on the composition of the electrolyte, pH, and nature and concentration of the supporting salts. For primary lithium air cells, LATP membranes demonstrate excellent stability in aqueous electrolyte based on NH_4Cl and LiCl (Fig. 6.9) or on NH_4NO_3 and $LiNO_3$. These electrolyte blends have the advantage of generating hygroscopic LiCl or $LiNO_3$ on cell discharge which scavenge H_2O from the external environment, alleviating the need to carry a stoichiometric quantity of water for the cell reaction [21–23]. In studies where we

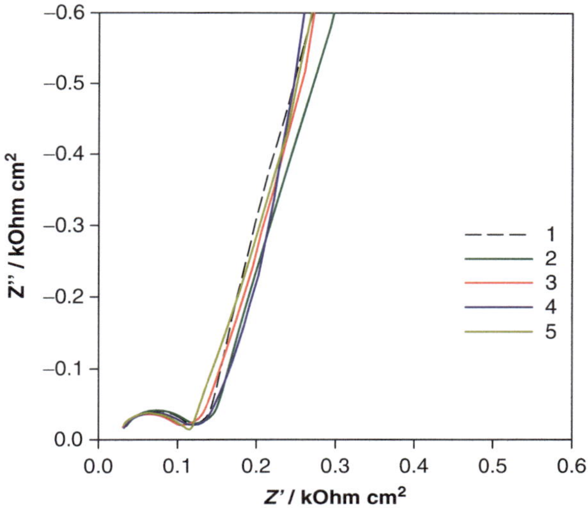

Fig. 6.10 Impedance spectra for solid electrolyte membrane immersed in aqueous $NH_4Cl/LiCl$ electrolyte for lithium air cell. Length of exposure: (*1*) less than 1 day, (*2*) 2 weeks, (*3*) 2 months, (*4*) 12 months, and (*5*) 21 months. Data courtesy of PolyPlus Battery Company

immersed LATP membranes in aqueous $NH_4Cl/LiCl$ (or $NH_4NO_3/LiNO_3$) electrolytes and determined the AC impedance of the membranes over time, we saw essentially no change in the impedance spectra for close to 2 years (Fig. 6.10). Since these electrolytes are slightly acidic prior to cell discharge, there is no uptake of CO_2 on exposure to ambient atmosphere. For rechargeable lithium air cells, PolyPlus developed polyprotic organic acid-based electrolytes [24]. Some of these polyprotic acids were found to degrade the performance of the LATP membranes as evidenced by a significant increase in the impedance arc corresponding to the total resistance of the membrane. Fortunately, we identified several polyprotic acid electrolytes in which LATP membranes exhibited excellent stability over time (see discussion below).

6.3.2.2 Primary Aqueous Lithium Air Batteries

Lithium air batteries share some common features with primary Zn-air batteries including the use of a GDE for oxygen reduction. The GDE comprises an intimate blend of carbon black and Teflon particles that create a tortuous path of gas and liquid channels to maximize three-phase boundaries in the air electrode. However, in a Zn-air cell the reaction products are stored in the anode compartment, leaving the GDE unperturbed as cell discharge progresses. In a lithium air battery the products of cell discharge are stored in the air electrode which means that the cathode compartment must accommodate a large volume of solid reaction product. Our air electrode design can accommodate a tremendous amount of solid discharge product as demonstrated by the data shown below. We have designed and fabricated lithium air cells with PLEs of approximately 10 Ah based on 50 mm × 50 mm solid electrolyte membranes and thick lithium foils (total thickness of 2–3 mm). In these cells the lithium capacity is matched with two air electrodes exposed to ambient air. One of the key features of the PLE is that it decouples the lithium electrode from the air electrode environment allowing complex tailoring of the aqueous electrolyte to enhance lithium air performance. This is illustrated clearly through the use of aqueous NH_4Cl electrolyte which on lithium air cell discharge liberates highly hygroscopic LiCl into the air electrode which then scavenges H_2O from ambient air. As the lithium air cell is discharged, a fairly complex reaction[1] takes place which can be viewed in its simplest case as such:

$$2Li + 2NH_4Cl(s) + \frac{1}{2}O_2 = 2NH_3 \uparrow + H_2O + 2LiCl \text{ (step 1)}$$
$$\text{and } 2Li + \frac{1}{2}O_2 + H_2O = 2LiOH \text{ (step 2)}$$

Since the first stage of discharge involves conversion of NH_4Cl to hygroscopic LiCl, the initial amount of water loaded into the cell prior to discharge can be significantly less than the amount of water required for a full discharge. The process of bringing water into the cathode is also affected by the presence of other species in solution. The rate at which water is scavenged from the external atmosphere is determined by a number of variables including depth of discharge, distribution, and morphology of the LiOH and LiCl hydrates (and/or complexes formed on cell discharge), structure and composition of the porous air electrode, and the water permeability of the gas diffusion membrane. We showed that for applications in ultralow-humidity environments, NH_4Br or NH_4I can be used to generate extremely hygroscopic LiBr or LiI salts. All of these factors must be considered when designing an air electrode for particular power requirements and environments. This approach is of course not feasible for nonaqueous lithium air chemistry highlighting a major shortcoming of the nonaqueous approach. Characteristic discharge curves for lithium air cells with PLEs and aqueous NH_4Cl/LiCl and NH_4NO_3/$LiNO_3$ electrolytes are shown in Fig. 6.11. These curves exhibit two typical regions: the initial discharge where the voltage decreases and then a long plateau where most of the capacity is delivered. For our cells, the major fraction (80–85 %) of discharge capacity is delivered in the two-phase (solid–liquid) region, corresponding to the long plateau on the discharge curves. At a current density of 0.5 mA/cm^2, these cells deliver over 200 mAh/cm^2 for a total capacity of 8,500 mAh at an energy density of more than 600 Wh/kg for fully engineered cells. To the best of our knowledge, no one to date has demonstrated a fully engineered nonaqueous lithium air cell with attractive energy density. In order to independently verify cell performance, several 10 Ah, 30 Wh primary aqueous lithium air cells were shipped to the US Army CERDEC at Aberdeen Proving Grounds (APG) in Aberdeen, MD. At APG the lithium air cells were tested on the benchtop in ambient conditions. While the cells tested at a current density of 0.5 mA/cm^2 delivered approximately 600 Wh/kg, lithium air cells tested at APG at a current density of 0.3 mA/cm^2 delivered a massive specific energy of 800 Wh/kg (Fig. 6.12); to the best of our knowledge the highest specific energy ever

[1] It is likely that a combination of LiOH and LiCl hydrates and/or complexes, as well as lithium peroxide, Li_2O_2, is formed on cell discharge.

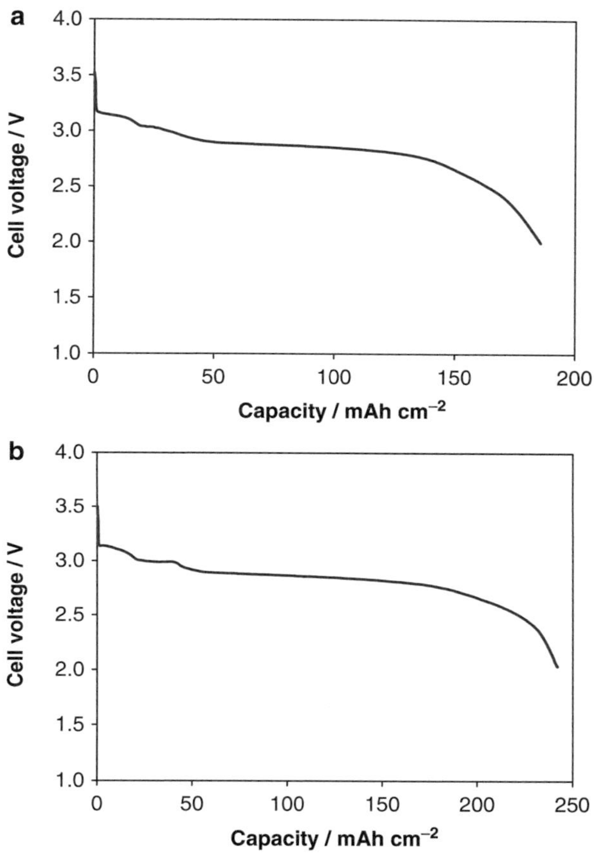

Fig. 6.11 Discharge performance of lithium air cells with aqueous electrolytes comprising (**a**) 2 M LiCl and 4 M NH_4Cl and (**b**) 2 M $LiNO_3$ and 4 M NH_4NO_3. Discharge was performed in air with 50 % relative humidity. Discharge current density was 0.5 mA/cm^2. Data courtesy of PolyPlus Battery Company

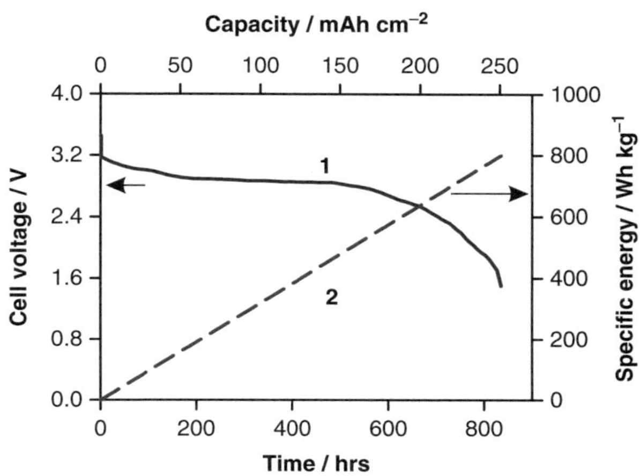

Fig. 6.12 Discharge curve and specific energy for PolyPlus 10 Ah lithium air cell with aqueous electrolyte. Discharge current density was 0.3 mA/cm^2. Discharge was performed in ambient air. Data courtesy of the US Army CERDEC

Table 6.3 Polyprotic acids for rechargeable aqueous lithium air batteries

	Solubility in water, mol/L	Melting point, °C	Decomposition temperature, °C
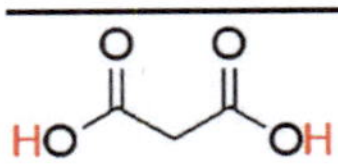 Malonic acid	7.2	136 (with decomposition)	136
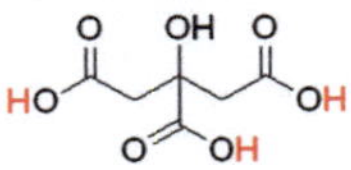 Citric acid	3.9	153	170

reported for an electrochemical device (other than our Li–water battery). These lithium air cells were stored for 50 days at APG prior to cell discharge with no deleterious effects on performance. Although the specific energy of these aqueous lithium air cells is remarkably high, at present the power density is relatively low. We intend to improve the rate capability of primary lithium air technology by rational design of the air electrode guided by porosimetry studies as described above and by moving to thinner, high surface area components.

6.3.2.3 Rechargeable Aqueous Lithium Air

Although we have demonstrated superb specific energy for primary lithium air cells using aqueous $NH_4Cl/LiCl$ electrolytes, these are not appropriate for secondary cells (partly due to the formation of NH_3 during discharge). In general strong inorganic acids are poor choices since they attack the LATP membranes and in some cases the corresponding salts are practically insoluble in water (e.g., Li_3PO_4) leading to rapid precipitation on discharge and/or limited reversibility. Accordingly, PolyPlus uses polyprotic organic acids in its rechargeable lithium air cells [24]. Two such electrolytes are based on citric and malonic acids (Table 6.3).

The lithium air cell reactions for citric and malonic acid electrolytes are shown below.

Malonic Acid
Anode:

$$Li \rightarrow Li^+ + e^-$$

Cathode:

1. $4H_2C_3H_2O_4 + O_2 + 4\,e^- \rightarrow 4HC_3H_2O_4^- + 2H_2O$
2. $4HC_3H_2O_4^- + O_2 + 4e^- \rightarrow 4C_3H_2O_4^{2-} + 2H_2O$

Citric Acid
Anode:

$$Li \rightarrow Li^+ + e^-$$

Cathode:

1. $4H_3C_6H_5O_7 + O_2 + 4e^- \rightarrow 4H_2C_6H_5O_7^- + 2H_2O$
2. $4H_2C_6H_5O_7^- + O_2 + 4e^- \rightarrow 4HC_6H_5O_7^{2-} + 2H_2O$
3. $4HC_6H_5O_7^{2-} + O_2 + 4e^- \rightarrow 4C_6H_5O_7^{3-} + 2H_2O$

Stability of LATP solid electrolyte membranes in triprotic citric acid is shown in Fig. 6.13. We do observe impedance rise at the solid electrolyte/aqueous electrolyte interface for the citric acid (which stabilizes after several days). After partial neutralization of citric acid with LiOH, the impedance practically does not rise and the solid electrolyte displays exceptionally stable impedance for more than 450 days. For the case of the electrolyte based on diprotic malonic acid, we did not see impedance rise and we observed exceptionally stable behavior for more than 260 days. A discharge curve for a lithium air cell having malonic acid electrolyte is shown in Fig. 6.14 overlaid with the experimental titration curve for the cell electrolyte. The titration curve for malonic acid (H_2Mal) traverses three distinct regions including that for diprotic acid (H_2Mal) and monoprotic salt (LiHMal), monoprotic salt (LiHMal), and neutral salt (Li_2Mal) and that for neutral salt (Li_2Mal) and lithium hydroxide. The use of polyprotic acid-based electrolytes can increase the discharge capacity and delay generation of insoluble discharge products compared to the behavior in basic electrolytes. Additionally, it allows for rejection of CO_2 at high states of cell charge. Using such polyprotic acid-based aqueous electrolytes, we have demonstrated excellent cycling for aqueous lithium air cells. Electrode processes for aqueous lithium air having malonic acid electrolyte are reasonably fast as can be seen from the high current pulse behavior shown in Fig. 6.15 (C rate). It was also important to determine that malonic acid was stable within the potential range used for lithium air cell cycling. Accordingly, on a glassy carbon electrode, at potentials exceeding 4 V (vs. Li/Li^+), we obtained the voltammetric curves for three electrolyte compositions: (1) malonic acid electrolyte, (2) partially neutralized malonic acid electrolyte having a pH of 2.7, and (3) that of a phosphate buffer solution that did not contain malonic acid and had the same pH. As can be seen in Fig. 6.16, the voltammetric curve for partially neutralized malonic acid almost overlaps that of the phosphate buffer having the same pH, indicating that the observed anodic currents are not due to oxidation of the malonic acid electrolyte. In Fig. 6.17 we show the long-term cycling behavior for lithium air cells with malonic acid electrolyte. We observed 77 cycles at 5 mAh/cm^2 of area specific capacity, a discharge current density of 1 mA/cm^2 (C/5), and a charge rate of 0.5 mA/cm^2, for a total cycled capacity of 385 mAh/cm^2 ($77 \times 5\ mAh/cm^2$). To the best of our knowledge, this is the largest cycled capacity ever reported for lithium air chemistry. However, in order to translate a cycled capacity of 5 mAh/cm^2 to the high specific energy possible with lithium air

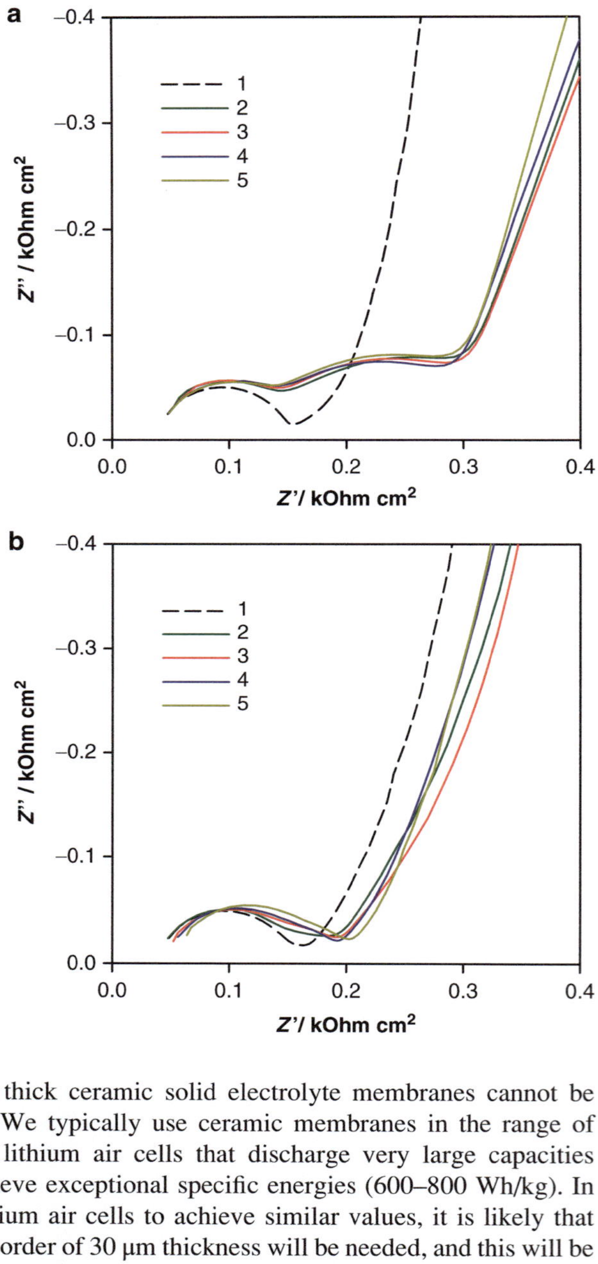

Fig. 6.13 (**a**) Impedance diagrams for LATP membrane exposed to 2.0 M citric acid electrolyte. Length of exposure: (*1*) less than 1 day, (*2*) 4 days, (*3*) 1 week, (*4*) 2 weeks, and (*5*) 3 weeks. Data courtesy of PolyPlus Battery Company. (**b**) Effect of partial acid neutralization (with LiOH) on the impedance behavior of LATP membranes exposed to citric acid. Length of exposure: (*1*) less than 1 day, (*2*) 1 month, (*3*) 6 months, (*4*) 9 months, and (*5*) 12 months. Data courtesy of PolyPlus Battery Company

chemistry, it is clear that thick ceramic solid electrolyte membranes cannot be used in cell construction. We typically use ceramic membranes in the range of 100–150 μm for primary lithium air cells that discharge very large capacities (>200 mAh/cm^2) and achieve exceptional specific energies (600–800 Wh/kg). In order for rechargeable lithium air cells to achieve similar values, it is likely that ceramic membranes on the order of 30 μm thickness will be needed, and this will be a challenge for membrane suppliers. We also believe that the present limitation to cycling of aqueous rechargeable lithium air might be mediated through better management of the water balance between the air electrode and the external environment.

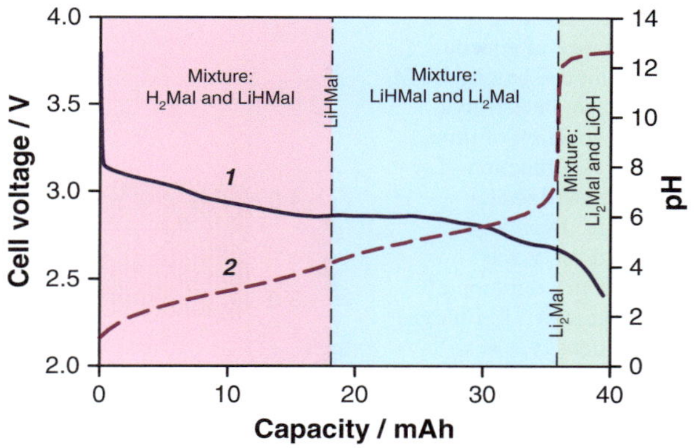

Fig. 6.14 (*1*) Discharge voltage profile of aqueous lithium air cell having an electrolyte based on malonic acid, (*2*) experimental titration curve for the same electrolyte plotted as a function of discharged capacity. Data courtesy of PolyPlus Battery Company

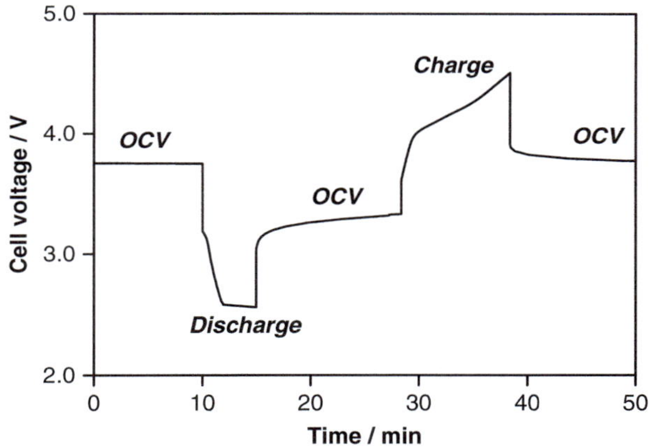

Fig. 6.15 High-rate pulse performance of rechargeable aqueous lithium air cell having an electrolyte based on malonic acid. Discharge current density was 6.0 mA/cm^2. Charge current density was 3.0 mA/cm^2. Duration of discharge was 5 min and duration of charge was 10 min. Data courtesy of PolyPlus Battery Company

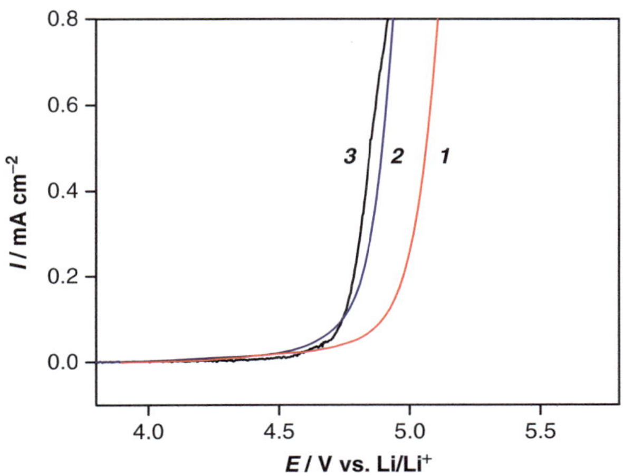

Fig. 6.16 Anodic stability of malonic acid-based electrolyte determined by cyclic voltammetry at a glassy carbon electrode. (*1*) 4 M malonic acid electrolyte, (*2*) partially neutralized malonic acid electrolyte having a pH of 2.7, and (*3*) a phosphate buffer solution (without malonic acid) having a pH of 2.7. Data courtesy of PolyPlus Battery Company

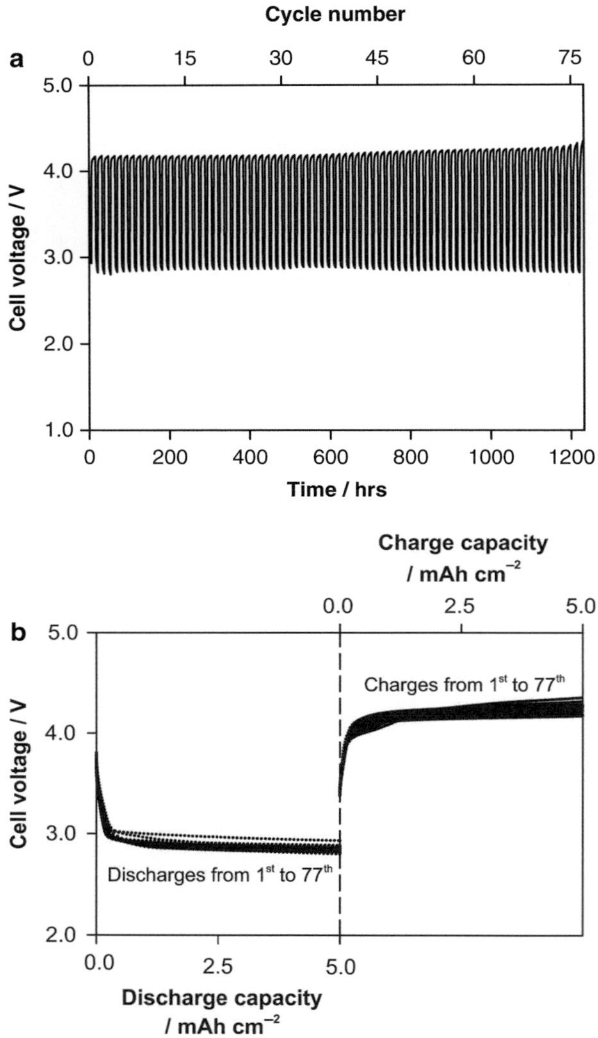

Fig. 6.17 Cycling performance of aqueous lithium air cell having 4.0 M malonic acid electrolyte (**a**) as a function of time and (**b**) as a function of discharge and charge capacities. Discharge current density was 1.0 mA/cm^2; duration of discharge was 5 h. Charge current density was 0.5 mA/cm^2; duration of charge was 10 h. Data courtesy of PolyPlus Battery Company

6.4 Conclusions

Lithium air chemistry offers the potential for unprecedented energy density and a step change relative to existing commercial batteries. In our opinion the use of a PLE is a basic requirement for both nonaqueous and aqueous approaches. In all cases, the use of a PLE leads to a self-discharge rate of effectively zero. For aqueous lithium air the need for a PLE is obvious. However, even for nonaqueous lithium air, it appears that the use of a PLE could be necessary to enable the use of electrolytes stable to Li_2O_2 but not lithium metal and to protect the negative electrode from reaction with moisture in ambient air. Nonaqueous lithium air chemistry is clearly plagued by a number of issues including reactivity of Li_2O_2 with many nonaqueous electrolytes leading to a high degree of irreversibility on cycling, irreversible loss of solvent from the air electrode to the atmosphere, the irreversible reaction of CO_2 with Li_2O_2, flooding of the air electrode leading to slow oxygen transport, and extremely limited solubility of Li_2O_2 leading to low electrode capacity and pore plugging. The situation with rechargeable aqueous lithium air cells is much better. We have demonstrated good cycling at reasonable current densities (C/5 discharge) and high current densities for short bursts (>C rate for 5 min). We have also demonstrated that primary aqueous lithium air batteries can achieve spectacular specific energies of over 800 Wh/kg and hope to demonstrate similarly high energy density for rechargeable lithium air soon. We have also demonstrated that porosimetry can guide the development of a model describing the behavior of air electrodes in lithium air cells and plan to use this capability for rational design of improved air electrodes for both primary and rechargeable lithium air technology.

Acknowledgements This work was funded in part by ARPA-E of the US Department of Energy and by the US Army CERDEC. The authors also acknowledge the participation of V. Loginova in the experimental work at PolyPlus, cell testing by Dr. Terrill Atwater at APG, and helpful discussions with Dr. P.N. Ross of the Lawrence Berkeley National Laboratory, in Berkeley, California.

References

1. Visco SJ, Nimon E, De Jonghe LC (2009) Lithium-air. In: Garche J (ed) Encyclopedia of electrochemical power sources. Elsevier, Amsterdam, p 376
2. Nimon YS, Visco SJ (2010) Active metal/aqueous electrochemical cells and systems. US Patent 7,645,543, 3 Feb 2004
3. Visco SJ, Katz BD, Nimon YS (2007) Protected active metal electrode and battery cell structures with non-aqueous interlayer architecture. US Patent 7,282,295, 14 Apr 2004
4. Visco SJ, Nimon YS, Katz BD, Petrov A (2010) Compliant seal structures for protected active metal anodes. US Patent 7,824,806, 8 Aug 2006
5. Bruce PG, Freunberger SA, Hardwick LJ, Tarascon JM (2012) Li-O_2 and Li-S batteries with high energy storage. Nat Mater 11:19
6. Christiansen J, Albertus P, Sanchez-Carrera RS, Lohmann T, Kozinsky B, Leidtke R, Ahmed J, Kojic A (2012) A critical review of Li/air batteries. J Electrochem Soc 159:R1–R30
7. Peng Z, Freunberger SA, Chen Y, Bruce PG (2012) A reversible and higher-rate Li-O_2 battery. Science 337:563–566
8. Walker W, Giordani V, Uddin J, Bryantsev VS, Chase GV, Addison D (2013) A rechargeable Li-O_2 battery using a lithium nitrate/N,N-dimethylacetamide electrolyte. J Am Chem Soc 135:2076–2079
9. Visco SJ, Nimon E, Katz BD, De Jonghe LC, Chu MY (2004) Lithium metal aqueous batteries. In: Abstracts of the International Meeting on Lithium Batteries, Nara, p 53
10. Visco SJ, Nimon E, Katz BD, De Jonghe LC, Chu MY (2004) Lithium fuel cells. In: Abstracts of the International Meeting on Lithium Batteries, Nara, p 396

11. Visco SJ, Nimon E (2004) Lithium air batteries. In: Abstracts of the International Meeting on Lithium Batteries, Nara, p 397
12. Abraham KM, Jiang Z (1996) A polymer electrolyte-based rechargeable lithium/oxygen battery. J Electrochem Soc 143:1–5
13. Aurbach D, Daroux M, Faguy P, Yeager E (1991) The electrochemistry of noble metal electrodes in aprotic organic solvents containing lithium salts. J Electroanal Chem 297:225–244
14. Mizuno F, Nakanishi S, Kotani Y, Yokoishi S, Iba H (2010) Rechargeable Li-air batteries with carbonate-based liquid electrolytes. Electrochemistry 78:403–405
15. Freunberger SA, Peng Z, Hardwick LJ, Chen Y, Barde F, Bruce PG (2010) Understanding the chemical reactions in the lithium-oxygen battery. In: Proceedings of the 218th Electrochemical Society Meeting, Las Vegas, NV
16. McCloskey BD, Bethune DS, Shelby RM, Girishkumar G, Luntz AC (2011) Solvents' critical role in nonaqueous lithium-oxygen battery electrochemistry. J Phys Chem Lett 2:1161–1166
17. Visco SJ, Nimon YS (2007) Li/air non-aqueous batteries. US Patent Application 20070117007, 22 Nov 2006
18. Xie B, Lee HS, Li H, Yang XQ, McBreen J, Chen LQ (2008) New electrolytes using Li_2O or Li_2O_2 oxides and tris(pentafluorophenyl) borane as boron based anion receptor for lithium batteries. Electrochem Comm 10:1195–1197
19. Jung HG, Hassoun J, Park JB, Sun YK, Scrosati B (2012) An improved high-performance lithium-air battery. Nat Chem 4:579–585
20. Nimon VY, Visco SJ, De Jonghe LC, Volfkovich YM, Bograchev DA (2013) Modeling and experimental study of porous carbon cathodes in Li-O_2 cells with non-aqueous electrolyte. ECS Electrochem Lett 2:A33–A35
21. Visco SJ, De Jonghe LC, Nimon YS, Petrov A, Pridatko K (2012) Catholytes for aqueous lithium/air battery cells. Cathodes and reservoirs for aqueous lithium/air battery cells. US Patent 8,323,820, 12 June 2009
22. Visco SJ, De Jonghe LC, Nimon YS, Petrov A, Pridatko K (2013) Cathodes and reservoirs for aqueous lithium/air battery cells. US Patent 8,455,131, 12 June 2009
23. Visco SJ, De Jonghe LC, Nimon YS, Petrov A, Pridatko K (2013) Hydrogels for aqueous lithium/air battery cells. US Patent 8,389,147, 12 June 2009
24. Visco SJ, Nimon YS, De Jonghe LC, Petrov A, Goncharenko N (2013) Aqueous lithium air batteries. US Patent Application 20130045428, 17 Aug 2012

Chapter 7
Air Electrodes for Aqueous Lithium Air Batteries

Philippe Stevens and Gwenaëlle Toussaint

7.1 Introduction

The aqueous rechargeable lithium battery such as the one developed by EDF and its partners, uses an aqueous electrolyte as opposed to the anhydrous lithium air concept which uses an organic electrolyte. Typical electrolytes used are saturated aqueous solutions of lithium hydroxide or lithium chloride. Since the negative electrode (lithium metal) is not stable in aqueous media, it needs to be separated from the positive aqueous compartment by using a watertight, lithium–ion conducting, separator such as a Lisicon ceramic membrane [1]. This separator is therefore the interface between the negative compartment and the aqueous electrolyte. The negative compartment needs to be hermetically sealed from the external environment, *that is*, both from the air and the aqueous electrolyte. The air electrode provides the other interface, between the aqueous electrolyte and the external environment. It must therefore be open to allow oxygen from the air to access the positive electrode. The air electrode not only needs to be sufficiently porous for the air to penetrate the electrode from outside the cell, but it also needs to act as a barrier to contain the aqueous electrolyte inside the cell, and prevent it from leaking through the electrode. This compromise between water tightness and porosity to air is typically obtained by the addition of a hydrophobic agent such as PTFE.

The principle of operation of an aqueous lithium air cell versus the anhydrous lithium air system has important implications on the air electrode. In the anhydrous lithium air concept, the air electrode serves both as an electrode and as a reservoir for the products of the discharge reaction, Li_2O_2, which is stored in the porosity of the electrode. The surface capacity of the battery (in mAh/cm^2) or areal capacity, will be determined by the quantity of Li_2O_2 that can be produced before either

P. Stevens (✉) • G. Toussaint
R&D Division, Les Renardières, Electricité de France (EDF),
77818 Moret sur Loing Cedex, France
e-mail: philippe.stevens@edf.fr; gwenaelle.toussaint@edf.fr

N. Imanishi et al. (eds.), *The Lithium Air Battery: Fundamentals*,
DOI 10.1007/978-1-4899-8062-5_7, © Springer Science+Business Media New York 2014

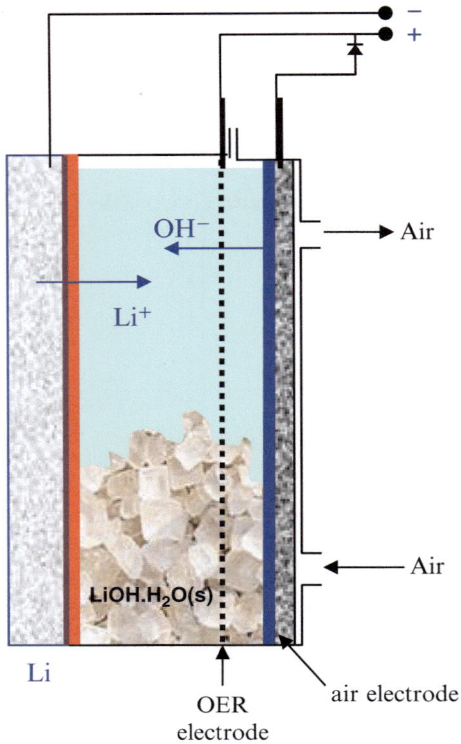

Fig. 7.1 Schematic diagram of the aqueous lithium air cell developed by EDF

filling and blocking the pores or by passivation due to the formation of an insulating layer on the electronic conductor of the electrode [2]. In the aqueous air electrode concept, the product of the reaction is stored as solid $LiOH{\cdot}H_2O$ in the aqueous electrolyte compartment and not in the air electrode (Fig. 7.1).

Unlike the anhydrous lithium air system, the product of the reaction is soluble in the aqueous electrolyte, up to 5 M at room temperature. This is equivalent to a capacity of 138 mAh/cm^3 of electrolyte, or more than 40 mAh/cm^2 for a 3-mm thick electrolyte compartment. Since $LiOH{\cdot}H_2O$ is soluble, it does not plate the electronic conductor to form an insulating layer in or on the air electrode. The precipitation reaction is slower and can be tailored to occur only at the bottom of the aqueous compartment [3]. By allowing the product of the reaction to precipitate in the positive compartment, much higher areal capacities can be obtained. The only limitation in areal capacity is then the amount of solid $LiOH{\cdot}H_2O$ which can be stored in the aqueous compartment. This capacity can be increased by simply increasing the thickness of the aqueous compartment or by increasing the volume fraction of the solid product in the saturated aqueous solution. For example, if 80 % of the aqueous compartment volume is taken up by $LiOH{\cdot}H_2O$, the capacity of the positive electrode can store up to 800 mAh/cm^3. A 3-mm thick aqueous compartment can therefore store up to 240 mAh/cm^2 on discharge. Volume fractions of solid $LiOH{\cdot}H_2O$ in the aqueous electrolyte up to 80 %

have been demonstrated without significantly impairing the performance of the positive electrode [4].

When the battery is charged, lithium ions in the aqueous solution are transferred to the negative compartment through the fast lithium–ion conducting ceramic membrane. The resulting drop in LiOH concentration from saturation will produce a shift in the LiOH dissolution/precipitation equilibrium and solid $LiOH{\cdot}H_2O$ will dissolve to saturate the solution again. The saturated lithium hydroxide solution is therefore a considerable storage buffer with more than 40 mAh/cm^2 of areal capacity for a 3-mm compartment that can be charged from the solution before running out of lithium ions. This buffer is more than sufficient to allow the solid $LiOH{\cdot}H_2O$ to redissolve in order to compensate for the consumption of lithium ions. For example, at a current density of 10 mA/cm^2, this gives 4 h for the precipitate to start redissolving.

7.2 Electrochemical Reactions at the Positive Electrode

During discharge:

$$O_2 + 2H_2O + 4e^- \rightarrow 4OH^-{}_{(aq)} \quad \text{for a } 4e^- \text{ reduction} \tag{7.1}$$

$$O_2 + H_2O + 2e^- \rightarrow HO_2{}^- + OH^-{}_{(aq)} \quad \text{for a } 2e^- \text{ reduction} \tag{7.2}$$

$$HO_2{}^- + H_2O + 2e^- \rightarrow 3OH^- \tag{7.3}$$

$$2HO_2{}^- \rightarrow 2OH^- + O_2 \quad \text{disproportionation reaction} \tag{7.4}$$

Once saturation is reached:

$$Li^+{}_{(aq)} + OH^-{}_{(aq)} + H_2O \rightarrow LiOH \cdot H_2O_{(s)} \downarrow \tag{7.5}$$

During charge:

$$4OH^-{}_{(aq)} \rightarrow O_2 \uparrow + 2H_2O + 4e^- \tag{7.6}$$

$$LiOH \cdot H_2O_{(s)} \rightarrow Li^+{}_{(aq)} + OH^-{}_{(aq)} + H_2O \tag{7.7}$$

7.3 Air Electrode in Charge/Discharge

The positive compartment of the aqueous lithium air cell can be compared to that of a fuel cell and the air electrode has similar characteristics during discharge. There are however important differences which need to be treated, namely the formation of a reaction product that is stored in the battery as opposed to evacuated in a fuel cell (as H_2O vapour), and the reversible function of the positive electrode which needs to work both in reduction and oxidation.

7.3.1 Discharge Reaction

During discharge, the positive electrode is used as an oxygen reduction electrode. Oxygen from the air is reduced to hydroxyl ions in the presence of water. Just as in acidic media, the oxygen reduction reaction (ORR) in alkaline electrolyte can proceed by two different pathways [5]: a direct and full four-electron reduction to produce hydroxyl ions only (Eq. 7.1), or it can proceed via an intermediate two-electron reaction to produce hydrogen peroxide (Eq. 7.2) which can then be further reduced to hydroxyl ions (Eq. 7.3). It can also decompose chemically to produce water and oxygen (Eq. 7.4). The reaction pathway will strongly depend on the catalyst used. For example, gold is known to favour the two-electron reduction [6].

The electrochemical ORR occurs at a triple-phase interface between a gas (oxygen from the air), a liquid (the electrolyte), and a solid (the electrode). Since the active material is a gas, itself diluted in the air, high surface area electrodes are required in order to reduce diffusion limitations at a higher current. Fuel cell air electrodes use high surface area activated carbon powders such as Vulcan XC72 as both the electronic conductor and support for the finely dispersed oxygen reduction catalyst. The use of an alkaline electrolyte as opposed to acidic electrolytes enables a much wider range of catalysts to be used, whereas platinum or platinum alloys are the only good catalysts in the latter case. The most widely used catalysts used in alkaline electrolyte are silver, manganese dioxide, pyrolysed iron(II) phthalocyanine, and cobalt tetra methoxyphenylporphyrin [7].

Water is consumed in this phase by two reactions: the ORR and the precipitation of $LiOH{\cdot}H_2O$ once saturation has been reached. This has the advantage of reducing liquid electrolyte volume variations due to the precipitation of $LiOH{\cdot}H_2O$. There is no *net* consumption of water since the same water molecules are released during charge. The only net loss or gain of water comes from evaporation or absorption of water from the air via air electrode.

7.3.2 Charge Reaction

During charge, the opposite reaction occurs and the positive electrode is used as an oxygen evolution electrode. Hydroxyl ions are oxidised to produce oxygen and water (Eq. 7.6). Water will also be released when $LiOH{\cdot}H_2O$ redissolves (Eq. 7.7).

The reaction interface on the positive electrode during charge is a two-phase interface (solid/liquid) as opposed to the three-phase interface in the discharge reaction. The concentration of active material (OH^-) in the discharge reaction is high and a very high surface area electrode is not necessary to reduce any diffusion limitations. On the contrary, the very porous, high surface area structure used in the air electrode for the ORR is not ideal for gas evolution from a liquid phase. The mechanical erosion resulting from the production of a gas causes a breakdown of

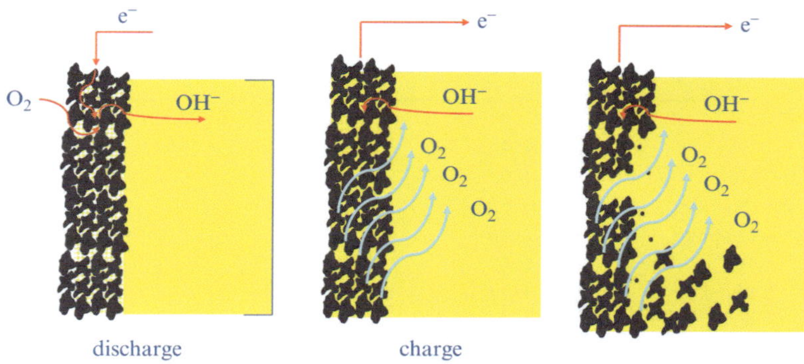

Fig. 7.2 Impact of oxygen evolution on a porous air electrode

the electrode and rapid failure (Fig. 7.2). In addition, the high OER potential causes accelerated corrosion of the carbon support and decomposition of the catalyst [8]. Thermal treatment of the carbon support to graphitise the surface can reduce carbon corrosion [9], and more stable bifunctional catalysts have been found in the past [10], but the lifetime of bifunctional air electrodes reported so far have been too low to satisfy the requirements of a battery for electric vehicles.

Another approach is to use a second positive electrode dedicated to the OER [8]. In this system, the oxygen evolution electrode is electrically isolated from the oxygen reduction air electrode (Fig. 7.1) and the latter is protected from the detrimental effect of the production of a gas and the high oxidation potential described above. This system also has the advantage that electrocatalyst can be tailored to the two different reactions. Low-cost and better performing electrodes can therefore be used. For example, air electrodes developed for primary zinc air cells or alkaline fuel cells can be used, with low-cost catalysts such as manganese dioxide. Because high surface area electrodes are not required for the two-phase OER, cheaper and more robust light metal grids can be used for example. Placed in the aqueous compartment between the negative electrode and the air electrode, the open mesh of a grid enables the electrolyte to penetrate the electrode and the ionic pathway between the negative electrode and the air electrode is not affected by the presence of the second electrode during discharge.

The oxygen evolution reaction comes with a severe overpotential penalty [11] which will have an impact on the energy efficiency of the lithium air battery. An overpotential of at least 500 mV is typically observed for this reaction, even on the best catalysts. Iron and cobalt and copper are the metals showing the lowest overpotentials, but they are not stable. Amongst the pure metals, nickel is one of the best candidates for an oxygen evolution electrode (Table 7.1).

Different transition metal oxide electrocatalysts for OER in alkaline electrolyte have been tried in the past, the best performing catalyst being Fe, Co, and/or Ni oxide based [12, 13]. Catalyst-coated electrodes have been reported using

Table 7.1 OER overpotential in 1 M KOH on different metals [23]

	Overpotential		
Electrode material	10 mA/cm^2	100 mA/cm^2	1,000 mA/cm^2
Cd	0.96	1.21	1.21
Co	0.46	0.54	0.60
Cu	0.66	0.73	0.77
Au	1.05	1.53	1.63
Graphite	0.96	1.12	2.20
Fe	0.48	0.56	0.63
Pb	0.97	1.02	1.04
Ni	0.75	0.91	1.04
Pd	1.01	1.12	1.28
Pt	1.32	1.50	1.55
Platinised Pt	0.66	0.89	1.14
Ag	0.71	0.94	1.06

techniques such as sol–gel, plasma spray, or pulsed laser ablation [14]; however, long-term lifetime tests have shown that they are not sufficiently stable with time in 5 M LiOH electrolyte [15]. One of the principal causes of failure is loss of the catalyst from the electrode surface provoked by the mechanical constraints resulting from the gas evolution. The most stable electrodes were found to be the ones in which the catalyst was produced by electroplating the electrode with the transition metal followed by thermal oxidation [16].

Surprisingly, a much simpler, cheaper, and more interesting OER electrode material for aqueous lithium air batteries is 316 L stainless steel. This alloy has demonstrated oxygen evolution performances in 5 M $LiOH_{(aq)}$ equal or better than catalyst coated electrodes. In addition, much longer lifetimes have been shown with an increase in activity with time [16]. This material forms a catalytic layer on the surface in situ during operation. A gradual increase in performance is observed during the first 250 h of operation in 5 M LiOH which has been attributed to the Ni enrichment of the surface to form a thin porous catalytic oxide layer. This layer appears to be composed to a large extent of nanocrystalline nickel hydroxides, with small amounts of iron and chromium.

7.4 Impact of the Electrolyte

The presence of lithium ions in the aqueous electrolyte has a negative impact on the performance of the positive electrode. A significant decrease in the catalytic activity is observed with the same electrodes when switching from 1 or 5 M KOH to 5 M LiOH aqueous solution.

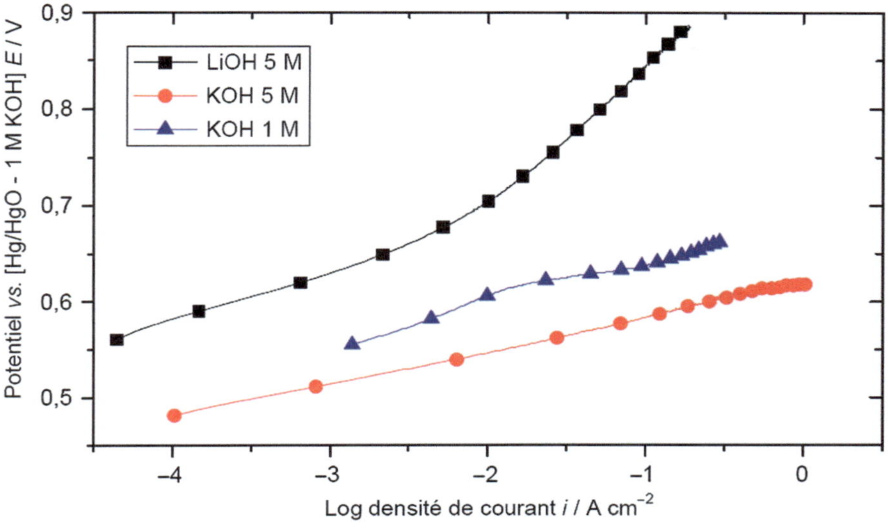

Fig. 7.3 OER on activated 316 L steel RDE; $T = 25\,°C$; 0.1 mV s^{-1} (IR drop compensated), from F. Moureaux [16]

7.4.1 Charge Reaction

The addition of LiOH to KOH aqueous electrolyte in NiMH batteries is known to increase its charging efficiency by increasing the OER overpotential [17]. Although this effect is beneficial to alkaline Ni batteries (Ni/Cd, NiMH, NiZn) because it delays the onset of oxygen evolution when charging these batteries, this results in higher charging voltages in aqueous lithium air batteries and therefore lower energy efficiency. A comparison of oxygen evolution on activated 315 L stainless steel electrodes in aqueous 1 and 5 M KOH vs. 5 M LiOH on a rotating disc electrode shows a significant increase in overpotential and in the Tafel slope [16] (Fig. 7.3).

One possible explanation for the poor OER activity in LiOH electrolyte could be the formation of $LiNiO_2$ which could be stable at the OER potentials. This would have the effect of stabilising Ni(3+) and prevent the formation of the more catalytically active Ni(4+).

The one OER mechanism proposed by Yeager [18] identifies the oxidation of the metal hydroxide on the surface as the rate determining step.

$$S^z + OH^- \rightarrow (SOH)^z + e^- \tag{7.8a}$$

$$(SOH)^z \rightarrow (SOH)^{z+1} + e^- \quad \textbf{RDS} \tag{7.8b}$$

$$\left[(SOH)^{z+1}\right]_2 + 2OH^- \rightarrow 2S^z + O_2 + 2H_2O \tag{7.8c}$$

By stabilising Ni^{3+} higher overpotentials are needed for this RDS.

However, an alternative mechanism has been proposed by Lyons [19] in which the oxidation state of Ni remains at (3+) and the RDS is the formation of a Ni superoxy species from the oxy-hydroxy species.

$$\left[\mathrm{Ni(III)O}_m(\mathrm{OH})_n\right]^{p-} + \mathrm{OH}^- \rightarrow \left[\mathrm{Ni(III)O}_{m+1}(\mathrm{OH})_{n-1}\right]^{p-} + \mathrm{H_2O} + e^- \quad (7.9a)$$

$$\left[\mathrm{Ni(III)O}_{m+1}(\mathrm{OH})_{n-1}\right]^{p-} \rightarrow \left[\mathrm{Ni(III)O}_m\mathbf{OOH}(\mathrm{OH})_{n-2}\right]^{(p-1)-} + \mathrm{H_2O} + e^- \quad \mathrm{RDS} \quad (7.9b)$$

$$\left[\mathrm{Ni(III)O}_m\mathbf{OOH}(\mathrm{OH})_{n-2}\right]^{(p-1)-} + 2\mathrm{OH}^- \rightarrow \left[\mathrm{Ni(III)O}_m\mathrm{O_2}(\mathrm{OH})_{n-1}\right]^{p-} + \mathrm{H_2O} + e^- \quad (7.9c)$$

$$\left[\mathrm{Ni(III)O}_m\mathrm{O_2}(\mathrm{OH})_{n-1}\right]^{p-} + \mathrm{OH}^- \rightarrow \left[\mathrm{Ni(III)O}_m(\mathrm{OH})_n\right]^{p-} + \mathrm{O_2} + e^- \quad (7.9d)$$

7.4.2 Discharge Reaction

The discharge reaction on the positive electrode of an aqueous lithium air battery involves both the consumption of oxygen and water (Eqs. 7.1 and 7.2). The activity of water will therefore have an impact on the air electrode potential. For example, manganese dioxide catalysed air electrodes have been shown to be very good in 1 M KOH electrolyte [20], but their performance is reduced in 5 M KOH electrolyte. The rate determining step of the ORR on MnO_x/C electrodes in 1 M KOH electrolyte is believed to proceed via proton insertion on the MnO_2 catalyst (Eq. 7.10). The lower proton concentration in high pH electrolyte therefore slows down the reactions.

$$\mathrm{MnO_2} + \mathrm{H_2O} + e^- \rightarrow \mathrm{MnOOH} + \mathrm{OH}^- \quad (7.10)$$

In addition to the impact of the electrolyte pH, it has been observed that the performance of the air electrode is further reduced when using 5 M LiOH electrolyte compared to 5 M KOH [21]. The presence of Li^+ ions further impedes the ORR. It has been proposed that Li^+ insertion in MnO_2 catalyst stabilises the Mn (III) intermediate due to the stronger Li–O bond. This prevents the catalyst from being regenerated to form Mn(IV).

7.4.3 Impact of CO_2

Carbon dioxide is known to be a poison for alkaline fuel cells. In fuel cells, high carbon dioxide levels can be present in the anode stream resulting from the use of natural gas or methanol as a fuel, and low levels of CO_2 are also present in the air

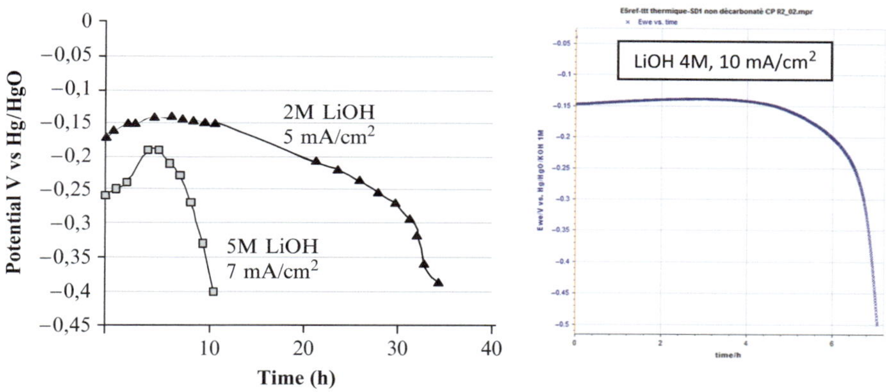

Fig. 7.4 Air electrodes running on untreated air

stream fed to the cathode. Alkaline fuel cells typically use a concentrated aqueous solution of KOH as an electrolyte. Even with the low concentrations of CO_2 present in the air, the degradation of performance and lifetime of alkaline fuel cells is significant when using untreated air. The poisoning mechanism has been attributed to the formation of K_2CO_3 and $KHCO_3$ which consumes the electrolyte in the long run but more immediately which will precipitate in the pores of the air electrode [22]. The precipitation of carbonates in the air electrode will modify the triple phase interface and will prevent access to electrolyte and to oxygen at the reaction sites. Metal–air batteries such as zinc air batteries which use KOH electrolyte also suffer from the same problem.

The solubility of Li_2CO_3 is very low (13 g/L) compared to K_2CO_3 (800 g/L) and the impact CO_2 on an air electrode running in untreated air is much more pronounced than in KOH electrolyte. Air electrodes running in LiOH electrolyte using untreated air will only last 10–50 h (Fig. 7.4) [3].

7.4.4 Impact of Discharge Product Precipitation

The air electrode of a lithium air battery is a porous electrode in which the porosity is partially filled with electrolyte. As solid $LiOH{\cdot}H_2O$ precipitates in the electrolyte during discharge, clogging of the electrode with precipitate is an important limiting factor which will limit the lifetime and performance of the electrode. For example, an air electrode running in saturated LiOH solution under CO_2 free air exhibits a rapid increase in polarisation after 5 h of operation when solid $LiOH{\cdot}H_2O$ precipitates inside the electrode (Fig. 7.5). A clear demonstration of oxygen diffusion limitation as a result of clogging of the electrode by the precipitate inside the electrode can be seen in Fig. 7.5. In this experiment, the

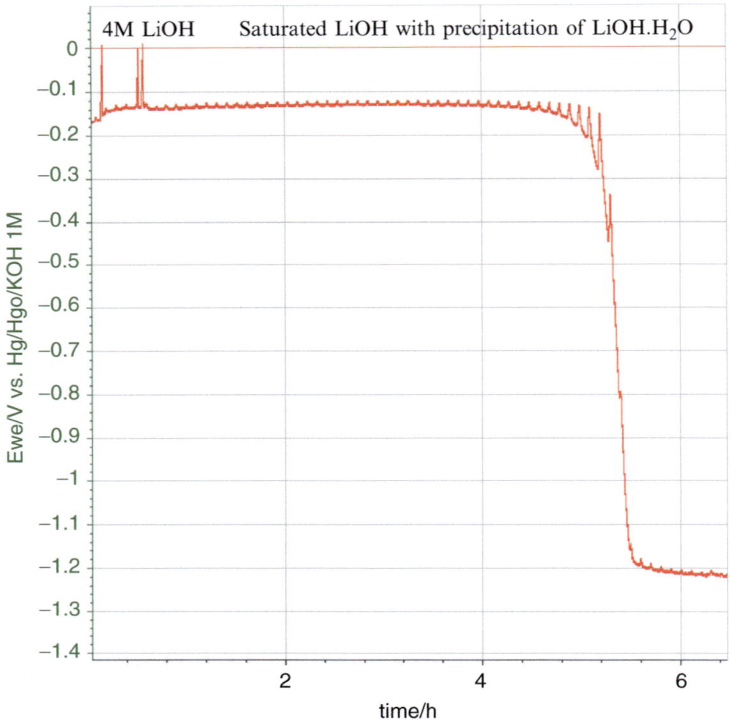

Fig. 7.5 Effect of $LiOH{\cdot}H_2O$ precipitation on the lifetime of an air electrode running in LiOH electrolyte in CO_2 free air

electrode was supplied with air which was periodically enriched with oxygen for short periods of time. The effect of oxygen enrichment before clogging of the electrode is mild, but it becomes much more important as it clogs up as demonstrated by the much larger variations in the air electrode potential when fed with enriched air.

7.4.5 Water Management

Water is consumed during discharge by both the ORR (Eq. 7.1) and by the precipitation of lithium hydroxide monohydrate (Eq. 7.5). The net consumption of water is therefore 1½ molecules of water per molecule of lithium oxidised,

$$Li^+ + \tfrac{1}{2}O_2 + 1\tfrac{1}{2}H_2O + e^- \rightarrow LiOH \cdot H_2O_{(s)} \qquad (7.11)$$

The consumption of water is reversible since on charge, the same quantity of water is produced. Fortunately, the variation in volume due to this water consumption is negligible since the molar volumes of water consumed and the $LiOH{\cdot}H_2O$ produced are very close (3 % difference).

The only water management issue to be concerned with is the transfer of water between the aqueous electrolyte and the air. This transfer can occur in both directions, *that is*, water can evaporate into the atmosphere via the air electrode or water can be absorbed by the electrolyte from the air via the air electrode.

7.5 Composite Air Electrode

The air electrode in an aqueous lithium air battery has three different limitations described above: precipitation of lithium carbonate, precipitation of lithium hydroxide inside the electrode, and water transfer between the electrolyte and the air fed to the air electrode. The use of a composite electrode composed of porous electrode and an anion conducting polymer electrolyte can significantly reduce these limitations. In this composite electrode, the electrochemical reaction occurs at polymer electrolyte/air electrode interface. The aqueous lithium hydroxide electrolyte provides the interface between the anion conducting polymer electrolyte and the anode compartment. The polymer electrolyte is not a cation conductor and prevents lithium–ion conduction in the air electrode. It therefore acts as an efficient barrier to the formation of lithium carbonate and lithium hydroxide inside the air electrode. It can also have a role in reducing the exchange of water between the aqueous lithium hydroxide solution and the air (Fig. 7.6).

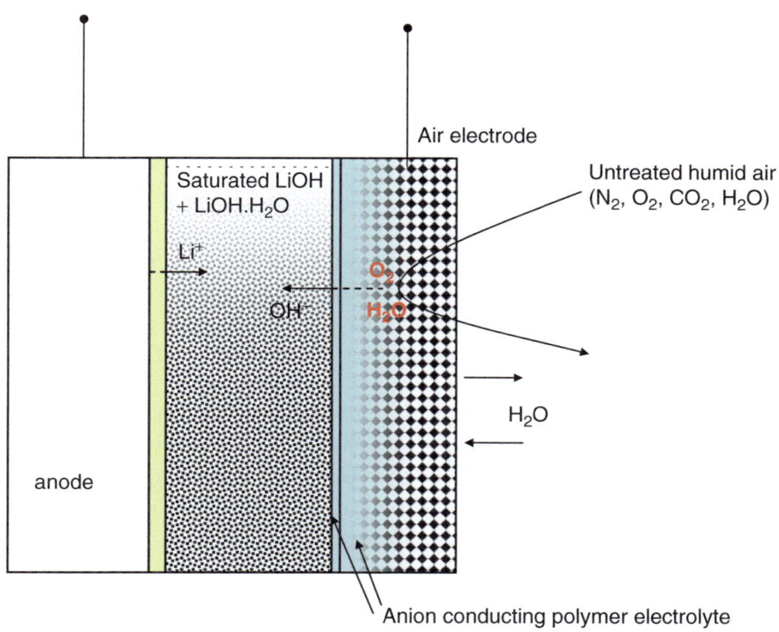

Fig. 7.6 Composite air electrode, a barrier to both CO_2 and LiOH

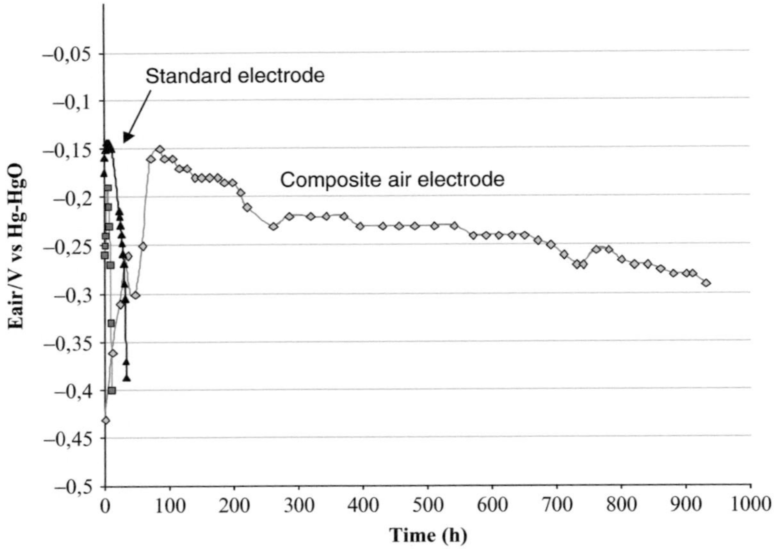

Fig. 7.7 Comparison between a standard air electrode and a composite air electrode in 2 M LiOH electrolyte using untreated air [3]

The polymer membrane will act as a barrier to CO_2 from the air to significantly reduce permeation of the gas into the aqueous electrolyte compartment. By using a composite electrode, the lifetime of air electrode running in untreated air has been increased from 10 h to nearly 1,000 h (Fig. 7.7).

7.6 Conclusion

The air electrode of an aqueous lithium air cell has many common features with a fuel cell air electrode, but it needs to be adapted to different environments. It needs to work both as an oxygen reduction electrode and an oxygen evolution electrode, it needs to be able to work in a saturated aqueous electrolyte and it needs to be CO_2 tolerant.

By working with a saturated electrolyte and by storing the discharge product as a solid, very high capacities for the positive electrode can be obtained. There is no theoretical limit to the capacity of this electrode, the only limit being the thickness of the positive compartment of the lithium air cell. Its capacity can therefore be tailored to match that of the negative compartment.

Lithium ions will increase the overpotential for oxygen evolution which will have a negative impact on round-trip energy efficiency of the battery by increasing the charge voltage necessary.

The aqueous lithium air electrode overcomes many of the problems encountered in the anhydrous lithium air concept such as low areal capacities, air sensitivity, electrolyte decomposition, and electrolyte evaporation.

References

1. Visco SJ, Nimon E, Katz B (2006) The development of high energy density lithium/air and lithium/water batteries with no self-discharge, abs. 389. In: Abstracts of the ECS 210th meeting, Cancun, Mexico
2. Albertus P, Girishkumar G, McCloskey B, Sanchez-Carrera RS, Kozinsky B, Christensen J, Luntz AC (2011) Identifying capacity limitations in the Li/oxygen battery using experiments and modeling. J Electrochem Soc 2011(158):A343–A351
3. Stevens P, Toussaint G, Caillon G, Viaud P, Vinatier P, Cantau C, Fichet O, Sarrazin C, Mallouki M (2010) Development of a lithium air rechargeable battery. ECS Trans 28(32):1–12
4. Stevens P, Toussaint G, Puech L, Vinatier P (2013) Very high specific area lithium-air battery. ECS Trans 50(25):1–11
5. Damjanovic A, Genshaw MA, Bockris JO'M (1967) Hydrogen peroxide formation in oxygen reduction at gold electrodes: II. Alkaline solution. J Electroanal Chem Interfacial Electrochem 15:173–180
6. Kinoshita K (1992) Electrochemical oxygen technology. Willey, New York, pp 19–29
7. Vielstich W, Gasteiger HA, Lamm A (eds) (2003) Handbook of fuel cells—fundamentals, technology and applications, vol 2 part 5. Wiley, Chichester
8. Stevens Ph, Toussaint G, Akrour L, Sarrazin C (2009) Development of an electrically rechargeable zinc air battery for electric vehicles. Abstracts of the 24th int. battery, hybrid and fuel cell electric vehicle symposium (EVS24), Stavanger, Norway, 13–16 May 2009
9. Ross PN, Sattler M (1988) The corrosion of carbon black anodes in alkaline electrolyte III. The effect of graphitization on the corrosion resistance of furnace blacks. J Electrochem Soc 135:1464
10. Muller S, Holzer F, Arai H, Haas O (1999) A study of carbon-catalyst interaction in bifunctional air electrodes for zinc-air batteries. J New Mater Electrochem Syst 2:227–232
11. O'Sulivan EJM, Calvo EJ (1988) Electrode kinetics: reactions. In: Compton RG (ed) Comprehensive chemical kinetics, vol 27. Elsevier, Amsterdam
12. Kannan AM, Shukla AK, Sathyanarayana S (1989) Oxide-based bifunctional oxygen electrode for rechargeable metal/air batteries. J Power Sources 25(2):141–150
13. Bockris JO, Otagawa T (1984) The electrocatalysis of oxygen evolution on perovskites. J Electrochem Soc 131(2):290–302
14. Lippert T et al (2007) Perovskite thin films deposited by pulsed laser ablation as model systems for electrochemical applications. Progr Solid State Chem 35:221–231
15. Moureaux F (2011) Etude des réactions mettant en jeu l'oxygène dans un système électrochimique lithium-air aqueux rechargeable électriquement. PhD thesis, Institut Polytechnique de Grenoble, France
16. Moureaux F, Stevens P, Toussaint G, Chatenet M (2013) Development of an oxygen-evolution electrode from 316L stainless steel: application to the oxygen evolution reaction in aqueous lithium–air batteries. J Power Sources 229:123–132
17. Zhang J, Zhang L, Liu H (eds) (2012) Electrochemical technologies for energy storage and conversion. Wiley, Weinheim
18. O'Grady W, Iwakura C, Huang J, Yeager E (1974) Ruthenium oxide catalyst for the oxygen electrode. In: Breiter MW (ed) Proceedings of the symposium on electrocatalysis, The Electrochemical Society Inc., p 286

19. Lyons MEG, Brandon MP (2010) A comparative study of the oxygen evolution reaction on oxidised nickel, cobalt and iron electrodes in base. J Electroanal Chem 641(1–2):119–130
20. Chainet IR, Chatenet M, Vondrak J (2007) Carbon-supported manganese oxide nanoparticles as electrocatalysts for the oxygen reduction reaction (ORR) in alkaline medium: physics characterizations and ORR mechanism. J Phys Chem C 111:1434–1443
21. Moureaux F, Stevens P, Chatenet M (2013) Effect of lithium and potassium cations on the electrocatalytic properties of carbon and manganese oxide electrocatalysts towards the oxygen reduction reaction in concentrated alkaline electrolyte. Electrocatalysis 4(3):123–133
22. Tewari A, Sambhy V, Urquidi Macdonald M, Sen A (2006) Quantification of carbon dioxide poisoning in air breathing alkaline fuel cells. J Power Sources 153:1
23. Hickling A, Hill S (1947) Oxygen overvoltage. Part I. The influence of electrode material, current density, and time in aqueous solution. Discuss Faraday Soc 1:236

Chapter 8
Solid Electrolytes for Aqueous Lithium Air Batteries

Nobuyuki Imanishi

8.1 Introduction

The lithium air battery system consists of a lithium metal anode, an electrolyte, and a carbon–air cathode. Two types of lithium air batteries have been developed based on the electrolyte type; nonaqueous and aqueous. The nonaqueous system with a lithium conducting organic electrolyte exhibits an extremely high theoretical energy density of 3,458 Wh kg^{-1}, based on the lithium metal weight and oxygen, and an open-circuit voltage (OCV) of 2.959 V. The cell reaction is

$$2Li + O_2 = Li_2O_2 \tag{8.1}$$

The nonaqueous system has some severe problems that must still be addressed, such as lithium metal corrosion by water from the air, electrolyte decomposition during the discharge process, and high polarization during the charge and discharge processes. The aqueous system was firstly proposed by Visco and Nimon [1] (see Chap. 6). This system consists of a lithium metal protected by a water-stable lithium conducting solid electrolyte, aqueous electrolyte, and a carbon–air electrode. The cell reaction is

$$4Li + 6H_2O + O_2 = 4(LiOH \cdot H_2O) \tag{8.2}$$

Water molecules are involved in the reaction and the theoretical cell voltage in an alkaline electrolyte is ca. 3.0 V, assuming the formation of hydrogen peroxide HO_2^-, which is dependent on the pH value of the electrolyte. The energy density of the aqueous system is lower than that of the nonaqueous system by 1,910 Wh kg^{-1} (including oxygen). However, the problems observed in the nonaqueous system are

N. Imanishi (✉)
Department of Chemistry, Mie University, Tsu, Japan
e-mail: imanishi@chem.mie-u.ac.jp

N. Imanishi et al. (eds.), *The Lithium Air Battery: Fundamentals*,
DOI 10.1007/978-1-4899-8062-5_8,

not present or have been solved in the aqueous system. The aqueous system also has advantages over the nonaqueous system such as the discharge products are soluble in aqueous electrolyte and oxygen diffusion in the electrolyte is much faster than in the nonaqueous electrolyte.

The key component of aqueous lithium air batteries is the water-stable lithium conducting solid electrolyte, due to the severe reaction of lithium metal with water. The aqueous-type system cannot operate without the solid electrolyte. At present, only two types of solid electrolytes, NASICON-type $Li_{1+x}Al_xTi_{2-x}(PO_4)_3$ (LATP) [2] and garnet-type $Li_7La_3Zr_2O_{12}$ (LLZ) [3], have been reported as water-stable lithium conducting solid electrolytes with high lithium ion conductivity at room temperature. The water-stable lithium electrode proposed by Imanishi and colleagues [4] consists of a lithium metal anode, a polyethylene oxide (PEO)-based lithium conducting polymer electrolyte of PEO-$Li(CF_3SO_2)_2N$ (LiTFSI), and LATP, where the polymer electrolyte is used to prevent direct contact between the lithium metal and LATP, because LATP is unstable in contact with lithium metal. The water-stable lithium ion conducting solid electrolytes should be stable in saturated LiOH aqueous solution, because the cell reaction product is a hydrated LiOH, as shown in reaction (Eq. 8.2). However, the water-stable LATP and LLZ solid electrolytes are unstable in saturated LiOH aqueous solution. Nevertheless, Imanishi et al. have reported that both LATP and LLZ are stable in saturated LiOH with saturated LiCl aqueous solution [5].

Lithium metal is the best anode candidate for high-energy density batteries, because it has a very high theoretical-specific capacity of 3,861 mAh g^{-1} and a high negative potential of -3.05 V vs. NHE. However, the formation of lithium dendrites during lithium deposition limits the use of lithium metal as the anode for lithium batteries. This phenomenon occurs even with polymer electrolytes, although to a lesser extent than that with liquid electrolytes. In this chapter, the electrical properties and stability of LATP and LLZ in aqueous electrolytes are introduced and the performance of lithium air cells with these water-stable lithium electrodes is discussed.

8.2 NASICON-Type Water-Stable Lithium Conducting Solid Electrolyte; LATP

Lithium ion conducting solid electrolytes have been extensively studied for the past half century and many solids with high lithium ion conductivity at room temperature, such as Li_3N, Li_2S-based glass, NASICON-type LATP, LISICON-type $Li_{3+x}Zn_{1-x}GeO_4$, and various polymer electrolytes, have been reported. Li_3N exhibits lithium ion conductivity as high as 10^{-3} S cm^{-1} at room temperature. Recently, Tatsumisago et al. [6] reported high lithium ion conductivity for the $Li_7P_3S_{11-z}$ glass ceramic, as high as 5×10^{-3} S cm^{-1} at room temperature. More recently, Kanno et al. [7] found a new lithium ion conducting solid of $Li_{10}GeP_2S_{12}$, which has a one-dimensional lithium diffusion pathway and an extremely high lithium ion conductivity of 1.2×10^{-2} S cm^{-1} at room temperature. The lithium ion conductivities of these sulfide-based solid electrolytes are comparable or higher to those of organic electrolytes used in lithium–ion batteries; however, these materials are quite hygroscopic and unstable in open air. The NASICON-type lithium ion conductors are attractive as electrolytes for batteries and electrochemical devices, because they exhibit a high conductivity of 7×10^{-4} S cm^{-1} at room temperature and are stable in an air atmosphere and in water [5]. Stability in the ambient atmosphere is useful for industrial applications. However, the LATP electrolyte is unstable in contact with the lithium metal.

The water-stable NASICON-type LATP lithium ion conducting solid electrolyte has a potential application for both aqueous lithium air and high-energy density nonaqueous lithium air rechargeable batteries, to protect the lithium metal anode from water and CO_2 in the air. The NASICON-type $Li_{1+x}M_{2-x}A_x$ $(PO_4)_3$ (A = Al, Sc, Y and M = Ti^{4+}, Ge^{4+}, Ta^{4+}) lithium ion conducting solid electrolytes have been reported by Aono et al. in 1989 [2, 8]. A high conductivity of 7×10^{-4} S cm^{-1} at 25 °C was reported for $Li_{1.3}Ti_{1.7}Al_{0.3}(PO_4)_3$ prepared by the conventional solid state reaction. The electrical conductivity of $Li_{1+x}Ti_{2-x}Al_x$ $(PO_4)_3$ has since been reported by many researchers. The conductivity is dependent on the preparation method and the x content in $Li_{1+x}Ti_{2-x}Al_x(PO_4)_3$. Fu [9] reported a higher conductivity of 1.3×10^{-3} S cm^{-1} at room temperature for the $Li_{1+x}Ti_{2-x}Al_x(PO_4)_3$ glass ceramic prepared by quenching from 1,450 °C and sintered at 550 °C. Impedance profiles of the glass ceramic indicated a bulk conductivity as high as 6×10^{-3} S cm^{-1}, which was dominated by the grain boundary resistance. Samples of this glass ceramic powder sintered at 1,000 °C for 12 h had a bulk conductivity of 9.2×10^{-4} S cm^{-1} and a grain boundary conductivity of 1.8×10^{-4} S cm^{-1} at room temperature [10]. Weppner et al. also reported a high bulk conductivity of 3×10^{-3} S cm^{-1} and a low-grain boundary conductivity of 3×10^{-4} S cm^{-1} at room temperature for $Li_{1.3}Al_{0.3}Ti_{1.7}(PO_4)_3$ [11]. The total conductivity of sintered LATP could be enhanced to over 10^{-3} S cm^{-1} at room temperature by reducing the grain boundary resistance. A conductivity of 10^{-3} S cm^{-1} is the target for the solid electrolyte of batteries for operation at high-current density. Figure 8.1 shows the temperature dependence of the electrical conductivity for the $Li_{1+x+y}(Ti,Ge)_{2-x}Al_xP_{3-y}Si_yO_{12}$ glass ceramic plate produced by Ohara Inc., Japan. The total conductivities were 1.05×10^{-4} S cm^{-1} at 25 °C and 4.42×10^{-4} S cm^{-1} at 60 °C. The stability of LATP in water was examined by Cretin et al. [12] and the weight loss for sample prepared by the sol–gel process after immersion for approximately 100 h in distilled water was ca. 0.9 %. Thokchom and Kumar [13] observed a slight buckling of the $Li_{1.3}Ti_{1.7}Al_{0.3}(PO_4)_3$ glass ceramic after 65 days immersion in tap water. Imanishi et al. [14, 15] reported the stability of a $Li_{1+x+y}Al_x(Ti,Ge)_{2-x}Si_yP_{3-y}O_{12}$ glass ceramic plate supplied by Ohara Inc. in aqueous solutions; Fig. 8.2 shows impedance spectra for the LATP glass ceramic plates measured at 25 °C after immersion in various aqueous solutions for 3 weeks or more at room temperature. The grain boundary resistance of the LATP plate increased from 144 to 242 Ω by immersion in distilled water at room temperature

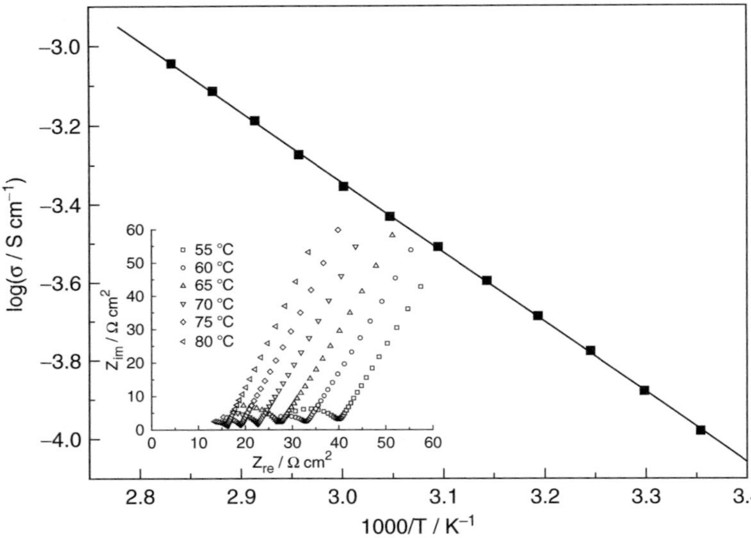

Fig. 8.1 Temperature dependence of electrical conductivity for the Ohara LATP plate

for 1 month. No increase in the grain boundary and bulk grain resistance was observed for LATP plate immersed in a mixed aqueous solution of 1 M LiCl and 1 M $LiNO_3$ for 3 weeks. The increase of grain boundary resistance by immersion in distilled water may be due to H^+ exchange for Li^+ [16]. However, the resistance of the LATP plates immersed in an aqueous solution of 1 M LiOH and 0.1 M HCl was significantly increased. The high resistances are due to decomposition of the LATP plate in aqueous LiOH and aqueous HCl, as confirmed by XRD analysis and scanning electron microscopy (SEM) observation [15].

According to cell reaction (Eq. 8.2) for the aqueous lithium air system, the discharge product is $LiOH{\cdot}H_2O$, and the OH^- concentration increases with increasing discharge depth. The electrolyte is saturated with LiOH (ca. 5 M) at only ca. 5 % discharge depth. Therefore, LATP cannot be used as the protective layer for the lithium metal electrode in the aqueous lithium air system, because LATP is unstable in aqueous LiOH. Imanishi et al. have examined the stability of LATP in weak acid solutions of CH_3COOH (HAc)–H_2O-saturated CH_3COOLi (LiAc) [10]. XRD results of LATP immersed in HAc (80 vol.%)-H_2O (20 vol.%)-saturated LiAc solution at pH 3.34 for 3 weeks at 50 °C showed no change. The impedance profile of LATP immersed in the HAc (80 vol.%)–H_2O (20 vol.%)-LiAc (saturated) for 3 weeks at 50 °C is shown in Fig. 8.3 along with that of pristine LATP. The impedance spectra show no significant change of the bulk resistance, but a reduction of the grain boundary resistance. The reduction of the grain boundary resistance could be explained by the removal of a resistive layer at the grain boundary by immersion in the acid solution. Thus, LATP could be used as a protective layer in acetic acid solution with saturated LiAc. The cell performance of a lithium air cell in acetic acid solution will be introduced later.

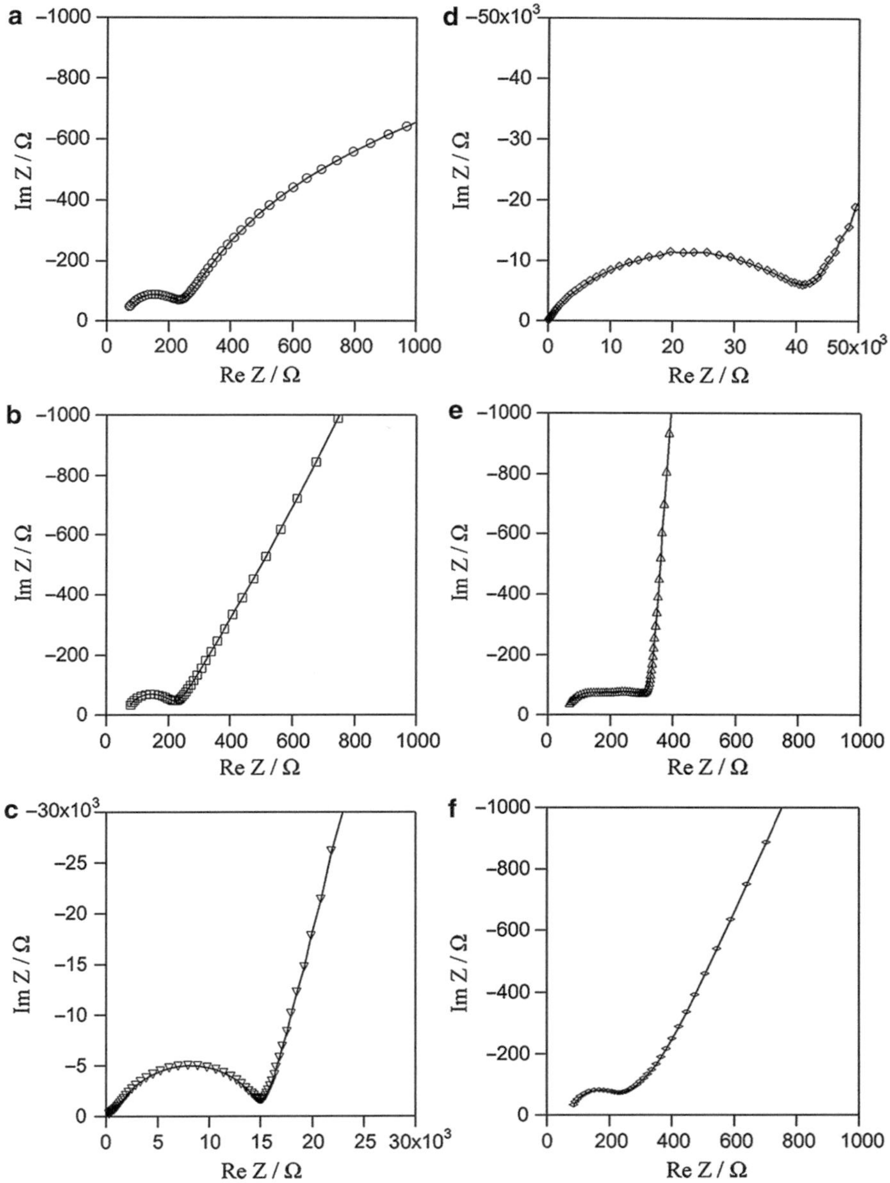

Fig. 8.2 Impedance profiles measured at 25 °C for (**a**) pristine LATP plate, and LATP plate immersed in (**b**) 1 M $LiNO_3$ for 3 weeks, (**c**) 0.1 M HCl for 3 weeks, (**d**) 1 M LiOH for 8 months, (**e**) distilled water for 1 month, and (**f**) 1 M LiCl for 3 weeks (from ref. 15)

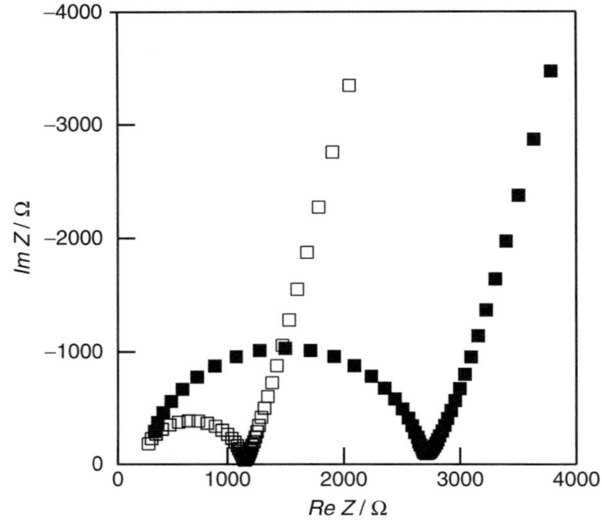

Fig. 8.3 Impedance spectra of the Au/LATP/Au cell at 25 °C before (*filled square*) and after (*open square*) immersion in HAc (80 vol.%)–H_2O (20 vol.%)-saturated LiAc at 50 °C for 3 weeks (from ref. 10)

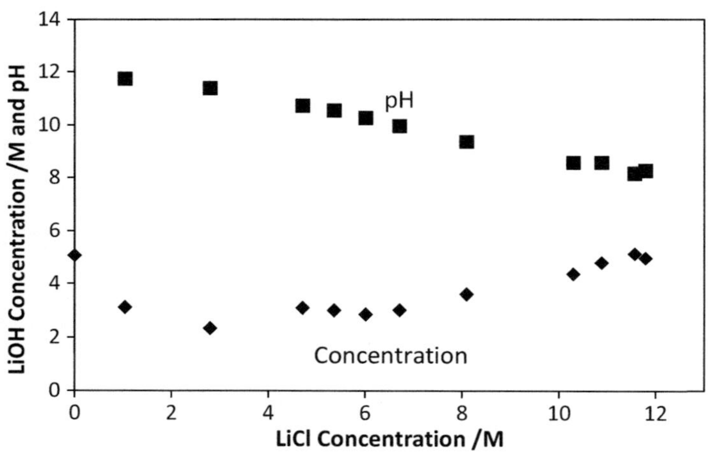

Fig. 8.4 Solubility and pH of the LiOH–LiCl–H_2O system at room temperature (from ref. 5)

More recently, Imanishi et al. [5] reported that LATP is stable in aqueous solutions saturated with LiOH and with a high content of LiCl, which suggests the possibility to develop a high-energy density lithium air rechargeable battery using an aqueous electrolyte. Figure 8.4 shows the experimental solubility of LiOH as a function of the LiCl concentration in LiOH–LiCl–H_2O at room temperature and the corresponding pH values. The solubility of LiOH decreased with increasing LiCl content in Li–Cl–LiOH–H_2O to ca. 2 M, and then increased. The solubility limit of LiOH in water and saturated LiCl aqueous solution at 25 °C are 5.06 and 5.12 M, respectively. However, reasons why the LiCl–LiOH–H_2O system exhibits

Table 8.1 Electrical conductivity of the LATP plate immersed in LiCl–LiOH–H_2O at 50 °C for 3 weeks

LiOH concentration (M)	LiCl concentration (M)	pH	Conductivity at 25 °C (S cm^{-1})
0.57	0	12.90	0.27×10^{-4}
0.057	0	12.78	0.58×10^{-4}
0.0057	0	11.91	1.7×10^{-4}
0.00057	0	10.91	2.0×10^{-4}
3.6	8.10	9.36	2.51×10^{-4}
5.0	11.8	8.25	2.61×10^{-4}
5.1	11.6	8.14	2.32×10^{-4}
2.3	10.4	7.62	2.55×10^{-4}
0.18	12.6	7.01	3.11×10^{-4}

such solubility behavior have yet to be clarified. The pH values of the LiOH–LiCl–H_2O system are also interesting. The pH of the 0.00057 M LiOH and 0.0057 M LiOH solutions were 10.91 and 11.91, respectively. The pH of the system is not dependent on the concentration of LiOH, but monotonically decreases with an increase of the LiCl content. The pH values of the 11.6 M LiCl–5.1 M LiOH–H_2O and 1.04 M LiCl–3.1 M LiOH–H_2O solutions were 8.14 and 11.7, respectively. These results suggest that a high content of Li^+ ions in LiOH–LiCl–H_2O reduces the dissociation of LiOH to Li^+ and OH^-. This is an important result when considering a lithium air rechargeable battery with an aqueous electrolyte and an LATP-protected lithium metal anode. The pH of the saturated LiCl aqueous solution increases slightly from 7.0 to 8.14 with the increase of the LiOH reaction product by the discharge process. Therefore, the water-stable lithium metal electrode with LATP will be stable under a deep discharge state, where the electrolyte solution is saturated with LiOH. According to cell reaction (Eq. 8.2), the concentration of LiOH at 50 % discharge depth is ca. 100 M. Therefore, almost all the LiOH is precipitated in the electrolyte solution, because the solubility limit of LiOH in water is only ca. 5 M. The solution will be saturated with LiOH at a discharge depth of a few percent. Table 8.1 shows the conductivity of the Ohara LATP plate measured at 25 °C after immersion in LiOH–LiCl–H_2O solution at 50 °C for 3 weeks, in addition to the pH values of the solutions. The electrical conductivity of the LATP plate is not changed significantly by immersion in aqueous solutions with pH lower than 10. Figure 8.5 shows impedance profiles of LATP plates immersed in 11.6 M LiOH–5.1 M LiCl–H_2O solution (pH = 8.14) at 50 °C as a function of the immersion period. The impedance profiles indicate no significant change after 3 months immersion; only the grain boundary resistance increases slightly and no change in the bulk conductivity is observed. The grain boundary conductivity decreased from 5.1×10^{-4} to 2.7×10^{-4} S cm^{-1} after 3 weeks immersion and then slightly increased after 3 months immersion. The impedance profile of LATP immersed in the LiCl and LiOH saturated aqueous solution shows no second deformed semicircle. The impedance profiles of LATP immersed in saturated LiCl-saturated LiOH–H_2O were similar to those of the LATP immersed in

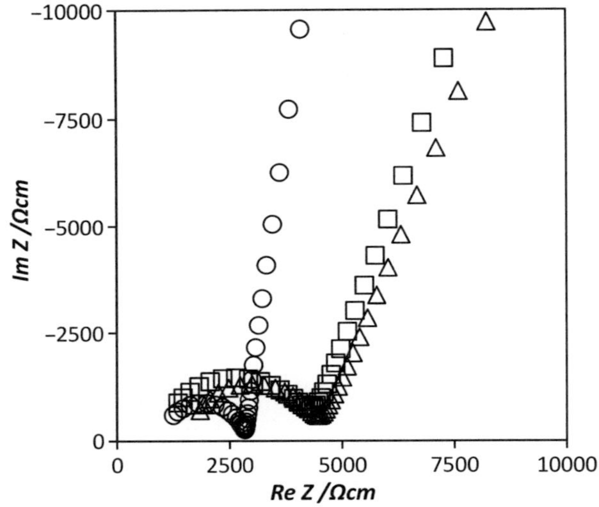

Fig. 8.5 Impedance spectra measured at 25 °C for (*open circle*) pristine LATP plate and LATP plate immersed in (*open square*) 11.6 M LiCl–5.1 M LiOH–H_2O at 50 °C for 3 weeks and (*open triangle*) 3 months (from ref. 5)

saturated $LiNO_3$–3 M LiOH–H_2O (pH = 9.79) at 50 °C for 3 weeks. These results suggest that the stability of LATP is dependent on the pH of the solution. It was concluded that LATP is stable in an aqueous solution with saturated LiCl and LiOH. Therefore, LATP plate could be used as the protective layer of the water-stable lithium electrode for aqueous lithium air rechargeable batteries.

8.3 Garnet-Type Water-Stable Lithium Ion Conducting Solid Electrolyte; $Li_7La_3Zr_2O_{12}$

The other candidate for the protective layer for the water-stable lithium electrode is the garnet-type lithium–ion conducting solid electrolyte $Li_7La_3Zr_2O_{12}$ (LLZ). This compound was reported by Weppner et al. in 2007 [3]. The ionic conductivity of a sintered pellet (92 % theoretical density) was reported to be 2.44×10^{-4} S cm^{-1} at 25 °C. This new garnet-type lithium conductor was found to be stable against molten lithium. Imanishi et al. reported the stability of LLZ in aqueous solutions [17]. XRD patterns of LLZ after immersion in aqueous solution of saturated LiCl, H_2O, 1 M LiOH and 0.1 M HCl at 50 °C for 1 week showed no significant change. SEM images of LLZ after immersion in 1 M LiOH and 0.1 M HCl aqueous solutions showed a significant change of the surface morphology due to reaction with these solutions. However, no significant change of the surface morphology was observed for LLZ immersed in saturated LiCl aqueous solution. Figure 8.6 shows impedance profiles for LLZ measured at 25 °C after immersion in various aqueous solutions at 50 °C for 1 week. A significant decrease in conductivity was observed after immersion in distilled water, and in aqueous solutions of 1 M HCl

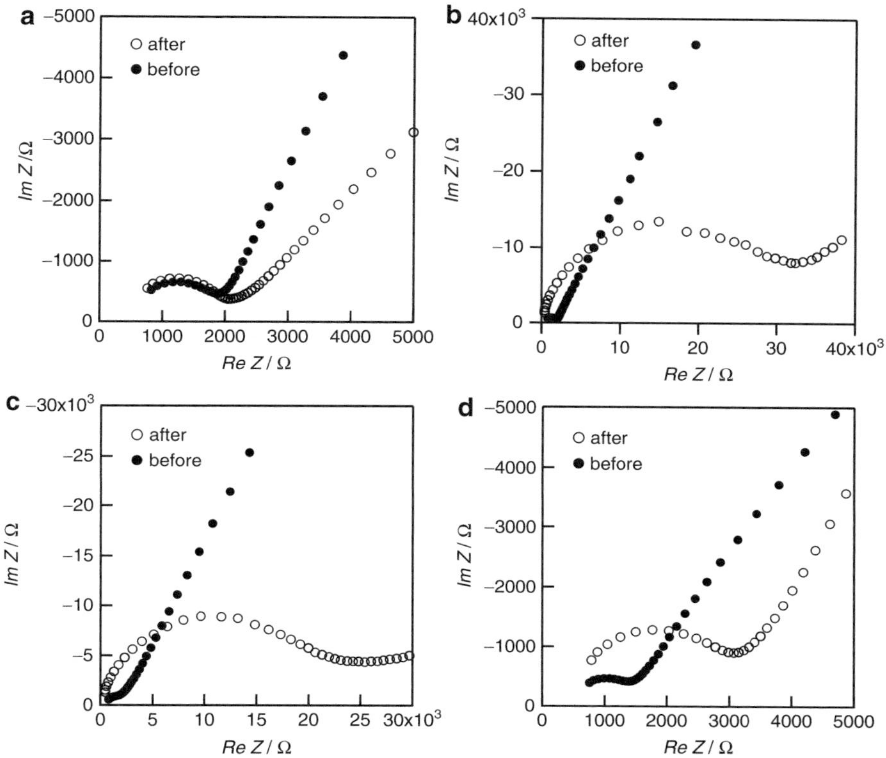

Fig. 8.6 Impedance profiles of LLZ measured at 25 °C after immersion in (**a**) saturated LiCl aqueous solution, (**b**) H_2O, (**c**) 0.1 M HCl and (**d**) 1 M LiOH aqueous solution at 50 °C for 1 week (from ref. 17)

and 0.1 M HCl at 50 °C for 1 week, but no conductivity change was evident after immersion in saturated LiCl aqueous solution. The bulk conductivity of LLZ after immersion in 1 M LiOH aqueous solution was estimated to be 5.8×10^{-4} S cm^{-1}, which is comparable to that of the pristine LLZ. However, the grain boundary conductivity decreased to 7.9×10^{-5} from the 1.39×10^{-4} S cm^{-1} for the pristine LLZ. LLZ immersed in 0.1 M aqueous HCl and in water also showed no change in the bulk conductivity, but a significant decrease in the grain boundary conductivity. The significant decrease in grain boundary conductivity by immersion in water may be due to the reaction of impurity phases such as La_2O_3 with water, because the impurity phase-free LLZ showed no such high decrease in conductivity after immersing in water. LLZ was determined to be stable in an aqueous solution saturated with LiOH and LiCl.

Sakamoto et al. [18] examined the effect of Al_2O_3 doping into LLZ and reported that $Li_{6.24}La_3Zr_2Al_{0.24}O_{11.98}$ had the total conductivity of 4.0×10^{-4} S cm^{-1} at room temperature, where the samples were hot-pressed at 1,000 °C and 40 MPa.

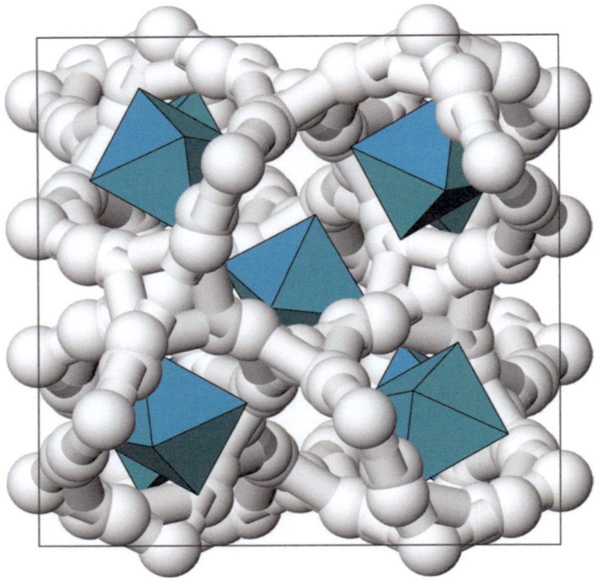

Fig. 8.7 Crystal structure model for cubic Al-containing $L_{7-x}La_3Zr_2O_{12-1/2x}$ projected on the (100) plane. Li sites are shown as *white spheres*, ZrO_6 octahedra are *blue*, and La atoms are omitted for clarity. Adjacent Li positions are connected via the white pathway that represent possible routes for Li diffusion (from ref. 22)

Ohta et al. [19] reported that Nb-doped LLZ had a high ionic conductivity of 8×10^{-4} S cm^{-1} at 25 °C, the conductivity value of which is comparable to that of LATP. These results suggest that LLZ is a promising candidate for water-stable protective layer of a water-stable lithium metal electrode for lithium air batteries, due to its high conductivity and stability with lithium metal and in a saturated aqueous solution of LiOH and LiCl. However, Kanamura et al. [20] reported that the Li/LLZ/Li cell exhibited unusual polarization behavior; the polarization potentials were steady with the polarization period up to a polarization current of 10 μA cm^{-2}, but at 20 μA cm^{-2}, the potential increased suddenly after polarization for 400 s. It is not clear why the polarization potential increased suddenly. Recently, Imanishi et al. [21] reported that the Li/Nb-doped LLZ/Li cell at 25 °C showed a short-circuit for 280 s polarization at 0.5 mA cm^{-1} and no short-circuit for 41 h at 0.1 mA cm^{-2}. The detailed polarization behavior between lithium metal and LLZ at high current density has not been sufficiently studied. Therefore, the interface between lithium metal and LLZ, and also between LLZ and aqueous electrolytes, should be systematically investigated.

Recently, Geiger et al. [22] determined the crystal structure of LLZ by X-ray single-crystal diffraction to be cubic with space group $Ia\overline{3}d$ in the temperature range of 100–500 K. Figure 8.7 shows a crystal structure model for cubic Al-containing $Li_{7-x}La_3Zr_2O_{12-0.5x}$. They reported that LLZ has isotropic three-dimensional Li-diffusion pathways, closely spaced Li sites, and delocalization that allow for easy and fast Li diffusion and low occupancy at Li sites, which may also be enhanced by the tetravalent substitution of Al^{3+} for Li^+. The obtained LLZ crystals had ca. 1.3 wt% Al_2O_3, which most likely originated from the ceramic crucibles.

The Al contamination may be the key aspect in stabilizing the cubic phase [22, 23]. Imanishi et al. [17] and Geiger et al. [22] reported the atomic composition of Li:La:Zr:Al to be 5.96:3.23:2.0:0.23 and 6.79:3.37:2.0:0.21, respectively. However, they did not measure the oxygen content. We would note that the exact stoichiometric formula of cubic LLZ has not yet been determined, because it is difficult to reproducibly prepare these samples due to an evaporation of lithium during sintering at a high temperature. LLZ is stable in direct contact with lithium metal and in saturated aqueous solution of LiOH and LiCl; therefore, a water-stable lithium electrode with a low interface resistance at room temperature could be prepared, because the main interface resistance of the electrodes when using LATP comes from the interface layer between lithium and the polymer electrolyte, and the ionic conductivity of the polymer electrolyte is too low at room temperature. The water table lithium metal electrode with LATP and the polymer electrolyte should be operated at over 60 °C [4]. The resistance of a water-stable lithium electrode with a 100-μm thick LLZ sheet, which has a high conductivity of 5×10^{-4} S cm^{-1}, is only 20 Ω cm^2, and the contribution of the IR-drop of LLZ at 10 mA cm^{-2} is 6.7 % of the OCV. This value is acceptable for high power batteries of EV applications. Important research targets for a water-stable lithium electrode using LLZ is to develop a water permeation free thin film with excellent mechanical properties and to prevent lithium dendrite formation between lithium and LLZ at high current density.

8.4 Preparation of Lithium Ion Conducting Thin Sheet Solid Electrolyte

The electrolyte plates of LATP were mainly supplied from Ohara Inc., Japan. The typical Ohara LATP plate has a thickness of 150–250 μm and ionic conductivities of 1 ~ 3.5×10^{-4} S cm^{-1} at 25 °C and 0.44 ~ 1.4×10^{-3} S cm^{-1} at 60 °C. The Ohara plates are prepared by quenching from the melt and are crystallized at high temperature [9]. The commercial LATP plates have good mechanical properties and are free from water permeation; however, they are expensive and their size is limited. Imanishi et al. prepared a water permeation-free LATP film with high ionic conductivity by a combination of tape-cast LATP with epoxy resin [24]. The LATP precursor of $Li_{1.4}$ $Al_{0.4}Ti_{1.6}$ $(PO_4)_3$ was prepared by the sol–gel method with citric acid. LATP powders with fine powders of TiO_2, ZrO_2, or Al_2O_3 as sintering additives were tape-cast using a doctor blade method. The green sheet was left to dry at room temperature for 24 h and sintered at 1,060 °C for 2 h. Figure 8.8 shows impedance profiles measured at 25 °C for as-sintered sheets of LATP-3 wt% TiO_3, LATP-3 wt% ZrO_2, and LATP-3 wt% Al_2O_3, and for these sheets annealed at 80 °C. The as-sintered LATP-3 wt% Al_2O_3 sample has a high ionic conductivity, but a significant thermal hysteresis for ionic conductivity, whereas the LATP-3 wt % TiO_3 sample did not exhibit thermal hysteresis. The LATP-3 wt% ZrO_2 has a low

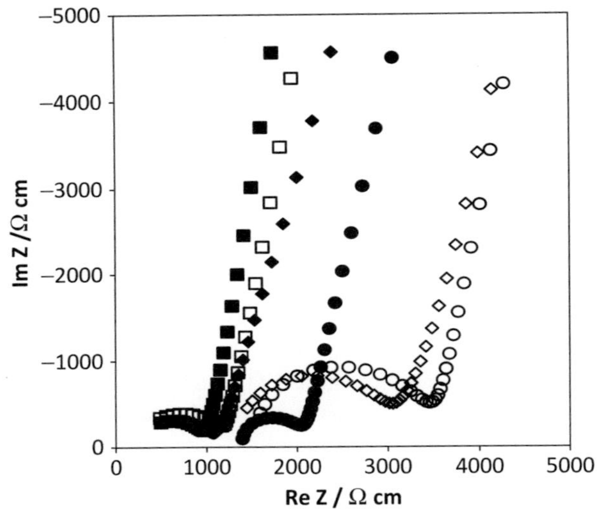

Fig. 8.8 Impedance profiles measured at 25 °C for the tape-cast LATP sheet sintered at 1,060 °C for 2 h with (*filled square*) as-prepared 3 wt% TiO_2 and (*open square*) 3 wt% TiO_2 after heating to 80 °C, (*filled diamond*) as-prepared 3 wt% Al_2O_3 and (*open diamond*) 3 wt% Al_2O_3 after heating to 80 °C, and (*filled circle*) as-prepared 3 wt% $ZrO_2$and (*open circle*) 3 wt% ZrO_2 after heating to 80 °C (from ref. 24)

initial conductivity and exhibits thermal hysteresis. The final conductivities of LATP-3 wt% Al_2O_3 and ZrO_2 are comparable to pristine LATP without these dopants. The highest stable conductivity of 7.6×10^{-4} S cm^{-1} was observed for LATP-3 wt% TiO_2. The conductivity of the tape-cast LATP-3 wt% TiO_2 was comparable to that of a pellet sample sintered at 950 °C using the LATP powder prepared by the sol–gel method. However, water permeation through the sintered LATP sheets was observed, probably due to the presence of open pores. Tape cast sheet is generally sintered at higher temperatures and for a longer period to obtain dense sheets. However, a dense LATP sheet could not be obtained by high-temperature sintering, due to the evaporation of lithium oxide at such temperatures over long periods. Therefore, to suppress water permeation through the sheet, the open pores in the sheet were filled with epoxy resin. Water permeation test results through the LATP-3 wt% TiO_2 sheet and LATP-3 wt% TiO_2–epoxy resin hybrid sheet are shown in Fig. 8.9, where the epoxy resin content was ca. 1.3 wt%. The water permeation test was conducted using an H-type cell. The water permeation rate through the LATP sheet was estimated from the change of the chlorine content in distilled water with time, measured using a chlorine meter. Figure 8.9a shows that the tape cast sheet of LATP-3 wt% TiO_2 quickly exhibits water permeation. In contrast, no chlorine was detected in the distilled water side of the H-type cell for the LATP-3 wt% TiO_2–epoxy resin after 17 days (Fig. 8.9b). Figure 8.10 shows impedance profiles of the LATP-3 wt% TiO_2–epoxy resin hybrid sheet measured at 25 °C. The total ionic conductivity of the hybrid sheet is 4×10^{-4} S cm^{-1} at 25 °C,

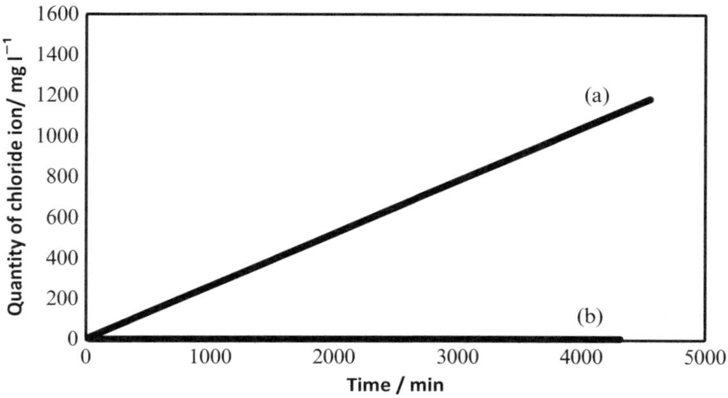

Fig. 8.9 Water permeation test for (**a**) tape-cast LATP-3 wt% TiO_2 and (**b**) tape-cast LATP-3 wt% TiO_2 with epoxy resin (from ref. 24)

Fig. 8.10 Impedance profiles measured at 25 °C for (*filled circle*) pristine tape-cast LATP-3 wt% TiO_2 with epoxy resin and the same immersed in (*open circle*) saturated LiCl aqueous solution at 50 °C for 1 week, and (*filled triangle*) in saturated LiCl and LiOH aqueous solution at 50 °C for 1 week (from ref. 24)

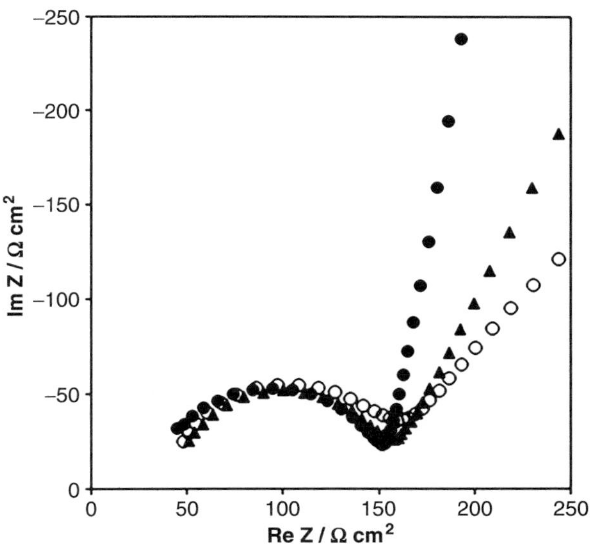

which is approximately half that of the LATP-3 wt% TiO_2 sheet without epoxy resin, and higher than that of the water permeation-free, dense-glass ceramic LATP plate supplied by Ohara (3,34 × 10^{-4} S cm^{-1} at 25 °C). The addition of epoxy resin resulted in a decrease in bulk conductivity from 5.5 × 10^{-3} to 2.5 × 10^{-3} S cm^{-1} and a decrease in the grain boundary conductivity from 7.7 × 10^{-4} to 4.7 × 10^{-4} S cm^{-1}. The stability of the LATP-3 wt% TiO_2–epoxy resin hybrid sheet in saturated LiCl aqueous solution and aqueous solution saturated with both LiCl and LiOH were examined, The impedance profiles of the hybrid sheet immersed in these solutions at 50 °C for 1 week are also shown in Fig. 8.10.

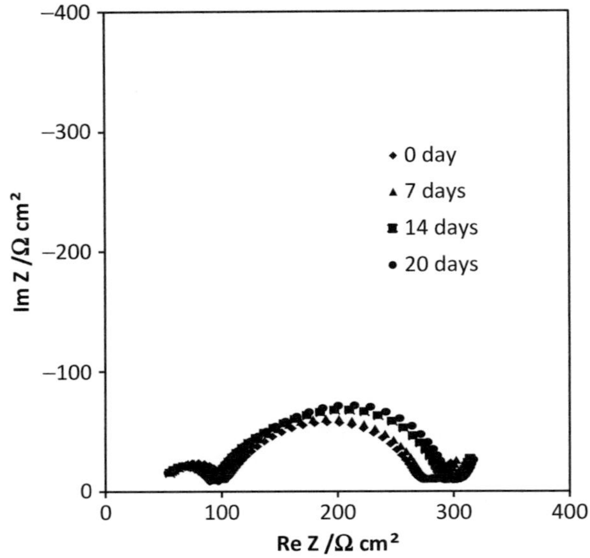

Fig. 8.11 Impedance profiles measured at 60 °C for the Li/PEO_{18}LiTFSI–10 wt% $BaTiO_3$/tape-cast LATP–3 wt% TiO_2–epoxy resin/saturated LiOH and LiCl aqueous solution/Pt, air cell as a function of the storage time (from ref. 24)

No significant change in the impedance profiles was evident. The similar water-stable lithium conducting hybrid sheets of $Li_{1.4}Al_{0.4}Ge_{1.6}(PO_4)_3$ (LAGP)-TiO_2 and epoxy resin were prepared by tape casting method [25]. The electrical conductivity of the water impermeable hybrid sheet with ca. 4 % epoxy resin was 4.19×10^{-4} S cm^{-1} at 25 °C. The electrical conductivity of the LAGP-TiO_2–epoxy hybrid sheet was comparable to that of LATP-TiO_2–epoxy hybrid sheet, but the three-point bending strength of the LAGP hybrid sheet (111 N mm^{-2}) was higher than that of the LATP hybrid sheet (52 N mm^{-2}).

The stability of an electrode fabricated with the LATP-3 wt% TiO_2–epoxy resin hybrid sheet was examined in an aqueous solution saturated with LiOH and LiCl. The test electrode consisted of lithium metal, PEO with LiTFSI and $BaTiO_3$, and the tape-cast LATP-3 wt% TiO_2–epoxy resin hybrid sheet. The polymer electrolyte was used to prevent direct contact between the lithium metal and the LATP hybrid sheet. The test electrode was immersed in saturated LiOH and LiCl aqueous solution at 60 °C. Figure 8.11 shows the impedance profiles of the Li/PEO_{18}LiTFSI-10 wt% $BaTiO_3$/LATP-3 wt% TiO_2–epoxy resin/saturated LiOH and LiCl aqueous solution/Pt, air cell as a function of the storage period at 60 °C. Platinum plate with a platinum black coating was used as a counter electrode. The impedance profiles show two semicircles at high and low frequency. The semicircle at high frequency corresponds to the resistance of the LATP sheet and polymer electrolyte, and that at low frequency corresponds to the interfacial layer between lithium metal and the polymer electrolyte, and the charge transfer resistance between the lithium electrode and solid electrolyte interface (SEI) prepared at the lithium metal/polymer electrolyte interface. The semicircle at high frequency did not change over time, while that at low frequency increased slightly after a storage

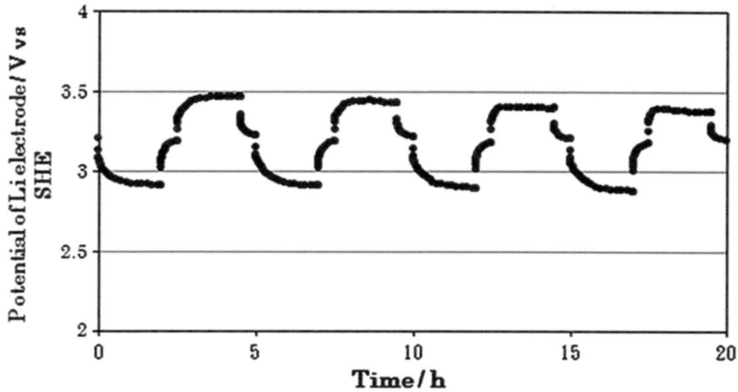

Fig. 8.12 Potential change of the lithium electrode at 0.5 mA cm^{-2} and 60 °C in the Li/PEO_{18}LiTFSI/tape-cast LATP–3 wt% TiO_2-epoxy hybrid sheet/saturated LiCl aqueous solution/Pt, air cell (from ref. 24)

period of 20 days. These results confirm that the tape-cast LATP-epoxy resin hybrid sheet is water permeation free; if a small amount of water permeates through the sheet, lithium metal reacts with water through the polymer electrolyte, which results in a change of the impedance profiles, because the polymer electrolyte is unstable in water.

The performance of the water-stable lithium electrode with the hybrid electrolyte sheet was examined. Figure 8.12 shows the typical lithium electrode potential change for the Li/PEO_{18}LiTFSI/LATP-3 wt% TiO_2-epoxy hybrid sheet/saturated LiCl aqueous solution/Pt, air cell at 60 °C, where the lithium electrode potentials were measured using a Ag/AgCl reference electrode. A stable cell OCV of 3.44 V at 60 °C was observed for 30 days, which was slightly lower than the theoretical OCV. Stable charge and discharge performance was obtained for the water-stable lithium metal electrode. The charge and discharge overpotential was 0.25 V at 0.5 mA cm^{-2}. The electrode performance indicated that this hybrid electrolyte sheet could potentially be used as the protective layer for the water-stable lithium metal electrode in high-energy density lithium air rechargeable batteries. No report for a thin sheet of LLZ has been published. However, in view of the energy and power densities of the aqueous lithium air batteries, water permeation-free thin sheet electrolytes with thicknesses less than 100 μm and excellent mechanical properties should be developed.

8.5 Cell Performance of Aqueous Lithium Air Rechargeable Cells

The electrode reaction of the aqueous lithium air system is different from that of the nonaqueous system. The aqueous electrolyte participates in the cathode reaction, as discussed in Sect. 6.1. The reaction product of LiOH is soluble in water, but

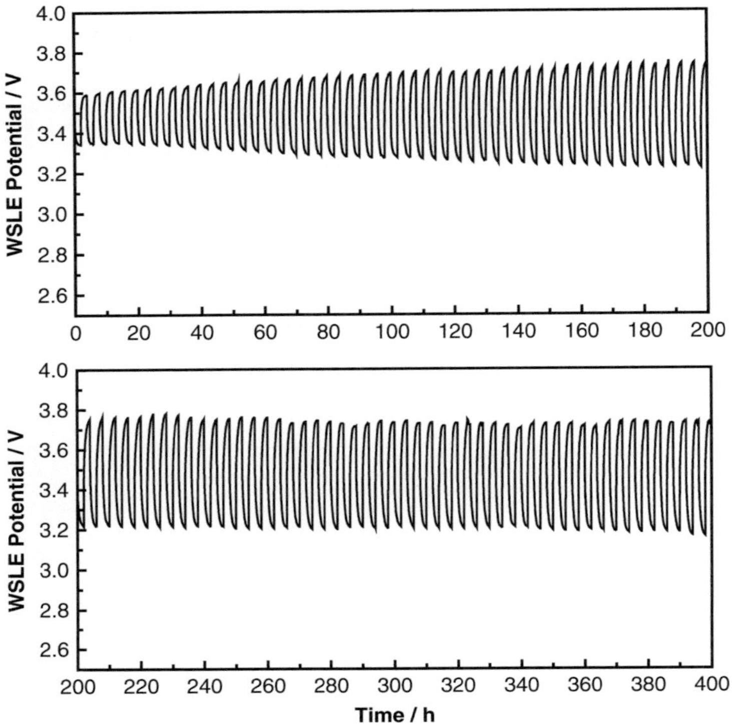

Fig. 8.13 Charge and discharge cycling performance for the Li/PEO_{18}LiTFSI-ionic liquid/LATP/aqueous 1 M LiCl/Pt, air cell at 0.3 mA cm^{-2} and 60 °C (from ref. 27)

becomes saturated in the electrolyte at only ca. 5 % discharge depth and is deposited in the electrolyte and/or on the air electrode. Stevens et al. [26] proposed the use of an anion-exchange membrane at the cathode and electrolyte interface, which would permit OH^- generated in the porous cathode to be transported out, while blocking Li^+ entering the electrode, so that LiOH precipitates outside the cathode (see Chap. 4). Such a cell has been demonstrated to operate for one hundred cycles without degradation in a narrow discharge depth (0.1 mAh cm^{-2}). The cyclic performance of a beaker-type cell with the water-stable lithium electrode and platinum air electrode was examined by Imanishi et al. [27]. Figure 8.13 shows the charge–discharge performance of the Li/PEO_{18}LiTFSI-ionic liquid/LATP/1 M LiCl aqueous solution/Pt, air cell at 0.3 mA cm^{-2} and 60 °C, where *N*-methyl-*N*-propylpiperidinium-bis(trifluoromethylenesulfonylimide (PP13TFSI) was used as the ionic liquid. The discharge potential is 3.2 V and charge potential is 3.75 V at 0.3 mA cm^{-2} after 100 cycles. These results were observed for a cell with an excess amount of water. However, the capability for high power density and extended deep cycling has not been shown for aqueous lithium air cells, except of that with acetic acid as the electrolyte.

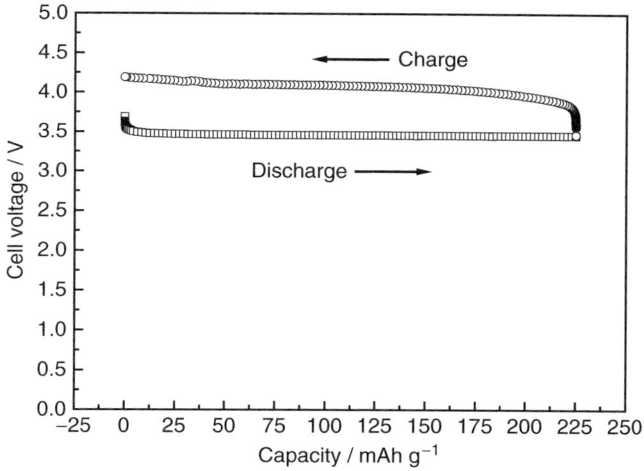

Fig. 8.14 Charge–discharge performance for the Li/PEO_{18}LiTFSI/HAc–H_2O–LiAc saturated/carbon, air cell at 0.5 mA cm^{-2} and 60 °C (from ref. 28)

The aqueous lithium air cell with an acetic acid solution electrolyte was proposed by Imanishi et al. [28]. The cell reaction of the proposed lithium air battery is:

$$2Li + 2CH_3COOH + 1/2O_2 = 2CH_3COOLi + H_2O \qquad (8.3)$$

In this system, the reaction products are CH_3COOLi and H_2O, and the pH in the solution is relatively constant during the charge and discharge process. The energy density of this system was calculated to be 1,320 Wh kg^{-1} (including oxygen) using an OCV of 3.69 V, which is ca. 70 % of that for the lithium/air cell with H_2O. LATP is unstable in strong acid and alkaline solutions, but Imanishi et al. reported that LATP was stable in a lithium acetate (LiAc) saturated acetic acid (HAc) aqueous solution [10]. Figure 8.14 shows the deep charge–discharge cycle performance of an aqueous lithium air pouch cell with an aqueous acetic acid electrolyte. The test cell consisted of a lithium metal anode, PEO_{18}LiTFSI interface, LATP plate, carbon black with a Pt catalyst, and the LiAc saturated-10 vol.% HAc-H_2O electrolyte. The OCV was 3.68 V at 60 °C and 3 atm of air. Flat discharge and charge potentials were obtained, with a discharge capacity of 225 mAh g^{-1}. The energy density of the aqueous lithium air cell including oxygen and lithium is calculated to be ca. 700 Wh kg^{-1}, which is much higher than the 387 Wh kg^{-1} for a lithium–ion battery with a carbon anode and a $LiCoO_2$ cathode, and the cell exhibited good cycling performance for 15 cycles.

The safety of lithium air batteries with a lithium anode is an important issue for EV applications. The large amount of lithium metal in the batteries cannot be made adequately safe, and the tendency of lithium ions to form dendrites during charge, especially at the high rates, results in penetration of the thin barrier layers to

Fig. 8.15 Log current density (J) vs. log dendrite formation onset time (t_o) for the Li/PEO_{18}LiTFSI–1.44PP13TFSI/Li cell at 60 °C (from ref. 31)

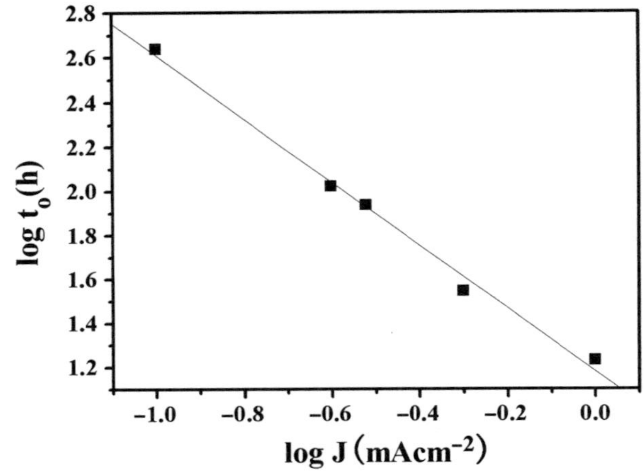

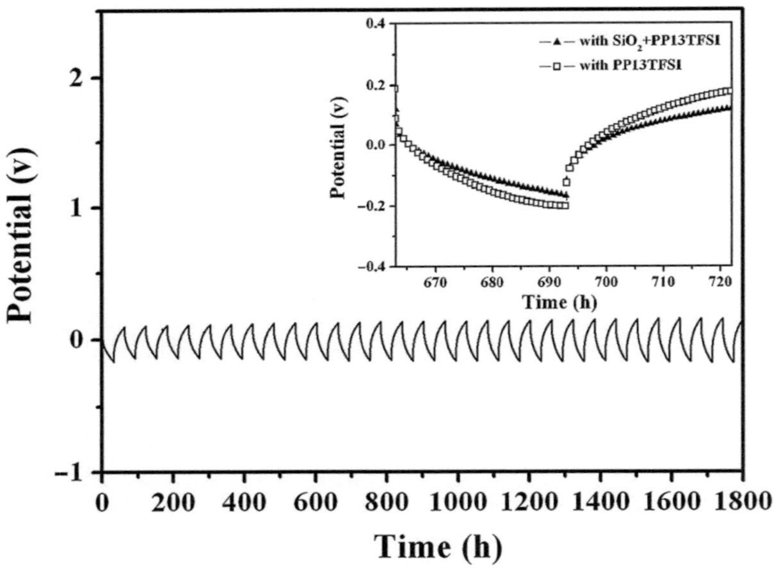

Fig. 8.16 Cycling performance for the Li/PEO_{18}LiTFSI–1.44PP13TFSI–10 wt% SiO_2/Li cell at 60 °C and at polarization of 0.3 mA cm^{-2} for 30 h. The inset shows a comparison of the 11th cycle for Li/PEO_{18}LiTFSI–1.44PP13TFSI/Li and Li/PEO_{18}LiTFSI–1.44 PP13TFSI–10 wt% SiO_2/Li cells (from ref. 30)

cause cell short-circuit with water. Lithium dendrite formation occurs even in polymer electrolytes, although to a lesser extent than that in liquid electrolytes [29]. The dendrite formation onset time is dependent on the current density and the resistance formed at the interface of lithium metal and the polymer electrolyte [30–32]. Figure 8.15 shows the onset time vs. current density curves for a

$Li/PEO_{18}LiTFSI–1.44PP13TFSI/Li$ cell at 60 °C. The dendrite formation onset time at 1.0 and 0.1 mA cm^{-2} were 17 and 434 h, respectively. The specific weight capacities of the lithium anode are 1,270 mAh g^{-1} at 1.0 mA cm^{-2} and 2,148 mAh g^{-1} at 0.1 mA cm^{-2}, where the capacities are calculated from the total weight including a 10-μm thick copper foil current collector and lithium metal deposited until dendrite formation. The specific weight capacity of the lithium electrode at 1.0 mA cm^{-2} is ca. 30 % of that for lithium metal and more than 3 times higher than that of a typical carbon anode in a conventional lithium–ion battery. Figure 8.16 shows the cycling performance for the lithium deposition and stripping process in the $Li/PEO_{18}LiTFSI–PP13TFSI–SiO_2/Li$ cell at 60 °C, where a constant current of 0.3 mA cm^{-2} was passed for 30 h during each process [30]. The area-specific capacity of 9 mAh cm^{-2} is a comparable typical area capacity of carbon–air electrode in nonaqueous lithium air cell (see Chap. 9). No significant change in cell voltage was observed during the cycles, and microscopic observation of the cell indicated no dendrite formation after the cycling test. Dendrite formation at the interface of lithium metal and the polymer electrolyte is suppressed by the addition of the ionic liquid and/or the oxide fillers into the polymer electrolyte; however, the complete suppression of lithium dendrite formation at a high current density of several milliamperes per square centimeter or more, as demanded for EV applications, is questionable. Thus, an alternative to lithium should be used in lithium air batteries for EV applications. Silicon and tin lithium alloys have been proposed as alternative anodes for lithium air rechargeable batteries. Therefore, the dendrite formation of these alloys should be investigated at high current densities, especially with EV applications in mind.

References

1. Visco SJ, Nimon B (2004) Lithium metal aqueous batteries. International meeting on lithium batteries, abs. # 53.
2. Aono H et al (1989) Ionic conductivity of the lithium titanium phosphate ($Li_{1+x}M_xTi_{2-x}$, M=Al, Sc, Y and La) systems. J Electrochem Soc 136:590–591
3. Murugan R, Thangadurai V, Weppner W (2007) Fast lithium ion conduction in garnet-type $Li_7La_3Zr_2O_{12}$. Angew Chem Int Ed 46:7778–7781
4. Zhang T et al (2008) Li/polymer electrolyte/water stable lithium-conducting glass ceramics composite for lithium-air secondary batteries with an aqueous electrolyte. J Electrochem Soc 155:A963–A969
5. Shimonishi Y et al (2011) A study on lithium/air secondary batteries—stability of the NASICON-type lithium ion conducting solid electrolyte in alkaline aqueous solutions. J Power Sources 196:5128–5132
6. Tatsumisago M et al (2002) New lithium ion conducting glass-ceramics prepared from mechanochemical Li_2S-P_2S_5 glasses. Solid State Ion 154–155:635–640
7. Kamaya N et al (2011) A lithium super conductor. Nat Mater 10:682–686
8. Aono H et al (1990) Ionic conductivity of solid electrolytes based on lithium titanium phosphate. J Electrochem Soc 137:1023–1027
9. Fu J (1997) Superionic conductivity of glass-ceramics in the system Li_2O-Al_2O_3-TiO_2-P_2O_3. J Power Sources 96:195–200

10. Shimonishi Y et al (2010) A study on lithium/air secondary batteries—stability of NASICON type glass ceramics in acid solution. J Power Sources 195:6187–6191
11. Brik P, Salman F, Doring S et al (1999) A first approach to a monolithic all solid state inorganic lithium battery. J Power Sources 119:149–157
12. Cretin M, Fabry F, Abello L (1995) Study of $Li_{1+x}Ti_{2-x}Al_x(PO_4)_3$ for Li^+ potentiometric sensor. J Eur Ceram Soc 15:1149–1156
13. Thockchom JS, Kumar B (2007) Water durable lithium ion conducting composite membranes from the Li_2O-Al_2O_3-TiO_2-P_2O_5 glass-ceramic. J Electrochem Soc 154:A331–A336
14. Imanishi N et al (2008) Lithium anode for lithium-air secondary batteries. J Power Sources 185:1392–1397
15. Hasegawa S et al (2009) Study on lithium-air batteries-Stability of NASICON-type lithium ion conducting glass-ceramics with water. J Power Sources 189:371–377
16. Bohnke O et al (2011) H^+/Li^+ exchange property of $Li_{3x}La_{2/3-x}TiO_3$ in water and in humid atmosphere. Solid State Ion 188:144–147
17. Shimonishi Y et al (2011) Synthesis of garnet-type $Li_{7-x}La_3Zr_2O_{12-x/2}$ and its stability in aqueous solution. Solid State Ion 183:48–53
18. Rangasamy E, Wolfenstine J, Sakamoto J (2012) The role of Al and Li concentration on the formation of cubic garnet solid electrolyte of nominal composition $Li_7La_3Zr_2O_{12}$. Solid State Ion 206:28–32
19. Ohta S, Kobayashi T, Asaoka T (2011) High lithium ion conductivity in the garnet-type oxide $Li_{7-x}La_3(Zr_{2-x}Nb_x)O_{12}$ (x=0-2). J Power Sources 196:3342–3345
20. Kotobuki M et al (2010) Conductivity of $Li_7La_3Zr_2O_{12}$ solid electrolyte to all-solid-state battery using Li metal anode. J Electrochem Soc 157:A1076–A1079
21. Ishiguro K et al (2013) Stability of Nb-doped cubic $Li_7La_3Zr_2O_{12}$ with lithium metal. J Electrochem Soc 160:A1684–A1689
22. Geiger CA et al (2011) Crystal chemistry and stability of "$Li_7La_3Zr_2O_{12}$" garnet. Inorg Chem 50:1089–1097
23. Jin Y, McGlnn PJ (2011) Al-doped $Li_7La_3Zr_2O_{12}$ synthesized by a polymerized complex method. J Power Sources 196:8683–8687
24. Takahashi K et al (2012) A water-stable high lithium ion conducting $Li_{1.4}Ti_{1.6}Al_{0.4}(PO_4)_3$-epoxy resin hybrid sheet. J Electrochem Soc 159:A1065–A1096
25. Zhang M et al (2013) Water-stable lithium anode with $Li_{1.4}Al_{0.4}Ge_{1.6}(PO_4)_3$-$TiO_2$ sheet prepared by tape casting method for lithium-air batteries. J Power Sources 235:117–121
26. Laurent P et al (2012) Elaboration and characterization of a free standing LiSICON membrane for aqueous lithium-air battery. J Power Sources 214:330–336
27. Zhang T et al (2011) Stability of Li/polymer electrolyte-ionic liquid composite/lithium conducting glass ceramics in an aqueous electrolyte. Electrochem Solid State Lett 14: A45–A48
28. Zhang T et al (2010) A novel high energy density rechargeable lithium/air battery. Chem Commun 46:1661–1663
29. Megahed S, Scrosati B (1995) Rechargeable nonaqueous batteries. Interface 4:34–37
30. Liu S et al (2011) Effect of co-doping nano-silica filler and N-propylpiperidinium bis(trifluoromethanesulfonyl)imide into polymer electrolyte on Li dendrite formation in Li/poly(ethylene oxide)-$Li(CF_3SO_2)_2N$/Li. J Power Sources 196:7681–7686
31. Liu S et al (2010) Lithium dendrite formation in Li/poly(ethylene oxide)-lithium bis(trifluoromethanesulfonyl)imide and N-propylpiperidinium bis(trifluoromethanesulfonyl)imide/Li cells. J Electrochem Soc 157:A1092–A1098
32. Liu S et al (2010) Effect of nano-silica in polymer electrolyte on Li dendrite formation in Li/poly(ethylene oxide)-$Li(CF_3SO_2)_2N$/Li. J Power Sources 195:6847–6853

Chapter 9
A Solid-State, Rechargeable Lithium Oxygen Battery

B. Kumar and J. Kumar

9.1 Introduction

The need and justification for developing environmentally sustainable and novel batteries were articulated by Armand and Tarascon [1]. In recent years, similar justifications have been echoed globally. Theoretically, $Li–O_2$ chemistry provides the most energy dense electrochemical power source. The concept of the cell was reduced to practice about 15 years ago by configuring a working $Li–O_2$ cell [2]. The $Li–O_2$ cell is capable of providing an open-circuit voltage (OCV) around 3.0 V and a theoretical specific energy of 5,200 Wh/kg if oxygen is contained in the cell. Oxygen can be accessed from the ambient atmosphere, and if such is the case, the theoretical specific energy of the $Li–O_2$ cell increases to 11,140 Wh/kg [2]. The $Li–O_2$ system has seized global attention in recent years as a potential power source for a range of small to large devices.

Since Abraham and Jiang [2] reported the successful demonstration of a nonaqueous rechargeable $Li–O_2$ battery cell, there was little activity for almost a decade. However, interest in $Li–O_2$ batteries has surged lately [3–9]. Girishkumar et al. [10] summarized the promises and challenges facing the development of practical $Li–O_2$ batteries. They opined that if successfully developed, $Li–O_2$ batteries could provide an energy source for electric vehicles rivaling that of gasoline in terms of energy density. Two review papers [11, 12] on $Li–O_2$ batteries also document recent advances and technical challenges. However, these reviews primarily discuss $Li–O_2$ cell chemistry based on nonaqueous and aqueous electrolytes.

A primarily solid-state $Li–O_2$ battery developed at the University of Dayton Research Institute (UDRI) addresses safety issues in addition to the high energy

B. Kumar (✉) • J. Kumar
Electrochemical Power Group, Energy Technology and Materials Division,
University of Dayton Research Institute, Dayton, OH 45469-0170, USA
e-mail: binod.kumar@udri.udayton.edu

N. Imanishi et al. (eds.), *The Lithium Air Battery: Fundamentals*,
DOI 10.1007/978-1-4899-8062-5_9, © Springer Science+Business Media New York 2014

density critical to the deployment of batteries for a number of different commercial applications. Since a solid-state Li–O_2 battery is primarily made of high-temperature ceramic materials, the probability of a runaway reaction leading to a catastrophic incident is minimized. The high energy density of the cell originates from two basic attributes of the cell design: first, the anode of the cell is comprised of lithium—the most energy dense solid metal—and second, O_2 is primarily contained in the cathode as an active oxidant. A proof of the concept of the solid-state Li–O_2 cell, along with associated attributes, has been reported in a recent publication [8].

This chapter enumerates electrochemical attributes and design features of a practical, solid-state Li–O_2 cell. The high rate of performance and long cycle life of a Li–O_2 cell can be achieved in these by the deployment of a membrane laminate (separator) exhibiting liquid-like conductivity at lower temperatures. The performance is further augmented by an oxygen electrode containing a glass–ceramic material that catalyzes oxygen reduction and evolution at low temperatures and without precious metals.

9.2 A Solid-State Cell

The schematic of a typical solid-state Li–O_2 cell is shown in Fig. 9.1. The cell consists of three primary components: anode, membrane laminate, and O_2 electrode. The three components are designed [13–16] to enhance electrode kinetics, to facilitate reversible and lithium-only ion transport (from anode to cathode to anode), and to provide interfaces with minimum resistances. The cell employs an O_2 cathode to achieve reversible discharge/charge cell reactions using O_2 either stored within the cathode compartment or harnessed from the ambient atmosphere. The cathode is a heterogeneous solid comprised of glass–ceramic, carbon, and polytetrafluoroethylene (PTFE).

9.3 Materials of Cell Components

The materials from which the solid-state Li–O_2 cell is constructed are identified in Fig. 9.1. The choice of the materials for these components depends upon the desired operating temperature range of the cell, energy density, and cycle life. The following paragraphs and subsections discuss the cell operation mechanism, requirements, and potential materials for components in a solid-state cell.

In an electrochemical cell the anode and cathode are spatially separated by a membrane (electrolyte). In a Li–O_2 cell, the oxidation reaction is carried out by an anode containing lithium. The oxidation of lithium produces electrons and lithium ions. The lithium ions are transported internally through the membrane and brought to the cathode. The electrons travel externally and arrive at the cathode,

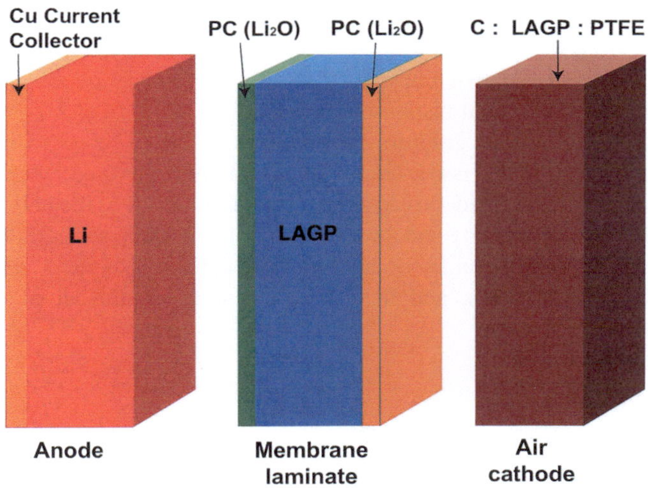

Fig. 9.1 A schematic of solid-state Li–O_2 cell and its component materials. The anodic component consists of lithium (active anode material) and copper (current collector) laminate. The membrane laminate, composed of PC(Li_2O), GC, and PC(Li_2O), couples the anode with the cathode. The current collector on the cathode side is a stainless steel casing, not shown in the schematic

where they reduce molecular oxygen. The reduced oxygen subsequently reacts with lithium to form the discharge product. A cathode, therefore, must have the ability to transport electrons and lithium ions, catalyze the reduction of oxygen, nucleate the discharge product, and provide space to accommodate a significant quantity of discharge product. All these requirements for the anode, membrane, and cathode provide challenges to the materials science and engineering community. To successfully develop Li–O_2 batteries, these challenges need to be successfully addressed.

9.3.1 Anode

Typical anodic materials that may be considered for a solid-state cell are lithium (Li) metal, alloyed lithium (Al–Li, Cu–Li), intercalated graphite (Li_xC_6), alloyed tin (Sn–Li), and germanium (Ge–Li). An efficient electrode reaction (oxidation), high conductivity (electronic and ionic), and a large potential difference with a compatible cathode are some of the primary requirements. From the capacity point of view, lithium metal is preferred; however, it is known that Li metal is chemically and morphologically unstable in most liquid electrolytes. Our understanding of the stability of Li metal with solid electrolytes is limited. The issue of lithium stability has been a dominant concern for the lithium battery community.

A reversible storage of energy requires that lithium be plated (charged) and stripped (discharged) from the surface of the anode with the highest Coulombic efficiency (CE). It may be argued that the use of a solid electrolyte could minimize the chemical and morphological instability of lithium anode.

Alloys and intercalated graphite can be used at the expense of the capacity and energy density. This issue could be rationally addressed once a suitable oxygen electrode has been identified and a structural design of the cell is finalized. Anodic and cathodic capacity should also match in an optimized cell design. Solid Li metal has been used as an anode in solid-state cells investigated and developed at UDRI for several years, and it will be discussed as the anode material in this chapter.

9.4 Electrolyte (Membrane)

The main purpose of the membrane is to ionically bridge and electronically insulate the anodic and cathodic compartments of an electrochemical cell. The purpose can be fulfilled by a variety of solid, dielectric materials, such as polymers, glasses, ceramics, and their combinations. The polymers are attractive because of their low density, manufacturability, and capacity to accommodate volume changes as compared to a solid, inorganic electrolyte. The subject of polymer electrolytes has been covered by review papers and a monograph [17, 18]. Polymer–ceramic composite electrolytes present additional advantages such as mechanical rigidity and high conductivity which have also received considerable attention in the last decade [19]. The doping of ceramic components into an ionically conducting polymer matrix leads to the formation and existence of a double layer (space charge). The double layer tends to be diffused at elevated temperatures due to the excessive thermal energy of ions and electrons. At lower temperatures, specifically around the operating temperature of conventional batteries, the space charge becomes effective and influences ion transport. The space charge-mediated transport of ions may lead to an enhanced ionic conductivity.

A glass–ceramic material formed by the conventional glassmaking process and subsequently nucleated and crystallized at appropriate temperatures has also drawn considerable attention in recent years. The processing and ionic conductivity of lithium aluminum titanium phosphate (LATP) and lithium aluminum germanium phosphate (LAGP) glass–ceramics have been reported [15, 20–22]. These glass–ceramics consist of a primarily crystalline phase that exhibits liquid-like conductivity. However, a glass–ceramic is a polycrystalline material. The conductivity is diminished by the existence of grain boundaries. Nonetheless, these glass–ceramics remain attractive for making electrochemical cells from the conductivity and its temperature dependence points of view.

A windable membrane comprised of polymer–ceramic (PEO:LiBETI-LAGP) electrolyte supported on fiberglass mat has been developed for Li–O_2 cells in the authors' laboratory. A mixed powder batch of LAGP (65 wt%) and polymer electrolyte (35 wt%) from the PEO:LiBETI (8.5:1) complex is spread onto a fiberglass mat and subsequently hot pressed at around 100 °C. The process yields a windable, mechanically strong, high-conductivity membrane.

9.5 Oxygen Electrode (Cathode)

Oxygen reduction is the fundamental electrochemical reaction at the electrode that is critical to the performance of $Li–O_2$ cells. It is also known that oxygen reduction is a sluggish reaction often requiring the use of the catalysts, such as precious metals or high temperatures. The catalysts and high temperatures often have their own set of limitations that may compromise the usefulness of electrochemical cells. An oxygen electrode is also a multifunctional component. The multifunctionality of the electrode is accomplished by a composite structure comprised of appropriate proportions of electro-catalytic, ionic conductive, and electronic conductive phases. Furthermore, the composite structure must also facilitate nucleation and accommodation of the discharge product and active material (oxygen).

9.5.1 Chemistry of Electro-Catalytic Phase

Ceramic materials, such as zeolites, mediate and enhance the rate of chemical reactions by accommodating large molecules and catalyzing the formation of intermediates and products. These are crystalline aluminum silicate ceramics composed of SiO_4 and AlO_4 tetrahedra arranged in various spatial configurations. The physical manifestations of these configurations are three-dimensional structures of polyhedral chains and layers. The zeolites are classic skeleton structures and used as molecular sieves. In nature, the zeolite structures are stabilized by water. Many natural zeolite minerals and a few synthetic varieties are known [23]. The zeolites have been widely used as catalysts for decades in industry. However, they have limited applicability for oxygen reduction because of the larger openings in the skeleton structure.

In 1976, Goodenough and Hong synthesized and reported a three-dimensional framework structure in the $Na_{1+x}Zr_2Si_xP_{3-x}O_{12}$ system which they named “skeleton” structure [24, 25]. Subsequently, the structure was called Na^+ super ionic conductor (Nasicon) and more recently as Li^+ super ionic conductor (Lisicon) when Na^+ was replaced by Li^+ in these framework structures. The Lisicon structures have high ionic conductivity (~10^{-3} Scm^{-1}) at ambient temperature, which is comparable to the conductivity of liquid electrolytes [20–22]. The zeolite, Nasicon, and Lisicon structures are similar with respect to their skeleton and rigid three-dimensional network; nonetheless, the sizes of the skeleton openings in these structures are different. The zeolites can accommodate and allow transport of larger molecules, hence being called a “molecular sieve.” The Nasicon and

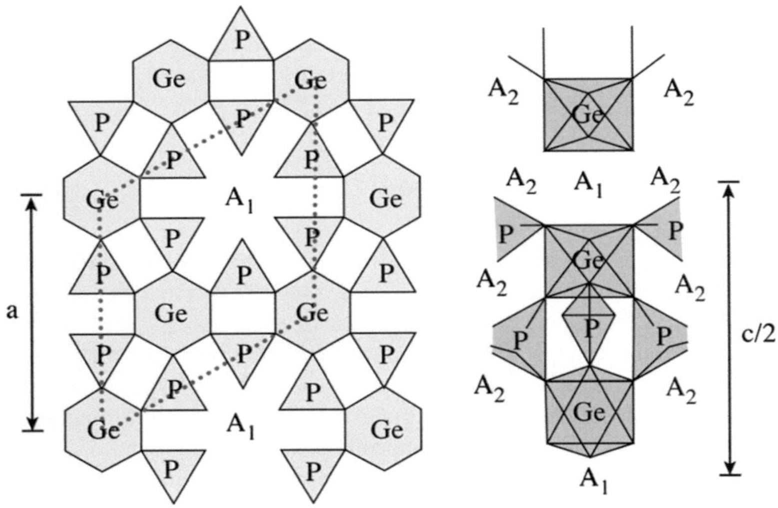

Fig. 9.2 Lisicon structure (not drawn to scale). The planar view on the *left* shows only A_1 sites. The three-dimensional view on the *right* shows both A_1 and A_2 sites. Reproduced with permission from Electrochem. Solid-State Lett. 14(6) A97–A99 (2011)

Lisicon structures are associated with smaller openings that can be stabilized by moving alkali ions.

LAGP glass–ceramic has received considerable attention during the last two decades as an electrolyte for lithium batteries [15]. The LAGP material belongs to the general class of the Lisicon structure. A Lisicon structure, $LiGe_2(PO_4)_3$, possesses a rhombohedral lattice (space group R3c) with an open three-dimensional framework of GeO_6 octahedra sharing all corners with PO_4 tetrahedra, such as shown in Fig. 9.2. The polyhedra are linked to their corners to form a $[Ge(PO_4)_3]^-$ rigid skeletal tunnel, and Li^+ ions migrate through the three-dimensional tunnel in the structure. Two different lithium ion sites (A_1 and A_2) exist in the $LiGe_2(PO_4)_3$ structure; however, only A_1 sites are occupied and A_2 sites are totally vacant. When a fraction of Ge is replaced by Al in the $LiGe_2(PO_4)_3$, a defect structure $Li_{1+x}Al_xGe_{2-x}(PO_4)_3$ results. In this case, some of the A_2 sites are occupied. At around $x = 0.5$, the defect structure provides the highest lithium ion conductivity [15]. The $Li_{1+x}Al_xGe_{2-x}(PO_4)_3$ material can be made either by the conventional solid-state reaction for ceramic processing or by the "glass–ceramic" route. The high lithium ion conductivity is a desirable property of the material that allows it to be used as a membrane. However, a fascinating attribute of the material—its ability to mediate oxygen reduction and evolution—has received little attention but is critical to the operation of solid-state $Li–O_2$ cells covered in this chapter.

The structure of LAGP in Fig. 9.2 clearly shows the existence of cavities and tunnels that can accommodate oxygen molecules. It is conceivable that in addition to the ability to transport Li^+, the Lisicon structure can also facilitate the catalysis

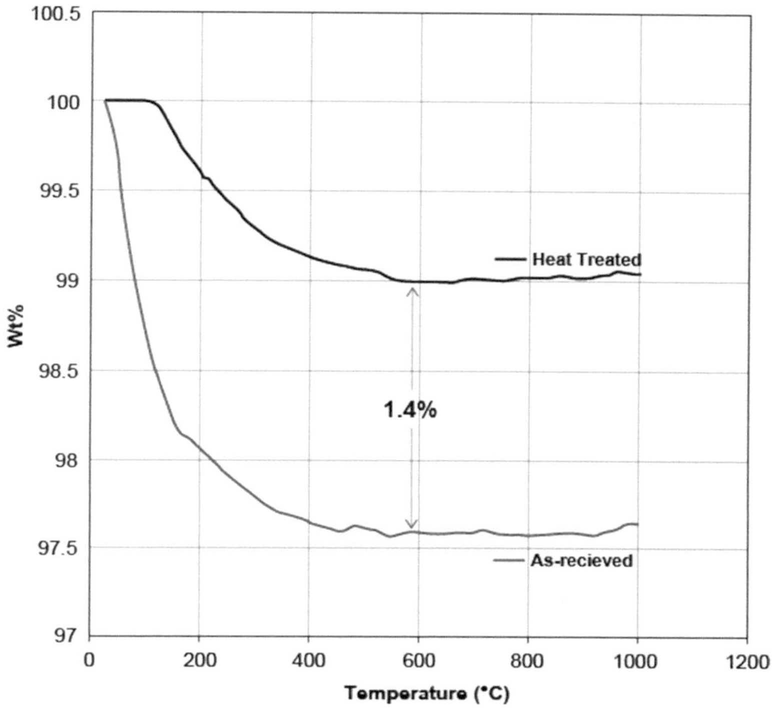

Fig. 9.3 Thermogravimetric analysis of as-received LAGP powder and heat-treated powder. Reproduced with permission from Electrochem. Solid-State Lett. 14(6) A97–A99 (2011)

of oxygen reduction and evolution. This conjecture was successfully verified and demonstrated by the use of a composite cathode made of LAGP, carbon, and PTFE in a Li–O_2 cell [8]. The experimental evidence of open space in the Lisicon structure is provided by the thermogravimetric analysis (TGA) and Brunauer–Emmett–Teller (BET) measurements. Figure 9.3 shows TGA measurements of two LAGP specimens—the first is as-prepared in the normal atmosphere and the second is as-prepared plus heat treated at 100 °C in a vacuum oven [26]. The first specimen was processed in the normal atmosphere, and therefore an interaction with or adsorption of atmospheric gases and a small concentration of impurity phase were expected. However, it is noted from Fig. 9.3 that the as-prepared specimen lost 2.4 wt% of weight when heated to 1,000 °C. Almost all the weight loss occurred below 500 °C. After the specimen was heat treated at 100 °C in a vacuum oven, the weight loss was reduced to 1.00 %. It is apparent that the second specimen lost 1.4 wt% by a heat treatment at 100 °C in a vacuum oven. The observations of Fig. 9.3 clearly point to the existence of free volume in the $Li_{1+x}Al_xGe_{2-x}$ $(PO_4)_3$ structure. Several BET measurements on the specimen revealed a pore size distribution from 0.90 to 2.80 nm and surface area of about 0.40 m^2/g.

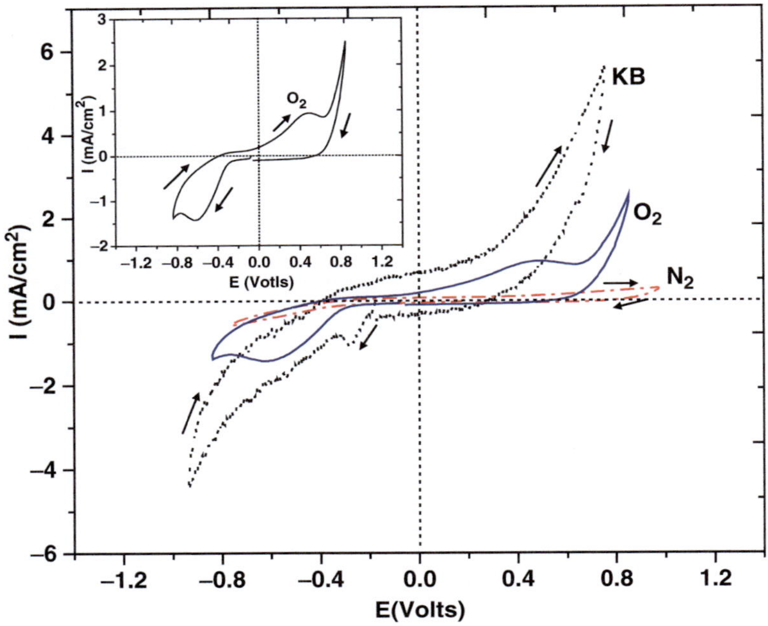

Fig. 9.4 Cyclic voltammogram from the LAGP (O_2 and N_2) and KB (O_2) specimens in 0.1 M KOH; sweep rate 5 mV/s, $T = 25$ °C. *Inset* shows the CV of LAGP (O_2). Reproduced with permission from Electrochem. Solid-State Lett. 14(6) A97–99 (2011)

If it is assumed that the weight loss determined by the TGA measurement arises from the adsorbed air (N_2, O_2, H_2O, CO_2), then the free volume is approximately 21 cc (0.024/0.00161[1]) per gram of the LAGP powder. This large free-volume LAGP powder can potentially be used to contain oxygen, catalyze its reduction, nucleate and accommodate the discharge product, and mediate oxygen evolution.

9.5.2 Electro-Catalysis by LAGP

Shown in Fig. 9.4 are cyclic voltammetry (CV) data of the LAGP and Ketjen black (KB) specimens with reference to the saturated calomel electrode (SCE, 0.242 V). In the case of LAGP specimen, 0.1 KOH electrolyte was bubbled with N_2 and O_2. No electrodic activity is noted when the KOH electrolyte is bubbled with N_2. Well-defined cathodic and anodic peaks located at −0.6 and 0.5 V, respectively,

[1] Density of air at 300 °K, Handbook of Physics and Chemistry.

were evident when the electrolyte was bubbled with O_2. It is noted that the KB curve exhibits a peak at 0.27 V. The LAGP peaks are distinctly different (inset, Fig. 9.4) in terms of their locations and significantly stronger in their intensities. These represent major electrodic activity corresponding to oxygen reduction and evolution.

The roles of LAGP material in cathodes of totally solid-state Li–O_2 cells were delineated as a function of cathode chemistry [16]. It was shown that the material possesses an inherent characteristic to adsorb oxygen molecules in its open structure. During discharge, the adsorption was followed by reduction of oxygen and subsequent reaction with lithium ion to form the discharge product.

It is justifiable to speculate on the nucleation site and accommodation of the discharge product in a cathode, which is primarily made of LAGP. In Fig. 9.2, two sites for lithium ions are shown. Even in highly conducting LAGP material, some A_2 sites are vacant in the skeleton structure. It is conceivable that the discharge product could be nucleated at these sites. The size of the mesopores in the LAGP ranges from 0.9 to 2.8 nm (BET measurement), which is adequate to accommodate the discharge product. During charge, the discharge product is dissociated to lithium and oxygen. Oxygen can be retained in the mesoporous structure, and lithium is transported under an influence of an electric field and plated on to the anode. The successful fabrication and characterization of more than 100 coin and pouch cells during a period of 5 years in the authors' laboratory justifies the proposed mechanism.

The aforementioned scenario for the nucleation and growth of the discharge product also illustrates challenges for its identification because it could be contained in the inaccessible mesopores of the LAGP. A positive identification of the product by commonly used analytical techniques such as X-ray diffraction and Raman spectroscopy has eluded our efforts.

9.6 Cell Reaction and Open-Circuit Voltage

In order to design a high-performance electrochemical cell based on Li–O_2 chemistry, one needs to have a basic understanding of the underlying reactions. The reversibility of reactions and the temperature range to carry out the charge and discharge must be established. The electro-catalyst facilitating reduction and evolution of oxygen at the cathode need to complement the cell discharge and charge reactions. The thermodynamic OCV characterizes a cell reaction. It is a fundamental parameter to define a cell operation and performance.

Three possible discharge products in the solid-state Li–O_2 cell can be foreseen: LiO_2, Li_2O_2, and Li_2O. The LiO_2 is an unstable oxide of lithium around the normal operating temperature range of the solid-state Li–O_2 cell; therefore it is of remote interest. However, it may be formed as an intermediate phase towards the formation of the final discharge product. The Li_2O_2 and Li_2O are of significant interest, and

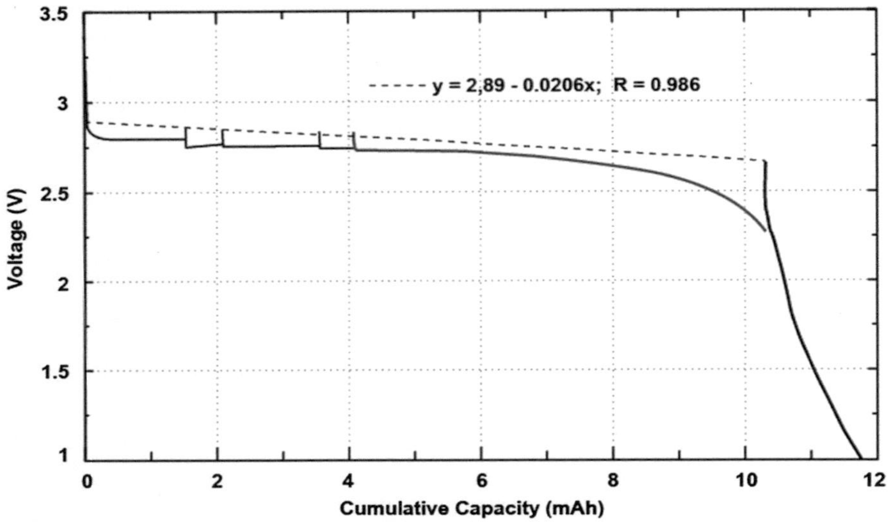

Fig. 9.5 Interrupted discharge curve of a typical Li–O_2 cell at 75 °C. The *curve* shows spikes after the discharge was interrupted. Extrapolation of the OCVs from the interrupted discharges yields a cell OCV of 2.89 V

both of these compounds have been suggested as the potential discharge product in the Li–O_2 cells [2].

The Li–O_2 cells fabricated according to the schematic of Fig. 9.1 have been characterized to delineate cell reaction and OCV. The OCV of the cells has been measured as a function of fabrication methods, thermal annealing, exposure to varying O_2 partial pressures, state of charge/discharge, and temperature. A typical cell subjected to an interrupted discharge with a current density of 0.05 mA/cm^2 is shown in Fig. 9.5. After fabrication, the OCV of the cells, both in a vacuum and at normal atmosphere, varied from 2.8 to 3.1 V. Annealing the cell at 75 °C and exposing it to varying O_2 partial pressures raised the OCV to 3.3 V. The OCVs of the typical cell at 75 °C after interrupted discharges show some spikes (Fig. 9.5) that correspond to relaxed cell voltages. The cell voltage dropped by about 0.5–0.6 V during the initial stage of discharge, which is attributed to the effects of impurities (H_2O, N_2) and electrode overpotential. After the interrupted discharges and cell relaxation, the OCVs exhibit a trend that may be characterized to be linear as a function of discharge capacity. The correlation coefficient of the linear regression for the relaxed OCVs is 0.986. An extrapolation of the relaxed OCVs to zero capacity yields a value of 2.89 V which is believed to be the equilibrium OCV representing the cell reaction. The OCV could represent the formation of either Li_2O or Li_2O_2 as the discharge product.

In an attempt to further characterize the discharge product, an open Li/PC(1 % Li_2O)/LAGP/PC(1 % Li_2O)/cathode coin cell was discharged with a current density of 0.1 mA/cm^2 at 75 °C in the oxygen atmosphere at normal pressure. The cathode

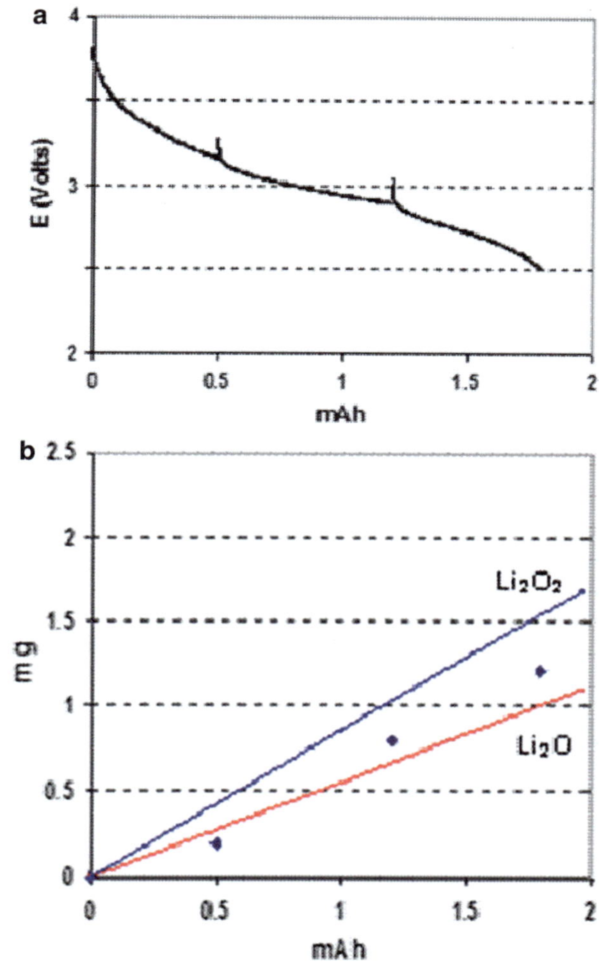

Fig. 9.6 Discharge (**a**) (0.1 mA/cm^2) profiles and (**b**) resulting weight gain of an open coin cell at 75 °C in oxygen atmosphere

of the cell contained LAGP (73.3 %), activated KB carbon (12.9 %), and PTFE (13.8 %) spread onto nickel foam substrate and subsequently sintered at 300 °C in an argon atmosphere. The cell was free from any liquid electrolyte or solvent. In effect, the cell was truly a solid-state cell. The discharge profile of the cell is shown in Fig. 9.6a. The spikes in the discharge curves correspond to the removal of the cell from the test fixture to measure its weight. Theoretical weight gains for Li_2O and Li_2O_2 as discharge reaction products are shown in Fig. 9.6b. The experimental weight gain of the cell resulting from the discharge is closer to the Li_2O rather than Li_2O_2. Initially, the weight gain of the cell lags behind the Li_2O curve. However, the cell weight exceeds the theoretical curve at later discharge stages. It is conceivable that a prolonged exposure of the cell to oxygen atmosphere may lead to physical adsorption of oxygen in the cell components and hence increased weight, but it remains to be analytically verified and proven. Based on the aforementioned interpretation, the remainder of the discussion will assume Li_2O as the discharge reaction product, though it remains a tentative identification. A positive identification of the discharged product by direct analytical methods has eluded our efforts. Concealment of the discharge product within the skeleton network of the LAGP is a probable explanation for the lack of positive identification.

9.7 A Practical Cell Design

The most energy dense reaction in the Li–O_2 system involves formation of Li_2O as the discharge product. The product Li_2O yields a specific capacity of 1,794 mAh/g. The theoretical energy density and specific energy of the Li_2O are 10.49 kWh/L and 5.22 kWh/kg, respectively. A practical cell design should therefore aim to accomplish the cell reaction that leads to the formation of Li_2O.

A source of lithium for the anode can be selected from the possible choices discussed earlier. From the energy density point of view, the choice could be narrowed down to lithium metal or an alloy containing very high concentrations of lithium. Its compatibility with membrane is critical. The high energy density requirement may be compromised, to some extent, to reconcile the compatibility issue. To fabricate a practical Li–O_2 cell, as will be discussed later, solid lithium metal is chosen as an anode.

The membrane is extremely important to keep the anodic and cathodic compartments ionically connected but electronically separated. The membrane should last for the entire life (years!) of the cell. This requirement is severe considering the chemistry of materials on both of its sides, electric field, and temperature excursions. The authors have mulled through this requirement in recent years. They have extensively evaluated glass–ceramic and polymer–ceramic materials for lithium battery applications. The research efforts may continue for a while before the final words are transcribed on membrane materials for Li–O_2 cells. At this stage of development, the choice has been narrowed down to LAGP and PEO:LiBETI (8.5:1)-LAGP (polymer–ceramic composite) supported onto a fiberglass mat.

To form the discharge product (Li_2O) in the cathode, a proper electrochemical environment must be maintained. For example, the nucleation sites for Li_2O need to be provided. If Li_2O nuclei are formed, they need to be grown and accommodated, and this will require free space. The capacity will depend upon the amount of Li_2O that can be accommodated within the cathode. Once the discharge is complete, the product should remain as Li_2O. Any parasitic reaction of the type shown by Eqs. (9.1) and (9.2) must be avoided because H_2O and CO_2 are atmospheric constituents and they can ingress into the cathode. During charge, oxygen gas evolves which may be allowed to escape to the ambient atmosphere, but this requires an open Li–O_2 cell configuration. The open configuration also

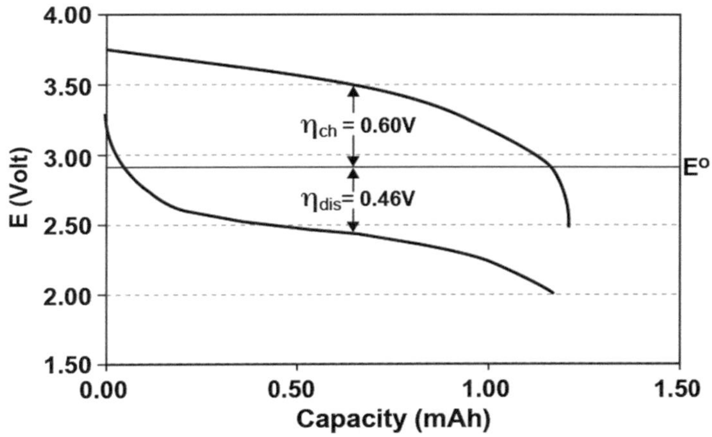

Fig. 9.7 Measured discharge–charge profiles of a sealed Li–O_2 cell operated at 75 °C with a current density of 0.5 mA/cm^2

allows an infinite source of reactants to carry out the parasitic reactions (Eqs. 9.1 and 9.2):

$$Li_2O + H_2O \rightarrow 2LiOH \quad (9.1)$$

$$Li_2O + CO_2 \rightarrow Li_2CO_3 \quad (9.2)$$

The lithium–O_2/air chemistry literature contains numerous proposals of an "open" Li–O_2 cell configuration. Investigations in the authors' laboratory have shown that reproducibility and cycle life are significantly compromised in such cells. In view of the limitations, the authors have pursued "sealed" Li–O_2 cell design, which requires oxygen containment commensurate with the capacity of the cell. The electrochemical performance data of such cells will be provided in the next section.

9.8 Electrochemical Performance of Sealed Cells

9.8.1 Electrical efficiency

A typical discharge–charge cycle of a sealed Li–O_2 cell is shown in Fig. 9.7. The cell consists of lithium metal foil as anode, a membrane laminate comprising of a tape cast, sintered LAGP encased with two polymer–ceramic layers (Fig. 9.1), and a porous cathode composed of LAGP, carbon, and PTFE composite spread onto nickel foam as the current collector. The initial open OCV of the cell is ≈3.3 V, which is significantly higher than the theoretical E^0 (2.89 V). This high OCV reflects mixed potential arising from impurities in the electrode. The high OCV drops rapidly as the cell is discharged. The working voltage of this cell during discharge is approximately in the range of 2.4–2.7 V, which is significantly less than the E^0 (≈2.89 V). The difference is the discharge overpotential, η_{dis}. The discharge capacity of the cell is approximately 1.2 mAh, which is primarily determined by the cathodic capacity for nucleating and accommodating the discharge product because the anode possessed excess capacity.

During constant current charging, the cell voltage increases to approximately 3.75 V (Fig. 9.7). The charge overpotential, η_{ch}, is significantly greater than the discharge overpotential, η_{dis}. The electrical efficiency for a discharge–charge cycle is 70 %. It is conceivable that further developments on cathode chemistry can reduce the overpotentials and increase the electrical efficiency. The LAGP material in the cathode possesses the electro-catalytic effect as discussed earlier. A systematic investigation and development of the Lisicon structures similar to LAGP may lead to further enhancement in the electrical efficiency.

9.8.2 Cell Cyclability

The component materials and physical constants of a typical coin cell #1 fabricated and characterized is presented in Table 9.1. Solid lithium metal was the anode. The membrane laminate consisted of three parts. The compositions of these parts are also presented in Table 9.1. The central part of the laminate is a fiberglass mat filled with a composite (LAGP and polymer mixture in powder form) and subsequently hot pressed at about 100 °C. The chemistry of the membrane laminate on the anode side is designed to passivate lithium and form a stable lithium–membrane interface. The polymer–ceramic layer on the cathode side is developed to facilitate the conduction of lithium ions and provide cathodic capacity in the cell through the use of 5 wt% of Li_2O. The Li_2O concentration is equivalent to a capacity of 4.70 mAh. The cathode is a composite of LAGP, carbon, and PTFE spread onto a nickel foam substrate and sintered at 320 °C in an argon atmosphere.

The cathode was saturated with a mixture of ethylene and propylene carbonate phase at 50 °C (liquid state). At ambient temperature, the mixed carbonate liquid phase (often used as solvents for lithium salts) transforms to a solid phase. It may be stated here that a cell without the carbonate phase in the cathode also functions as discussed in Fig. 9.6a, b but at a higher temperature or lower rate. The purpose of the mixed carbonate phase in the cathode was to enhance low temperature discharge and charge performance (high rate) of the cell.

Figure 9.8 shows the first cycle of charge and discharge profiles of the cell at 65 °C. The cell was initially charged with a current of 0.11 mA, and subsequently the charge current was increased to 1 mA to expedite the charging process. The cumulative charge capacity is approximately 2.8 mAh. There is a relatively linear and flat charge voltage which ranges from 3.4 to 3.6 V and it depends upon the

Table 9.1 Components, materials, and physical constants of a coin cell (area = 2 cm^2)

Cell #1						Cell #2			
Component		Thickness (μm)		Weight (mg)		Thickness (μm)		Weight (mg)	
Anode (lithium)		380		40		380		40	
Membrane laminate	P* (1 wt% Li_2O)	140	600	52	162	200	640	58	238
	Fiberglass mat supported 65 wt% LAGP + 35 wt% P*	320		158		310		122	
	P^* (5 wt% Li_2O)	140		52		P^*(30 wt% Li_2O) = 200		58	
Cathode	LAGP	410		61.6	170		460	67.4	172
	C			10.8				11.87	
	PTFE			11.6				12.7	
	$(EC + PC)^{\#}$			60.0				56.0	
	Nickel foam			80.0				80.0	

P^* = PEO:LiBETI (8.5:1); $(EC + PC)^{\#}$ = 90 wt% ethylene carbonate + 10 wt% propylene carbonate

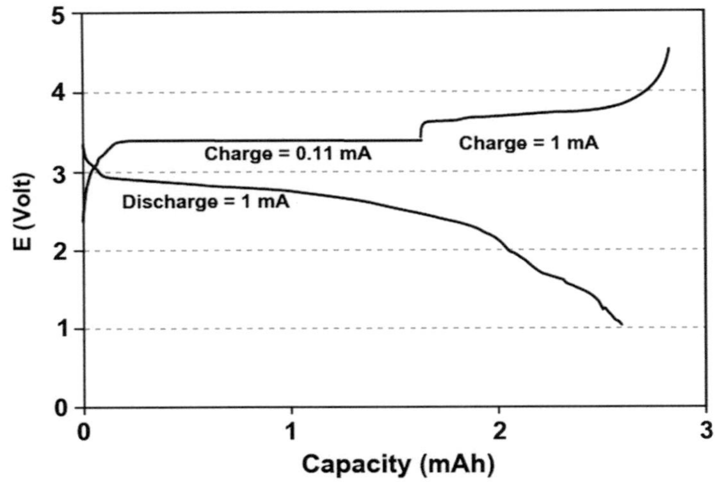

Fig. 9.8 First charge and discharge profiles of a coin cell at 65 °C

charge current. Towards the end of the charge, voltage increases rapidly and the charging is terminated at 4.5 V. It is apparent that the charge capacity is less than one would expect from the concentration of Li_2O on the cathode side (4.70 mAh). Only about 60 % of Li_2O is active during the charge process. The remaining 40 % Li_2O remains isolated and inactive. The discharge curve of Fig. 9.8 exhibits a major segment of the discharge taking place around 2.5 V. The curve drops towards the end of discharge and exhibits a moderate slope down to 1 V. The discharge capacity, approximately 2.7 mAh, is comparable to the charge capacity. The discharge capacity primarily depends upon the concentration of active material, O_2, on the cathode side in the sealed configuration.

9.8.3 Effect of Temperature

Figure 9.9 shows the discharge and charge profiles of the coin cell #1 at 45, 55, and 65 °C. The cell was discharged and charged with current densities of 1 and 0.5 mA/cm^2, respectively. It is apparent from Fig. 9.9 that the cell capacity decreases as the temperature is lowered. The decrease is primarily related to the Ohmic loss as evidenced by a significant increase in the slope of the discharge curves above 2 V. The capacity of the cell could not be reversed with a charge current density greater than 0.5 mA/cm^2.

At 45 °C, the cell was discharged with 1.0 mA/cm^2 to an approximate capacity of 1.5 mAh as shown in Fig. 9.10. Subsequently, the cell was subjected to ten discharge and charge cycles at 80 % of the initial capacity. The voltage loss between the first and tenth cycle of discharge is significant. The loss is more

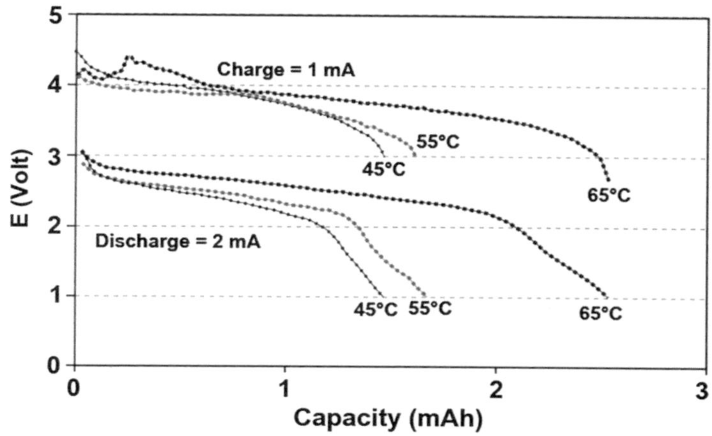

Fig. 9.9 Discharge and charge profiles of a solid-state, rechargeable coin cell at 45, 55, and 65 °C. The discharge and charge current densities were 1 and 0.5 mA/cm^2, respectively

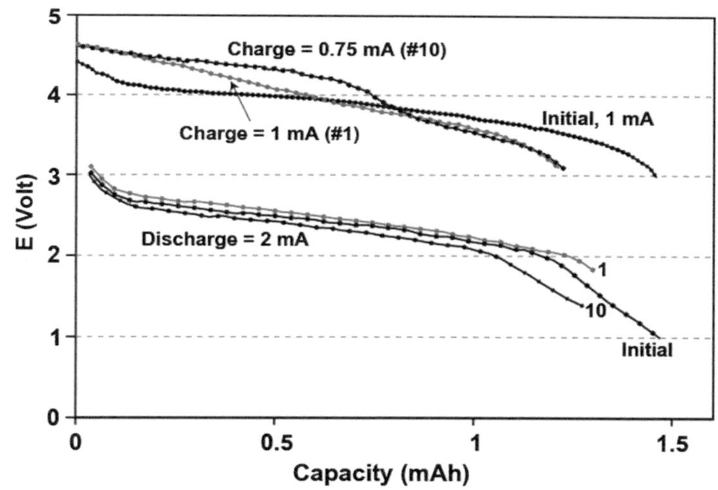

Fig. 9.10 Discharge and charge profiles of a solid-state, sealed coin cell at 45 °C. The cell was discharged with 1 mA/cm^2 and charged with currents from 0.375 to 0.5 mA/cm^2

significant at lower voltage (<2 V) discharges. A similar trend is reflected in the charge curves. The first charge was accomplished with 0.5 mA/cm^2, whereas the tenth charge used a current of 0.375 mA/cm^2. The voltage loss during charge reflects the declining Coulombic efficiency of the cell.

The cumulative cycle life of the coin cell #1 at 45, 55, and 65 °C with a varying discharge current was determined to be about 500 cycles. The cell degradation is noted during cycling. The nature of the parasitic electrochemical reaction leading to

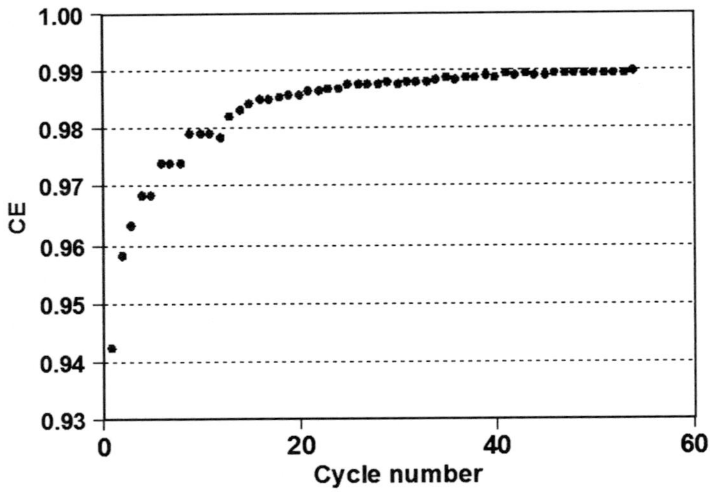

Fig. 9.11 Coulombic efficiency of a solid-state, rechargeable Li–O_2 cell at 65 °C when cycled at 80 % of its initial capacity

the degradation remains to be investigated. It is conceivable that a cycle life of hundreds could be extended (perhaps thousands) if the degradation mechanism could be delineated and corrective measures implemented.

9.8.4 Coulombic Efficiency (CE)

The CE of a cell is defined as the ratio of present discharge capacity to prior charge capacity. After the cell #2 (Table 9.1) was stabilized at 65 °C, its capacity was measured with a discharge current density of 0.5 mA/cm^2. Subsequently, the cell was cycled at 80 % of the initial capacity using both discharge and charge current density of 0.5 mA/cm^2. The current density corresponded to approximately a 0.4 C rate. The measured CE as a function of cycle number is shown in Fig. 9.11. The CE increases and stabilizes to about 0.99 with increasing cycle number up to 53. The CE of Fig. 9.11 reflects a loss in capacity attributed to parasitic reactions and the error arising from the experimental limitations. As such, the CE is potentially attractive for an experimental cell considering the fact that no special effort was made to purify the cell materials and components.

9.9 Prospect and Future Trends

The emerging fields of renewable energy technology such as solar and wind energies require efficient energy storage batteries. The batteries can be deployed in applications ranging from small, portable electronic devices to large transportation vehicles.

The battery is a bottleneck in the widespread, consumer-based uses of the renewable energy.

The Li–O_2 battery is an attractive technology to harness and expand the use of renewable energy. The electrochemical data of solid-state, rechargeable Li–O_2 cells presented and discussed in this chapter have highlighted its potential and justify future development. While aqueous and nonaqueous Li–O_2 chemistries are also being developed simultaneously, the solid-state battery presents advantages in terms of safety and life.

The Li metal of the solid-state cell is one of the most challenging components. Prior studies with liquid electrolytes have suggested that it tends to roughen and develop dendrites with cycling. The use of a dense glass–ceramic membrane addresses the issue to a large extent. The chemical compatibility of the glass–ceramic membrane with lithium is improved by laminating it with a polymer–ceramic film.

A reversible cell reaction in the Li–O_2 chemistry has been demonstrated. Understanding the formation and distribution of the discharge product in the cathode (most likely within the mesopores of LAGP) is critical for achieving high capacity, energy density, rate capability, and long cycle life. It will be illuminating to observe and understand the morphology of the discharge product during discharge and charge cycles. The microstructure and morphology of the discharge product must exhibit reversibility for a long cycle life cell. A morphology suggesting blocking of the lithium conduction path and diffusion of molecular and ionic oxygen may limit the cell performance.

The role of Lisicon-type ceramics and glass–ceramic in oxygen reduction and evolution deserves additional focus and attention for future research. Such an effort may lead to more efficient crystal structures for oxygen electrodes and improved electrical efficiency.

Acknowledgments The authors express sincere gratitude to the University of Dayton Research Institute and Air Force Research Laboratory (AFRL), Dayton, for sustained support over several years.

References

1. Armand M, Tarascon J-M (2008) Building better batteries. Nature 451:652–657
2. Abraham KM, Jiang Z (1996) A polymer electrolyte-based rechargeable lithium/oxygen battery. J Electrochem Soc 143(1):1–5
3. Dobley A, Morein C, Roark R, Abraham KM (2006) 42nd power sources conference, Philadelphia, 13 June 2006
4. Kuboki T, Okayama T, Ohsaki T et al (2005) Lithium-air batteries using hydrophobic room temperature ionic liquid electrolyte. J Power Sources 146:766–769
5. Beattie SD, Manolescu DM, Blair SL (2009) High-capacity lithium–air cathodes. J Electrochem Soc 156(1):A44–A47
6. Ogasawara T, De Bart A, Holzapfel M et al (2006) Rechargeable Li_2O_2 electrode for lithium batteries. J Am Chem Soc 128:1390–1393

7. Visco SJ, Nimon E, Katz B et al (2006) 42nd power sources conference, Philadelphia, 13 June 2006
8. Kumar B, Kumar J, Leese RL et al (2010) A solid state, rechargeable, long cycle life lithium–air battery. J Electrochem Soc 157:A50–A54
9. Wang D, Xiao J, Xu W et al (2010) High capacity pouch-type Li–Air batteries. J Electrochem Soc 157:A760–A764
10. Girishkumar G, McClosky B, Luntz AC et al (2010) Lithium–air battery: promise and challenges. J Phys Chem Lett 1:2193–2203
11. Chistensen J, Albertus P, Sanchez-Carrera RS (2012) A critical review of Li/air batteries. J Electrochem Soc 159(2):R1–R30
12. Bruce PG, Freunberger SA, Hardwick LJ et al (2012) Li-O_2 and Li-S batteries with high energy storage. Nat Mater 11:19–29
13. Kumar J, Rodrigues SJ, Kumar B (2010) Interface-mediated electrochemical effects in lithium/polymer-ceramic cells. J Power Sources 195:327–334
14. Kumar J, Kumar B (2009) Development of membranes and a study of their interfaces for rechargeable lithium-air battery. J Power Sources 194:1113–1119
15. Kumar B, Thomas D, Kumar J (2009) Space-charge-mediated superionic transport in lithium ion conducting glass-ceramics. J Electrochem Soc 156(7):A506–A513
16. Kumar B, Kumar J (2010) Cathodes for solid-state lithium–oxygen cells: roles of Nasicon glass-ceramics. J Electrochem Soc 157(5):A611–A616
17. MacCallum JR, Vincent CA (eds) (1987) Polymer electrolyte review 1 and 2. Elsevier, London
18. Gray FM (1997) Polymer electrolytes. The Royal Society of Chemistry, Cambridge
19. Kumar B, Rodrigues SJ, Scalon LG (2001) Ionic conductivity of polymer-ceramic composites. J Electrochem Soc 148(10):A1191–A1195
20. Fu J (1997) Superionic conductivity of glass-ceramics in the system Li_2O-Al_2O_3-TiO_2-P_2O_5. Solid State Ion 96:195–200
21. Fu J (1997) Fast Li+ ion conducting glass-ceramics in the system Li_2O–Al_2O_3–GeO_2–P_2O_5. Solid State Ion 104:191–194
22. Thokchom JS, Kumar B (2006) Microstructural effects on the superionic conductivity of a lithiated glass-ceramic. Solid State Ion 177:727–732
23. Venuto PB (1968) Advances in catalysis, vol 18. Academic, New York
24. Goodenough JN, Hong HY-P, Katalas JA (1976) Fast Na^+ ion transport in skeleton structures. Mat Res Bull 11:203–220
25. Hong HY-P (1976) Crystal structures and crystal chemistry in the system $Na_{1+x}Zr_2Si_xP_{3-x}O_{12}$. Mat Res Bull 11:173–182
26. Kumar B, Kichambre P, Rodrigues SJ et al (2011) Lisicon glass-ceramics mediated catalysis of oxygen reduction. Electrochem Solid State Lett 14(6):A97–A99

Chapter 10
Primary Lithium Air Batteries

Ji-Guang Zhang, Jie Xiao, and Wu Xu

10.1 Background

Abraham and Jiang [1, 2] first reported a lithium (Li) air battery based on a nonaqueous electrolyte in 1996. Because of various barriers in lithium air batteries, the interests and works in this field were very limited in the first decade following their discovery [3–7]. However, in recent years, the increased need for high-energy batteries for electric vehicles as well as the rapid advances in electrode materials and electrolytes has renewed intensive efforts in the development of lithium air batteries, and we have witnessed a rapid increase in the publications in this field [8–38]. For rechargeable lithium air batteries, electrolyte decomposition and the associated limited reversibility is one of the most important barriers, as has been discussed in other chapters of this book. This chapter will focus mainly on the basic principles and designs of primary lithium air batteries. The main barriers in primary lithium air batteries will be discussed, including their sensitivity to water, low power rate associated with air electrode blocking, and various side reactions. It should be recognized that all of the barriers in primary lithium air batteries also exist in rechargeable lithium air batteries; therefore, most of the fundamental principles and solutions discussed in this chapter can also be applied to rechargeable lithium air batteries.

Lithium air batteries have much higher specific energies than most currently available primary and rechargeable batteries [1, 2, 39–41]. In contrast with most other batteries that must carry both anode and cathode inside a storage system, lithium air batteries are unique in that the active cathode material (oxygen) is not stored in the battery. Instead, oxygen can be absorbed from the environment and reduced at catalytic surfaces inside the air electrode. When only Li is considered (oxygen will be absorbed from the air outside of the battery), it has a theoretical

J.-G. Zhang (✉) • J. Xiao • W. Xu
Pacific Northwest National Laboratory, Richland, WA 99352, USA
e-mail: jiguang.zhang@pnnl.gov

N. Imanishi et al. (eds.), *The Lithium Air Battery: Fundamentals*,
DOI 10.1007/978-1-4899-8062-5_10, © Springer Science+Business Media New York 2014

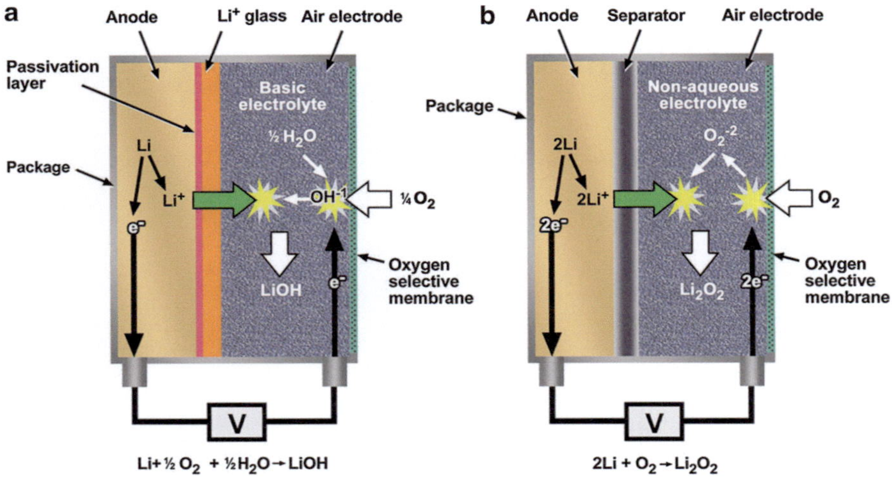

Fig. 10.1 Schematics of reaction processes in lithium air batteries. (**a**) Aqueous lithium air batteries with a Li-ion conducting glass and an intermediate layer that is stable with lithium. (**b**) Nonaqueous lithium air battery

specific energy of 11,972 Wh kg^{-1}. One of the most significant barriers to the development of lithium air batteries is the sensitivity of lithium anode to water. Two strategies have been developed to overcome this barrier. One is to use a nonaqueous electrolyte [1], and the other is to use a combination of an aqueous electrolyte and a water-impermeable Li-ion conducting glass [3]. The typical reaction processes in aqueous and nonaqueous-electrolyte-based lithium air batteries are shown in Fig. 10.1a, b, respectively. For lithium air batteries using an aqueous electrolyte, the electrolyte can be alkaline, acid, or neutral. For simplicity, only the process associated with alkaline electrolyte is illustrated in Fig. 10.1a. One important feature in a lithium air battery is that Li^+ is the ion carrier transferred from the anode to the air electrode. As a result, the reaction products are accumulated on the air electrode (cathode) side rather than on the anode side of the cell (as in the case of Zn–air batteries). This feature poses one of the most significant challenges for lithium air batteries, i.e., the blocking of the air cathode during the discharge process. This phenomenon increases with increasing current density and largely limits the power rate of lithium air batteries. Because of the water sensitivity of lithium air batteries, especially nonaqueous lithium air batteries, most current research on lithium air batteries is actually performed in pure oxygen environments, and they should be called Li–O_2 batteries. In this chapter, the words "Li–O_2 battery" and "lithium air battery" will be used interchangeably based on the convention used in the literature. "Lithium air battery" will be used exclusively when we discuss the lithium air batteries operated in ambient conditions.

Another approach to protect the Li electrode is to use a glass electrolyte (such as LISICON glass ($Li_{1+x+y}Al_xTi_{2-x}Si_yP_{3-y}O_{12}$) made by Ohara Inc. of Japan).

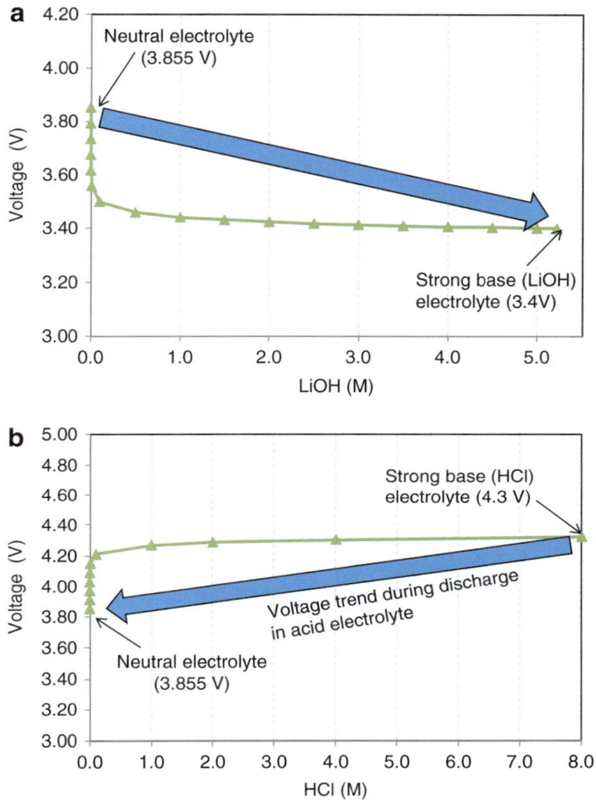

Fig. 10.2 Voltage of aqueous-electrolyte-based lithium air batteries as a function of (**a**) LiOH concentration in basic electrolyte and (**b**) HCl concentration in acidic electrolyte $\{V(x) = 4.268 - 0.059\,\text{pH (V)}\}$ [42]

However, this glass is not stable when in contact with Li metal. Visco et al. [3] of PolyPlus Battery Company first solved this problem by depositing a solid-state interfacial layer (e.g., Cu_3N, LiPON, etc.) between the Li metal and the LISICON glass, thus forming a protected Li electrode. Lithium air batteries using this protected Li electrode can be operated in both aqueous and nonaqueous electrolytes. Several groups (Read and colleagues [31], Zhou and colleagues [32], Yamamoto and colleagues [33]) also used an organic electrolyte as the interfacial layer between the Li metal and the LISICON glass, thereby forming a triple-electrolyte structure (organic electrolyte/LISICON glass/aqueous electrolyte). In the case of aqueous electrolytes, the voltage of the lithium air battery is a function of pH value of the electrolyte as shown in Fig. 10.2 $\{V(x) = 4.268 - 0.059\,\text{pH (V)}\}$ [42]. In the case of neutral electrolytes, the theoretical voltage is 3.86 V; in the case of very strong acid, the voltage can be as high as 4.44 V; in the case of strong base, the theoretical voltage can be as low as 3.40 V. Figure 10.2a shows the voltage of basic-electrolyte-based lithium air batteries as a function of the concentration of LiOH (which is formed during discharge). Figure 10.2b shows the voltage of acidic-electrolyte-based lithium air batteries as a function of the concentration of HCl (which is consumed during discharge). Most groups used basic solutions as the electrolytes in aqueous lithium air batteries. Figure 10.2a shows that the voltage of these batteries drops quickly during discharge, and the average voltage of these batteries is ~3.44 V. Similarly, the average voltage of acidic-based lithium air batteries is ~4.27 V. In practice, the operation of aqueous lithium air batteries is greatly limited by the stability of the glass electrolyte used in these batteries. Yamamoto and colleagues [33] investigated the stability of the Ohara glass in aqueous solution and they found that this glass is not stable when the pH value of the aqueous electrolyte is less than 4 or larger than 10. To operate aqueous lithium air batteries using a glass electrolyte, significant work has been done to stabilize the pH value of the aqueous electrolyte [33]. These efforts are discussed in more detail in another chapter of this book. Therefore, we will focus further discussion on the operation of nonaqueous primary lithium air batteries in this chapter.

Although lithium air batteries have great theoretical energy densities when only Li is considered, the practical specific energies of lithium air batteries are greatly reduced when other components of the batteries (e.g., the electrolyte and packaging) are considered, even if oxygen can still be absorbed from the environment. Zheng et al. [43] simulated both aqueous- and nonaqueous-electrolyte-based lithium air batteries in which the total weight of the Li, the carbon-based air electrode, and the electrolyte were considered. Their analysis showed that the theoretical maximum energy densities of lithium air batteries are ~1,300 and ~1,400 Wh kg^{-1} for lithium air batteries with basic and acidic electrolytes, respectively. For lithium air batteries with nonaqueous-based electrolytes, the estimate is ~2,790 Wh kg^{-1}. The main difference between these two kinds of lithium air batteries originates from the fact that the solvent is consumed in aqueous-electrolyte-based lithium air batteries, but it is not consumed in the nonaqueous-electrolyte-based lithium air batteries. For lithium air batteries, it is ideal to have the reaction products fully dissolved in the electrolytes, so the air electrodes will not be blocked. If the solubility of the reaction products (such as LiOH and LiCl) is also considered, the theoretical specific energy of the battery will be further reduced [43]. Table 10.1 summarizes the theoretical specific energy of lithium air batteries using different electrolytes. When more components other than Li and oxygen are included in the calculations, a more realistic view of the specific energy density of lithium air batteries can be obtained.

In general, the most critical components in a lithium air battery are its air electrode and electrolyte. The overall design of the battery also plays an important role in the energy density of the full battery. These three elements will be discussed in detail in this chapter.

10.2 Air Electrodes

The performance of lithium air batteries is influenced by many factors. First, the carbon materials that are used to construct the air electrode have different morphologies and physical properties which directly determine the oxygen

Table 10.1 Specific energy of Li-O_2 couple in different electrolytes saturated with reaction products

Li/O_2 reaction in different electrolytes	Electrolyte/ components included	Theoretical voltage (V)	Theoretical specific energy based on metal ($Wh\ kg^{-1}$)	Theoretical specific energy based on reactants (excluding O_2) ($Wh\ kg^{-1}$)	Theoretical specific energy based on reaction products ($Wh\ kg^{-1}$)	Specific energy based on full reaction ($Wh\ kg^{-1}$)	Comments
$Li + \frac{1}{2}O_2 \leftrightarrow \frac{1}{2}Li_2O_2$	Li/carbon electrode/ non-aqueous electrolyte	2.96	11,431	11,431	3,458	2,790	Carbon-based air electrode with a porosity of 70 %, filled with electrolyte
$Li + 0.25O_2 + 1.5H_2O \leftrightarrow LiOH{\cdot}H_2O$	Li/carbon electrode/ basic electrolyte with one bonded water	3.44	13,285	2,717	2,198	1,500	LiOH solubility is 5.3 M; used voltage value at 1 M LiOH, pH = 14
$Li + 0.25O_2 + HCl \leftrightarrow LiCl + 0.5H_2O$	Li/carbon electrode/ basic electrolyte with one bonded water	4.27	16,491	2,637	2,227	1,607	LiCl solubility is 23.2 M; used voltage value at 1 M LiCl, pH = 0

transportation and thus the deliverable capacity from the cell [4, 10, 13]. Second, even with the same type of carbon, the final air electrode design can also vary greatly depending on the carbon loading [10, 22] and the lamination approach (porosity) [43]; thus, the electrochemical behaviors of the air electrodes consisting of the same carbon are sometimes quite different. In addition, the testing conditions such as the oxygen partial pressure [24, 44] and humidity [43] also have significant effects on the energy densities obtained from the lithium air batteries. For example, Read et al. [44] found that the capacity of a Li–O_2 cell increases with the oxygen partial pressure only when the oxygen pressure is less than 1 atm. Above 1 atm the discharge capacity delivered from the cell does not show any significant changes, indicating a "saturation" is reached. Xia et al. [24] reported a capacity increase of 40 % when oxygen pressure was increased from 1 to 10 atm. According to the Young–Laplace equation, in order to take full advantage of the as-formed different diffusion tunnels, the oxygen pressure should be controlled between $2\sigma \cdot \cos\theta / r_{small} > P > 2\sigma \cdot \cos\theta / r_{large}$, where σ is the surface tension, θ is the wetting angle, and r_{small} and r_{large} are the effective radius of the small and large capillaries/channels, respectively. If pressure $P < 2\sigma \cdot \cos\theta / r_{large}$, all large channels will be flooded by the electrolyte, slowing down the oxygen-diffusion rate since the solubility of O_2 in the organic electrolyte is very limited. If $P > 2\sigma \cdot \cos\theta / r_{small}$, oxygen will enter the small pores, reducing the number of tri-phase (solid–liquid–gas) reaction regions. While it is difficult to quantitatively determine the optimum pore radius, the different results from the literature about the relationship between oxygen partial pressure and the delivered capacity can be partially explained by the various microstructures of the carbon-air electrode used in different groups.

Because oxygen has to be supplied to the interior carbon during discharge, the design of carbon electrodes with appropriate microstructures becomes critical in this system. The air electrode usually consists of carbon and a small amount of binder such as polytetrafluoroethylene (PTFE). For several years following Abraham's discovery [1, 2], work continued to focus on the primary nonaqueous lithium air batteries. Read [6, 7] investigated the effects of electrolyte and air cathode formulation on discharge capacity and rate capability. They found that electrolyte formulation has the largest influence on cell performance. Electrolytes with high oxygen solubility result in cells with high discharge capacity. The ability of an electrolyte to wet a particular carbon is also an important factor in determining cell capacity. Scanning electron microscopy (SEM) micrographs of air cathodes (see Fig. 10.3) reveal that the formation of reaction products can quickly block the oxygen-diffusion path, especially in high current-density conditions. Substantial improvement in the cell performance was achieved through electrolyte optimization. Increasing the oxygen partial pressure up to 10 atm also increases the discharge capacity, as shown by recent results from Xia and colleagues [24]. Apparently, the performance of lithium air batteries in ambient conditions (with an oxygen partial pressure of only 0.21 atm) needs to be investigated because the capacity of these batteries is only about one-fifth of that obtained in a pure oxygen environment [7].

Fig. 10.3 SEM micrographs of PTFE/Super P® air cathodes (air side of electrode). (**a**) Undischarged, discharged at (**b**) 0.05 mA cm^{-2}, (**c**) 0.2 mA cm^{-2}, and (**d**) 1.0 mA cm^{-2} (Reproduced from Read [7])

10.2.1 Surface Area and Pore Size/Volume of Different Carbon Candidates

The porous carbon electrode not only supplies electrons continuously for the electrochemical process, it also provides the required solid–liquid–gas tri-phase regions for the oxygen reduction. The pores and the surface of the carbon electrode are the hosts of the Li_2O_2 discharge product [13]; thus, they play a key role in determining the overall electrochemical performance of the Li–O_2 batteries and will be discussed in detail. The carbon surface area is related to the performance of Li–O_2 batteries because a larger surface area means more active sites for electrochemical reactions and also more catalysts can be loaded onto the porous carbon. However, detailed studies on the relationship between surface area and the capacity delivered from Li–O_2 batteries reveal that higher surface area does not always lead to a higher capacity. Instead, the discharge capacity is largely determined by the pore volume of the carbon, especially the amount of mesopores [10]. For example, in a 1 atm pure oxygen environment, Super P with only 62 m^2 g^{-1} surface area demonstrated a high capacity of more than 3,000 mAh g^{-1} carbon [45]. In the same pure oxygen environment, activated carbon, which has very high surface area (2,100 m^2 g^{-1}), exhibited much lower capacity than Super P, which has a large

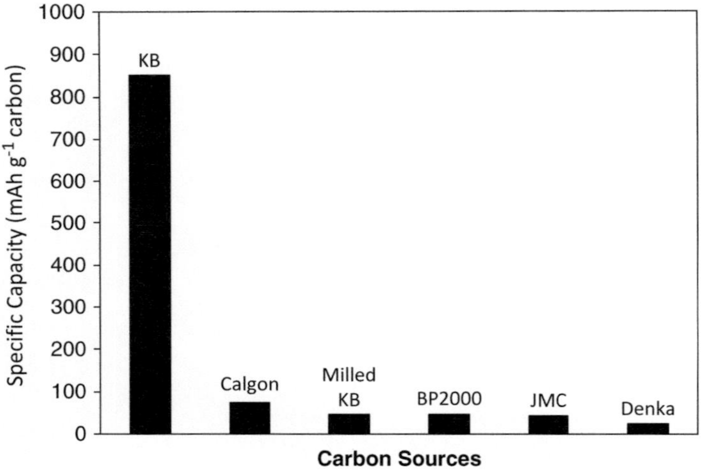

Fig. 10.4 Comparison of the discharge capacities of lithium air batteries using different carbons. All the coin cells are discharged to 2.0 V at 0.05 mA cm^{-2} and then held at 2.0 V until current (I) is less than $I/5 = 0.01$ mA cm^{-2} (Reproduced from Xiao [10])

pore diameter of 50 nm. This result indicates that the main factor that influences the capacity of the air electrode is related to the pore volume rather than the surface area of carbon used in the air electrode. A comparison of different commercial carbons along with lab-made mesoporous carbon (named JMC) with controlled mesopore size further revealed that the mesopore volume is the most critical factor directly influencing the capacity of lithium air cells [10]. This is consistent with the fact that KB has the highest total pore volume of 4.95 cc g^{-1}, most of which was contributed by mesopores. The highest discharge capacity was obtained from a lithium air battery with a KB air electrode, confirming the relationship between the mesopore volume and the capacity.

The discharge capacities of lithium air batteries based on different carbon sources are compared in Fig. 10.4. The specific capacities (mAh g^{-1} carbon) for each cell have the following order from high to low: KB > Calgon > ball-milled KB ≈ BP2000 > JMC > Denka. The excellent performance of KB-based lithium air batteries can be attributed to its high mesopore volume. After ball milling, the pore volume (0.4334 cc g^{-1}), specific capacity (47 mAh g^{-1}), and surface area (342.4 m^2 g^{-1}) of KB were reduced to 5.6 %, 5.5 %, and 12.8 % of their original values, respectively. This result clearly indicates a direct correlation between the mesopore volume of carbon and the specific capacity of lithium air batteries, especially when there is a large difference in pore volume between the carbon sources. It was found that although Denka black carbon has a larger pore volume (0.5355 cc g^{-1}) than that of JMC carbon (0.2376 cc g^{-1}), the lithium air batteries that used Denka carbon-based air electrodes exhibit a capacity much smaller than those that use JMC carbon-based air electrodes as shown in Fig. 10.3. It is expected that pore sizes that are either too large or too small will lead to less efficient use of

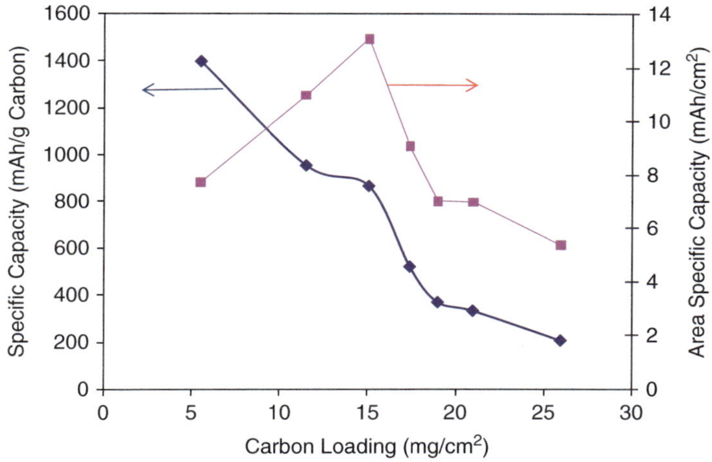

Fig. 10.5 Relationship between capacity and carbon loading. The left axis shows the specific capacity (mAh g^{-1} carbon) of the air electrode; the right axis shows the area-specific capacity (mAh cm^{-2}) of the same air electrode. All cells are discharged to 2.0 V at 0.05 mA cm^{-2} and then held at 2.0 V until the current density is less than or equal to 0.01 mA cm^{-2} in a dry air environment (Reproduced from Xiao [10])

mesopore volumes. Therefore, JMC carbon (which has a narrow distribution of mesopore size centered on ~3.5 nm) could be used more efficiently than Denka black carbon (which has a wide distribution of pore sizes).

It is well known that the specific capacity of a lithium air electrode depends strongly on the porosity of the electrode. The KB-based electrode was prepared by a dry rolling method in this aforementioned work. Because KB is a highly porous material with a chainlike structure, the thickness of the electrode can expand after it has been compressed under higher pressure. Therefore, no significant change in its porosity was observed for the electrode prepared under different loadings and different pressures (a constant ratio of KB:PTFE [85:15] was used in this work). The porosity of the KB-based electrode is determined mainly by its mesopore volume and highly springlike structure, especially after it is soaked with electrolyte, as we discuss later.

Most previous efforts have reported the performance of lithium air batteries in terms of mAh g^{-1} carbon. Although this value may increase with decreasing carbon loading, too small a carbon loading per unit area will lead to a small capacity per unit area of a practical lithium air battery. Considering the fact that the capacity of a lithium air battery is always proportional to the area of the air electrode exposed to the air, the capacity of the lithium air cells is proportional to both the specific capacity per gram carbon (mAh g^{-1}) and carbon loading per unit area (g cm^{-2}). Therefore, the product of these two parameters (i.e., the area-specific capacity [mAh cm^{-2}]) can be used as a more practical parameter to optimize the performance of the complete air electrode used in practical lithium air cells. Figure 10.5

compares the plots of both specific capacity (mAh g^{-1} carbon) and area-specific capacity (mAh cm^{-2}) of lithium air cells as a function of carbon loading (mg cm^{-2}). Although the specific capacity of the cells decreases monotonically with increasing carbon loading, a maximum area-specific capacity of 13.1 mAh cm^{-2} is found at a carbon loading of 15.1 mg cm^{-2}. When the carbon loading is less than 15.1 mg cm^{-2}, although a higher specific capacity (mAh g^{-1} carbon) can be obtained, the capacity per unit area of the lithium air cell decreases because less active material per unit area is used. When the carbon loading exceeds 15.1 mg cm^{-2}, the area-specific capacity of lithium air cell also decreases because a thicker electrode leads to a low utilization of the active carbon.

In fact, the importance of the mesopore volume of carbon in lithium air batteries can be well explained from the kinetic process including oxygen diffusion and reduction. During discharge, oxygen transports into the carbon electrode either through open channels in gaseous phase or through the electrolyte as dissolved oxygen [13]. However, by whichever route oxygen penetrates, the successful formation of Li_2O_2 only occurs on the triple junctions where electrolyte, carbon, and oxygen coexist. In other words, the more tri-phase regions, the more Li_2O_2 will be produced and therefore the higher the capacity obtained from the cell. If the pore size is too small and within the micro size range, the entrances of the micropores will quickly be blocked by either the electrolyte or Li_2O_2 preventing further access to the interior carbon surface. On the other hand, carbons with large macropores are not suitable for Li–O_2 batteries because those large pores are easily flooded by the electrolyte, reducing the generation of triple junctions. Furthermore, the volumetric energy density of the air electrode will also decrease by using macroporous carbons [8, 46]. From this point of view, KB carbon satisfies most of the requirements for the air electrode. Based on the chainlike structure [47] of KB, it was postulated that the KB expands significantly like a "spring" when soaked in the organic electrolyte, which causes the increased porosity of the wetted air electrode. Compared with other carbons, more tri-phase reaction points were generated during the expansion process of the air electrode, facilitating the reaction between Li^+ and O_2 on those triple junctions. Meanwhile, the newly formed pores can hold more discharge products (Li_2O_2). Carbon aerogels with different controlled porosities have also been prepared by changing the molar ratio of resorcinol to catalyst during gel preparation and pyrolysis conditions [48]. The corresponding capacities from those carbon aerogels demonstrate again that the mesopore volume is the key factor in determining the obtainable capacity from lithium air cells.

10.2.2 Microstructures of Air Electrode

The pore sizes/volume and porosity discussed above are the intrinsic properties of different carbons and can be tuned by various synthesis approaches [49]. However, the final pore size/volume and porosity of the air electrode can be very different from those of the staring carbon materials because of the lamination process and the addition of binder. Appropriate manipulation of the air electrode can even significantly increase the mesopore volume and porosity of the air electrode compared with their carbon precursors and thus should be discussed as well. Freestanding carbon films consist of nanotube/nanofiber have been used as an air electrode without any binder [50]. The discharge capacity is found to be strongly dependent on the thickness (carbon loading) as well as the current density, consistent with other reports on lithium air batteries using different carbons. The higher the current density, the faster the deposition of discharge product on the air electrode surface that is exposed to the air. The formation of this insulating layer not only increases the resistance of the whole cell, but more importantly it blocks the oxygen-diffusion path to the interior carbon, leading to incomplete utilization of the carbon electrode. This also explains why a thinner air electrode or a lower carbon loading has higher capacity since the specific capacity is calculated based on the weight of the total carbon material on the electrode.

A more recent work uses hollow carbon fibers grown on a porous ceramic substrate as the oxygen electrode [51]. The energy density from this lithium air battery is four times greater than that of the state-of-the-art lithium intercalation materials. The low carbon packing in the grown carbon-fiber electrodes yields a highly efficient utilization rate of the carbon and the void volume for Li_2O_2 formation. The interesting design of the nanofiber structure allows the direct observation of Li_2O_2 growth and disappearance on the sidewalls of the aligned carbon fibers, providing valuable information for understanding the intrinsic challenges existing in Li–O_2 batteries. Dual-layer carbon electrodes, which are widely used in fuel cells, were also built for Li–O_2 batteries. An active layer containing 2.5 % α-MnO_2 catalyst was stacked with another diffusion layer consisting mainly of carbon. In a pure oxygen environment the discharge capacity reached 3,100 mAh g^{-1} at a relatively high current density of 0.18 mA cm^{-2} because of the effective transport of oxygen through the diffusion layer to the interior active layer. The cell resistance from the as-designed Li–O_2 battery is recorded by impedance spectra, which indicate a gradual increase at different states of charge (SOC) due to the accumulation of reaction products. A similar hybrid air electrode is also designed by mechanically mixing V_2O_5, MnO_2, or CF_x with carbon in order to improve the rate capability of the cell [28].

Another popular carbon material, graphene, was recently reported by several groups as the air electrode for lithium air batteries. Although graphene sheets and composites have been extensively studied as electrode materials for lithium-ion batteries [52, 53], they typically form a two-dimensional (2D) layered structure that is not suitable for lithium air battery application. For example, even though a metal-free graphene-based air electrode showed a high discharge voltage comparable with that of a 20 wt% Pt catalyzed carbon electrode in an aqueous lithium air system, the reversible capacity is too low as a metal–air cell [54]. The reason for the very low capacity in the graphene air electrode can be attributed to the fact that most of the graphene sheets are stacked together without forming any pores, which are critical for oxygen diffusion. By reducing the graphene loading to only 0.3 mg in the air electrode, the capacity reaches 9,000 mAh g^{-1} based on the carbon weight [55]. However, it is

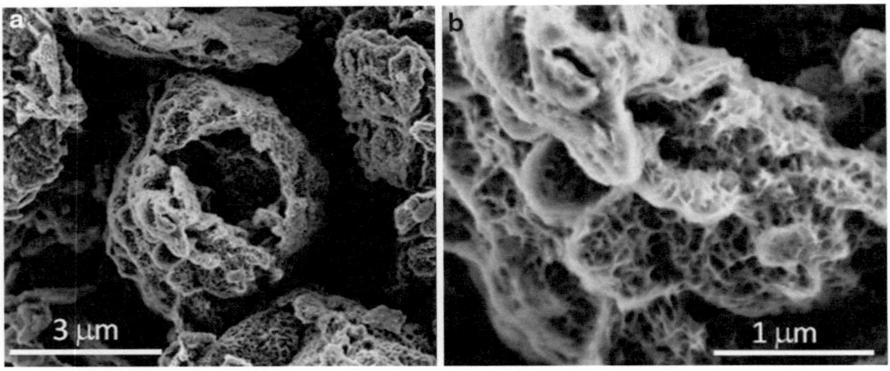

Fig. 10.6 Morphologies of the graphene-based air electrode. (**a**, **b**) SEM images of as-prepared functionalized graphene sheet (C/O = 14) air electrodes at different magnifications (Reproduced from Xiao et al. [56])

well known that the capacity of a lithium air cell is strongly dependent on the carbon loading [22, 24]. This is because the air permeability of the air electrode decreases with increasing carbon loading. Besides, as discussed earlier, because Li_2O_2 will be first deposited on the outermost side of the carbon electrode, the deposited Li_2O_2 will limit further oxygen penetration into the interior of the air electrode. Therefore, some of carbon in the air electrode cannot be accessed and fully used during discharge. The capacity based on the total weight of carbon will decrease with increasing carbon content and increasing current density.

In a recent work, Xiao et al. [56] demonstrated that through appropriate manipulation the graphene air electrode generated self-assembled, highly porous, hierarchical architecture of the air electrodes with interconnected pore channels as shown in Fig. 10.6 [56]. These architectures are not only useful for lithium air batteries but also important for many other energy-storage, conversion, and catalytic applications. Figure 10.6 shows that the functionalized graphene nanosheets aggregate into loosely packed, “broken egg” structures leaving large interconnected tunnels that continue through the entire electrode depth. These tunnels can function as numerous arteries that continuously supply oxygen into the interior of the electrode during the discharge process. This unique morphology of graphene air electrode is an ideal design for an air electrode. That is, during discharge the robust large tunnels can function as “highways” to supply oxygen to the interior parts of the air electrode, while the small pores on the walls are the “exits” which provide tri-phase regions required for oxygen reduction and provide a large space to store the reaction product without blocking the gas transport “highways.” This unique electrode structure greatly improves the capacity of the air electrode. Figure 10.7a shows the discharge curve of a Li–O_2 cell utilizing an air electrode (carbon loading: 2 mg cm^{-2}) prepared with functionalized graphene sheets (C/O = 14). The cell was first tested in pure oxygen at an initial pressure of ~2 atm. The discharge capacity reaches 15,000 mAh g^{-1} carbon with a plateau at around 2.7 V. The corresponding specific

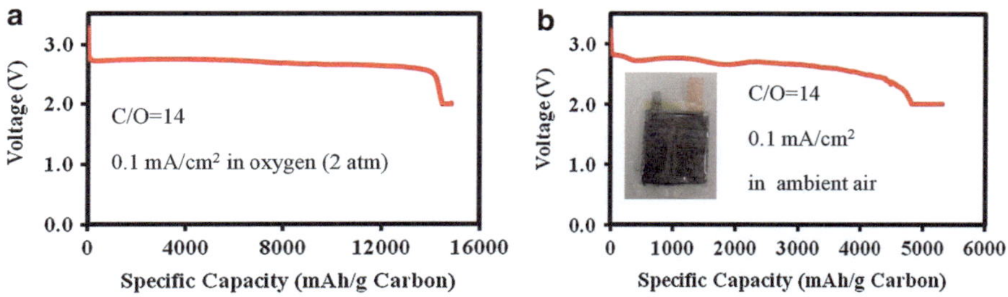

Fig. 10.7 Discharge curves of lithium air cells using graphene as the air electrode in (**a**) pure oxygen and (**b**) ambient environment (Reproduced from Xiao et al. [56])

energy is 39,714 Wh kg^{-1} carbon, giving an average voltage of 2.65 V. This is the highest capacity reported to date for nonaqueous Li–O_2 batteries tested under the same conditions. These graphene-based air electrodes were also tested in a pouch-type cell configuration (see inset of Fig. 10.7b) in which the polymeric pouch was used to regulate O_2 and H_2O ingress [9]. The pouch cell was operated in the ambient environment with an oxygen partial pressure of 0.21 atm and 20 % relative humidity. The resulting discharge curve for the cell is shown in Fig. 10.7b. At the same current density of 0.1 mA cm^{-2}, the specific capacity based on the carbon weight still exceeded 5,000 mAh g^{-1} which is double that of a KB-based air electrode tested in the same conditions.

It was further revealed, experimentally and theoretically, that Li_2O_2 prefers to nucleate and grow near the functionalized lattice defect sites on graphene with functional groups as a result of the stronger interaction between the deposited Li_2O_2 monomer at the 5-8-5 defects. More importantly, the aggregation of Li_2O_2 clusters in the vicinity of those defective sites is energetically unfavorable; therefore, the deposited Li_2O_2 would form isolated nano-sized "islands" on graphene, further enhancing smooth oxygen transport during the discharge process. Limited size or thickness of the reaction products, Li_2O_2, with preferred growth points is important for rechargeable lithium air batteries because it prevents continuous increase in electrode impedance and improves the efficiency of the cell. Those defects and/or functional groups also provide ideal locations to anchor the catalysts, which need to keep close contact with Li_2O_2 through repeated cycling.

In general, a good carbon candidate for lithium air or Li–O_2 batteries should have well-controlled pore sizes/volume, which can be performed during synthesis. The surface area of the ideal carbon should also be tailored to suppress electrolyte decomposition but promote reversible Li–O_2 reactions. On the other hand, the porous carbon can also be "produced" from nonporous two-dimensional carbons such as graphene nanosheets. The functional groups and/or defects on those carbons [57] (not only in graphene) can be modified to satisfy different energy requirements, and their interactions with Li_2O_2 are worth further investigation.

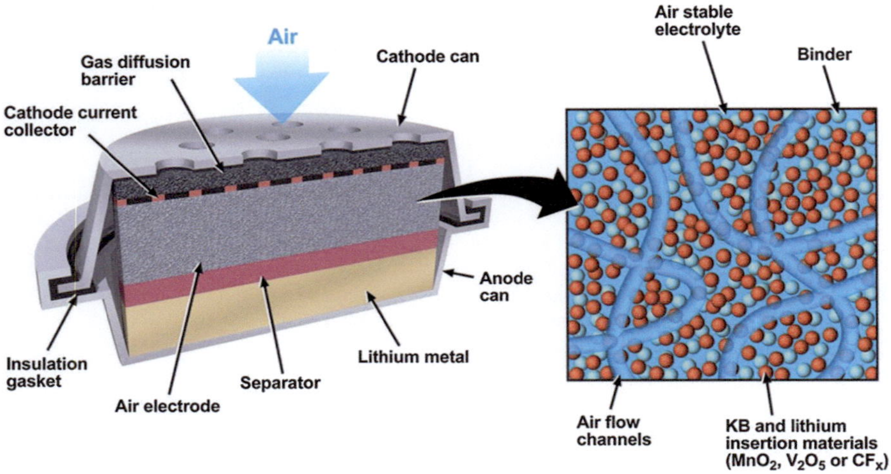

Fig. 10.8 Schematic of a lithium air hybrid battery based on coin-cell design

10.2.3 Modification of Air Electrode with Improved Energy/Power Density

As discussed earlier, the rate capability of a lithium air cell is limited due to the quick blocking of the carbon surface by the accumulation of Li_2O_2 particles. This leads directly to a low power density of the lithium air system and thus limits its practical use. In zinc air batteries, a hybrid power source consisting of a high-energy-density but low-power-density component (zinc air battery) is combined with another low-energy but high-power-density portion (Li-ion battery), which compromises the energy and power requirements. Similarly, this idea can be applied in lithium air batteries [28]. Figure 10.8 is a schematic of a lithium air hybrid battery. Besides KB carbon, another electrochemically active material such as MnO_2, V_2O_5, or CF_x was incorporated in the air electrode. For the MnO_2/KB air electrode, the catalyst effect from the commercial MnO_2 is not significant, and MnO_2 mainly works as a lithium insertion material with a discharge plateau similar to that of the lithium air battery. The lithium air battery using the V_2O_5/KB-based hybrid air electrode exhibits several distinct voltage plateaus and has a specific capacity greater than 800 mAh g^{-1} at a current density of 0.1 mA cm^{-2}, which is twice as large as those of pure KB-based lithium air batteries tested under similar conditions. The advantage of this hybrid battery design is more pronounced in CF_x/KB hybrid cells in which high discharge capacities of 950 and 521 mAh g^{-1} can be achieved at current densities of 0.1 and 0.2 mA cm^{-2}, respectively, as shown in Fig. 10.9. The high theoretical capacity, rate capability, and hydrophobic property of CF_x enhance the electrochemical performance of a conventional air electrode based on pure active carbon; the combination with KB, on the other hand, also

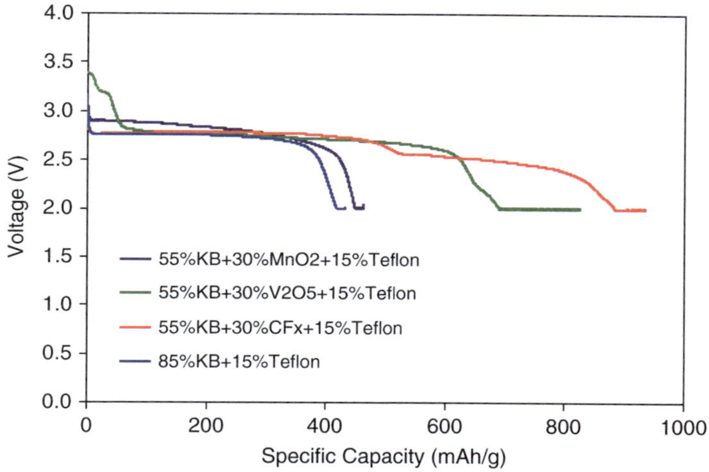

Fig. 10.9 Hybrid lithium air batteries discharged at 0.1 mA cm^{-2}. All the batteries were tested in dry air at ambient pressure (oxygen partial pressure = 0.21 atm). Capacities are expressed as per gram of total active mixtures in the air electrode (Reproduced from Xiao et al. [28])

improves the conductivity and utilization of CF_x. The CF_x:KB weight ratio in the air electrode can be tailored to satisfy the specific capacity and energy density requirements for different applications. When compared with pure carbon-based lithium air batteries, the batteries using hybrid air electrodes demonstrate significantly improved specific capacities and energies, especially for the CF_x-based hybrid lithium air batteries. Extremely hydrophobic CF_x also facilitates the formation of airflow channels in the carbon matrix and alleviates the air-electrode blocking problem during the discharge process. These hybrid air electrodes provide a promising approach to improve the energy density of lithium air batteries. Another interesting hybrid structure is the lithium/sulfur–oxygen battery which can operate either in the air or in a closed system [58]. The cathode is made of sulfur–carbon, within which the oxygen reduction reaction and the reaction between lithium and sulfur occur continuously. The capacity is improved from 650 mAh g^{-1} in a pure lithium air cell to 857 mAh g^{-1} in a Li–S/O_2 hybrid cell at 0.2 mA cm^{-2}. The self-discharge issue in Li–S batteries is also alleviated in the hybrid cell.

The most effective approach to improve the energy and power density is to identify appropriate catalysts that can accelerate the oxygen reduction and evolution reactions. However, due to the complicated oxygen reduction and evolution processes, the exploration of lithium air catalysts is still in its fundamental research phase. Unlike in the case of fuel cells where liquid water is the main reaction product and can be easily drained out, the discharge product (Li_2O_2) in lithium air batteries is in the solid state; therefore, it is much more difficult to identify and to use a dual-functional catalyst during repeated cycling. The presence of the catalyst in any reaction only changes the reaction pathway with lower activation energy; the overall thermodynamics are the same with or without a catalyst in the reaction.

Therefore, a catalyst that can promote the oxygen reduction reaction should also facilitate the oxygen evolution. However, because carbon (which is the dominant component in the air electrode) significantly narrows the electrochemical window of the electrolyte, especially in the presence of oxygen [59], the electrolyte decomposition during charging may occur concurrently or even before the oxidation of Li_2O_2 starts; therefore, the catalytic effect is usually not easy to observe in the charge process unless the oxygen evolution voltage can be reduced to be less than 4.0 V.

Several efficient methodologies have been established to systematically investigate the oxygen reduction/evolution reactions. For example, by using rotating disk electrode measurement [60, 61], the quantitative evaluation of the electrocatalytic activity of different catalysts is made feasible for Li–O_2 batteries. Based on the rotating disk electrode techniques, it was reported that Pt–Au nanoparticles worked as a highly active bifunctional electrocatalyst for rechargeable Li–O_2 batteries [61]. The cation of the conducting salts used in the electrolyte is also reported to strongly influence the reduction and subsequent oxidation of oxygen according to the results obtained from the complementary cyclic voltammogram and rotating disk electrode techniques. It was then expected that the mixture of Li and K and/or tetrabutylammonium salts as the supporting electrolytes may increase the solubility of the oxygen reduction product and thus promote the rechargeability of Li–O_2 cells [38].

Although an appropriate carbon material itself can function well in a lithium air battery at low current densities without catalyst, the improvement of the energy/power densities along with the development of truly rechargeable Li–O_2 cells still makes it urgent to explore or design different catalysts that can enhance the electrochemical behaviors of Li–O_2 batteries, especially for their high-power applications. Dobley et al. [25] and Debart et al. [34] investigated the effects of various metal catalysts on the performance of lithium air batteries. They showed that catalysts enhance the performance. Among the different metal catalysts studied, the manganese-catalyzed air cathodes gave the highest specific energy density. Shiga et al. [62] reported a new catalyst based on a donor–acceptor molecule in which an electron donor, D, having a porphyrin ring is connected to an electron acceptor, A, composed of a fullerene derivative with a conductive spacer between. An example of the donor–acceptor molecule is the triphenylporphyrinyl bithienyl *N*-methylpyrrolidino fullerene. A nonaqueous lithium air battery using this catalyst has been claimed to demonstrate a specific capacity of 14,560 mAh g^{-1} carbon [62]. However, the mechanism and the origin of such an exceptional capacity are not yet understood.

In summary, the performance of lithium air batteries strongly depends on both the meso- and macrostructure of its air electrode. The capacity of the air electrode increases with increasing mesopore volume of the carbon material used in the air electrode. In addition, the macrostructure of the air electrode has to be well designed and processed so both oxygen and lithium ions can have easy access through the entire electrode at different depths of discharge. A good catalyst is helpful for high rate operation of these batteries. Lastly, the gas transport and the Li–O_2 reaction rate in an air electrode depend not only on the electrode structure but also on the interfacial properties (or contact angle) between the air electrode and liquid electrode. This phenomenon will be discussed in the next section.

10.3 Electrolytes

Electrolyte is an indispensable component in any type of lithium air batteries. It plays a critical role in the reduction and oxidation reactions that occur at both cathode and anode and in the performance of the battery. Like in other electrochemical cells, electrolytes for lithium air batteries can also be divided into different categories by their components, states, and functions. Generally they can be classified as aqueous electrolytes and nonaqueous electrolytes. Although many nonaqueous electrolytes have been investigated and used in Li batteries in the last 40 years, most of these studies are of performance in an oxygen-starved environment. The chemical and electrochemical stability of these electrolytes need to be reinvestigated when they are used in an oxygen-rich environment, such as in lithium air batteries.

Among nonaqueous electrolytes, there are liquid electrolytes and solid electrolytes, or organic electrolytes and inorganic electrolytes. The currently accepted divisions of nonaqueous electrolytes are organic liquid electrolytes (including protic electrolytes and aprotic electrolytes), polymer electrolytes (including all-solid-state polymer electrolytes, polymer gel electrolytes, and polymer–inorganic composite electrolytes), and all-solid-state inorganic electrolytes. In recent years, ionic liquids (or room-temperature molten salts) have been studied as electrolyte solvents for lithium air batteries, but they can be included in the category of organic liquid electrolytes.

The electrolyte, especially the liquid electrolyte in an electrochemical device, is the only component that contacts every other component inside the device (such as cathode, anode, separator, cathode current collector and tab, anode current collector and tab, and the container or package). Therefore, the electrolyte must meet some basic requirements that are listed below:

- High chemical and electrochemical stability to both anode and cathode active materials, conductive carbons, binders, current collectors, separators, and packaging materials
- Wide voltage range
- High ionic conductivity
- High fluidity (i.e., low viscosity) if it is a liquid electrolyte
- Wide temperature range
- Safety
- Low cost
- Low toxicity and environmental benignity

In lithium air batteries, because the active cathode material, oxygen, is not stored inside the battery when the battery is assembled but drawn from the ambient air during discharge, lithium air batteries are open systems rather than sealed as are many other battery systems; therefore, the nonaqueous electrolytes used in lithium air batteries should have the following additional properties:

- Stable in oxygen-rich environment

- Stable with the discharged intermediates and products (i.e., O_2^-,$^1LiO_2$, Li_2O_2 and/or Li_2O for nonaqueous electrolytes, LiOH in base and neutral aqueous electrolytes, and Li salts in acidic aqueous electrolytes)
- Low volatility (or high boiling point)
- High oxygen solubility and oxygen diffusivity
- Capable of dissolving the cell discharge products

Various aqueous electrolytes, nonaqueous liquid electrolytes, and polymer gel electrolytes have been used for lithium air batteries in recent years. Lithium air batteries with nonaqueous electrolytes exhibit limited power capability because their discharge products have very limited solubility in the electrolytes even when some additives are used. These reaction products (mainly Li_2O_2, Li_2CO_3, and alkylcarbonates) will deposit on the surface of the air electrode and quickly block the pathways of oxygen diffusion deeper into the interior of the air electrode, especially at high discharge current rates. On the other hand, lithium air batteries with aqueous electrolytes may have much higher power capability because their discharge products are much more soluble in the aqueous solutions. For example, the solubilities of LiOH and LiCl are 5.3 and 23.2 M in aqueous solutions, respectively. Aqueous- and nonaqueous-electrolyte-based lithium air batteries have their own advantages and disadvantages. Use of aqueous electrolytes in lithium air batteries has been described in detail in other chapters of this book. In this chapter, we will focus more on nonaqueous electrolytes but will also cover the application of aqueous electrolytes in primary lithium air batteries and discuss methods to improve battery capacities by choosing appropriate electrolytes for different applications.

10.3.1 Nonaqueous Liquid Electrolytes

Read et al. [44] studied the effect of oxygen transport properties of several organic electrolytes on the performance of Li–O_2 batteries using Super P carbon-based air electrodes. In studies of electrolyte properties on the discharge performance of Li–O_2 batteries by Read and colleagues [7, 31], they also found that the electrolyte formulations had a large influence on the discharge capacity and rate capacity. Electrolytes with higher oxygen solubility and lower viscosity resulted in higher discharge capacity and better rate performance. Ether-based electrolytes had relatively lower viscosity than carbonate-based electrolytes, but both electrolytes had similar oxygen solubility. In ether-based electrolytes, the lithium salt with the larger anion showed slightly higher oxygen solubility than the salt with smaller anion due to the increase of ion pairing in the latter solution, which increased the electrolyte's dielectric constant. Although the salt with the larger anion had higher ionic conductivity because it dissociated more easily, it still resulted in higher electrolyte viscosity and showed much lower discharge capacity. When the electrolyte viscosity increases, the oxygen diffusivity decreases. Thus, it seems that oxygen solubility

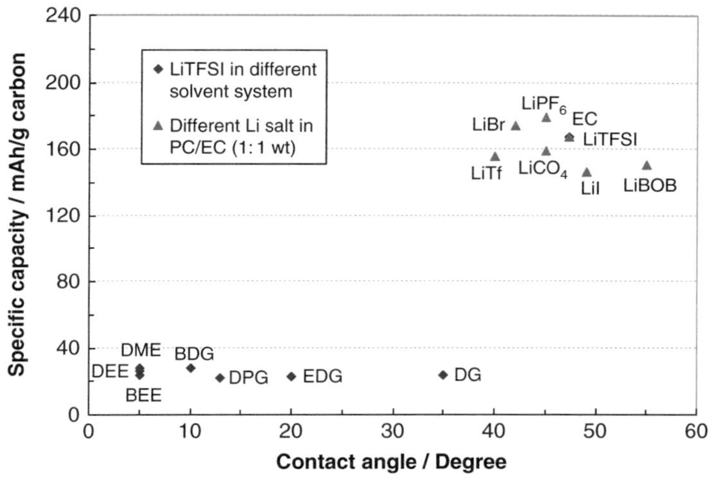

Fig. 10.10 Effect of the contact angle of electrolyte at the carbon surface of the air electrode on the discharge capacity of lithium air batteries tested in dry air. *Diamonds*: 1.0-M LiTFSI in different solvent mixtures of PC with another cosolvent at 1:1 weight ratio. *Triangles*: various 1.0-M lithium salts in PC-EC (1:1 wt) (Reproduced from Xu et al. [30])

is no longer an important factor in determining the discharge capacity of a Li–O_2 battery once it reaches a certain level, after which the electrolyte viscosity becomes the dominant factor.

Xu et al. reported on the selection and optimization of nonaqueous electrolytes for practical primary lithium air batteries [29, 30]. They compared a series of properties that included evaporation rate, moisture absorption rate by solvents and electrolytes, chemical stability of solvents and electrolytes with Li metal, viscosity, ionic conductivity, oxygen solubility in the electrolytes, and contact angle of the electrolyte on the air electrode. In their studies, various air-stable electrolytes, especially the electrolytes with low vapor pressure and low moisture sensitivity, have been investigated. They found that diethylene glycol dibutyl ether (i.e., butyl diglyme, BDG) and propylene carbonate (PC) lost less than 5 % weight in 2 months when tested inside the glove box at 29 °C. Among the various solvent systems investigated, a mixture of PC and ethylene carbonate (EC) was found to be the most effective solvent system and lithium *bis*(trifluoromethanesulfonyl)imide (LiTFSI) the best salt for ambient operation of lithium air batteries using Darco® G-60 activated carbon for the air electrode, the performance of which was relatively insensitive to PC:EC ratio and salt concentration. The performance of the lithium air battery was determined by the number of tri-phase regions formed by oxygen, electrolyte, and active carbons (with catalyst). The oxygen-diffusion rate (strongly related to the polarity of the electrolyte) through the open channels was several orders of magnitude larger than that through the liquid electrolyte; therefore, they found that the electrolyte polarity, i.e., the wettability of the electrolyte to the carbon-based air electrode, was the most important factor on the performance of lithium air batteries as compared with other properties of the electrolyte, such as oxygen solubility, viscosity, and ionic conductivity. Figure 10.10 shows the effect of contact angle of the electrolyte at the electrode surface on the discharge capacity of lithium air batteries when they were tested in dry air. All of the samples that exhibit a contact angle (between electrolyte and air electrode) of less than 35° show very limited specific capacities. On the other hand, all of the electrolytes with a contact angle between 40° and 60° have demonstrated significantly improved specific capacities. Therefore, the electrolyte polarity (or its wettability to the air electrode) is an important factor in the performance of lithium air batteries.

Xu et al. also reported that the discharge performance of a lithium air battery could also be improved by the addition of certain amounts of organic additives or cosolvents [63]. Crown ethers could significantly affect the discharge performance of nonaqueous electrolytes in lithium air batteries. A small amount ($<$10 wt%) of crown ethers reduced the battery performance, and a minimum discharge capacity appeared at ca. 5 wt% of crown ether in the electrolytes. When the content was increased to 15 wt%, both 12-crown-4 and 15-crown-5 improved the capacity of lithium air cells by about 28 % and 16 %, respectively. The 15-crown-5-based electrolytes even showed a maximum discharge capacity in the crown ether content range of 10–15 wt%. However, increasing the amount of 18-crown-6 in the electrolytes continuously lowered the cell performance. Besides the crown ethers, *tris*(pentafluorophenyl)borane (TPFPB) was also studied as a functional additive and cosolvent in electrolytes. It was reported that although TPFPB could help partially dissolve Li_2O and/or Li_2O_2 formed during battery discharge, which was beneficial for increasing the discharge capacity, this addition also greatly reduced the contact angle and increased the electrolyte viscosity, thus decreasing the overall discharge capacity of the lithium air battery.

Zhang et al. [64, 65] reported using partially fluorinated solvents as a cosolvent in the nonaqueous electrolytes for lithium air batteries. Three fluorinated solvents were tested in the mixtures with PC and $LiSO_3CF_3$ in the ambient environment, including methyl nonafluorobutyl ether (MFE), *tris*(2,2,2-trifluoroethyl) phosphate (TTFP), and *tris*(2,2,2-trifluoroethyl) phosphate (TFP). Results showed that at certain cosolvent:PC ratios, all three partially fluorinated solvents were able to increase the specific capacity of carbon in Li–O_2 and lithium air batteries. With TTFP, they found that addition of 30 wt% (vs. solvent) into a 0.2 M (molality) $LiSO_3CF_3$-PC electrolyte could significantly improve the discharge performance of lithium air batteries. With TFP, the best discharge performance was found at 3:2 (by weight) PC/TFP for lithium air batteries but 7:3 (by weight) PC/TFP for Li–O_2 batteries, where the salt concentration was still 0.2 M $LiSO_3CF_3$. These optimized electrolytes had significantly higher performance in specific capacity and rate capability than the control PC electrolytes. They also reported that the discharge performance of the lithium air batteries with the partially fluorinated electrolytes increased with increasing discharge current density. Such improvement was believed to be due to the increased dissolution kinetics and oxygen solubility in these partially fluorinated solvents and electrolytes. The low volatility of the partially fluorinated phosphate solvents was also favored for the long-term operation of lithium air batteries in dry ambient environments.

Crowther et al. [66] reported that the addition of methoxybenzene at 33 % by volume as a cosolvent to the electrolyte of 1.0 M $LiBF_4$ in PC-dimethyl carbonate (DMC) (1:1 by volume), i.e., PC:DMC:methoxybenzene of 1:1:1 by volume, could significantly improve the discharge capacity of pouch-type lithium air batteries with KB EC-660G air electrodes. When discharged to 1.5 V at 0.2 mA cm^{-2} in a pure O_2 atmosphere at 101 kPa (i.e., 1 atm), the discharge capacity was over 3,158 mAh g^{-1} of carbon for the Li–O_2 cell with a 100 cm^2 cathode and methoxybenzene, an increase of 46.6 % over the cell without methoxybenzene. When the cathode area was reduced to 10 cm^2, the discharge capacity of the Li–O_2 cell with methoxybenzene was about 4,680 mAh g^{-1} and had an increase of 45.3 % compared to the one without methoxybenzene. The increase in discharge capacity was due to the increase in solubility of the discharge product (Li_2O_2 and/or Li_2O) in methoxybenzene.

Zhang et al. [67] also studied the effect of quaternary ammonium as an additive or a co-salt on discharge characteristics of Li–O_2 batteries. They chose tetrabutylammonium triflate ($NBu_4SO_3CF_3$) as the quaternary ammonium salt, a Super P carbon for the air cathode, and a solution of 0.2 mol kg^{-1} $LiSO_3CF_3$ in PC/dimethyl ether (DME) (1:3 by wt) as the baseline electrolyte. The results showed that addition of $NBu_4SO_3CF_3$ increased the discharge capacity of the Li–O_2 battery only in the concentration range from 5 to 50 mol% vs. the total of Li and ammonium mixed salts. The optimum concentration was about 5 mol%, and this amount in the baseline electrolyte could increase the discharge capacity of the Li–O_2 battery from 732 to 1,068 mAh g^{-1} based on the weight of Super P carbon and at a discharge current of 0.2 mA cm^{-2}. However, the addition of this quaternary ammonium salt led to slightly lower discharge voltage compared with the baseline electrolyte because the larger molar volume of quaternary ammonium cations led to less deposition of oxygen reduction products on the carbon surface. Wang et al. [68] reported that the discharge rate capability of a lithium air battery could be greatly improved by adding perfluorinated compounds commonly used as blood substitutes as oxygen carriers. Perfluorotributylamine (FTBA) as an additive in $LiPF_6$–PC electrolyte. 0.5 % of FTBA in the control electrolyte could enhance the O_2 solubility in nonaqueous electrolytes, thus substantially improving the discharge rate capability of the gas diffusion electrode for lithium air batteries. Discharge current rates of 1 and 5 mA cm^{-2} were demonstrated in a lab-scale Li–O_2 battery. Although such perfluorinated compounds have very low solubility in regular electrolytes, this study does open a new approach to improve the dynamic performance of lithium air batteries, which can make the lithium air chemistry feasible in practical applications. Besides the above nonaqueous electrolytes, the electrolytes reported for rechargeable lithium air or Li–O_2 batteries can also be used in primary lithium air batteries, such as tetraethylene glycol dimethyl ether (TEGDME) with $LiPF_6$ reported by Laoire et al. [38] and oligoether-functionalized silane electrolytes reported by Zhang et al. [69]. The latter electrolyte based on tri(ethylene glycol)-substituted trimethylsilane (1NM3) showed a slightly lower discharge capacity than the PC-based electrolyte.

10.3.2 Polymer and Solid Electrolytes

In 1996, Abraham and Jiang reported the first rechargeable Li–O_2 battery based on a polymer electrolyte [1]. The Li^+ conductive polymer electrolyte membrane was prepared from the polymer gel electrolyte comprising polyacrylonitrile (PAN), EC, PC, and $LiPF_6$ with a weight ratio of 12:40:40:8. The Li–O_2 battery using this PAN-based polymer gel electrolyte had a discharge capacity of 1,410 mAh g^{-1} of carbon at 0.1 mA cm^{-2} discharge current and a discharge voltage of about 2.5 V. Abraham et al. [23] also reported Li–O_2 batteries using other polymer electrolytes based on poly(vinylidene fluoride)-hexafluoropropene (PVdF-HFP) copolymer and a liquid electrolyte of different lithium salts (such as $LiSO_3CF_3$, LiTFSI, or $LiPF_6$) in oligomeric poly(ethylene glycol) dimethyl ethers (PEGDME) with molecular weight from 250 to 500. A Li–O_2 battery containing the polymer electrolyte of 30 % PVdF 2822, 60 % PEGDME 400, and 10 % LiTFSI had an open-circuit voltage of 2.95 V and an average discharge voltage of ca. 2.5 V when discharged at 0.1 mA cm^{-2}. The discharge capacity of 1,262 mAh g^{-1} of carbon was obtained at the voltage cutoff of 2.0 V.

Mohamed et al. [70] reported epoxidized natural rubber-based polymer gel electrolytes for lithium air batteries. The polymer gel electrolytes, composed of 50 % epoxidized natural rubber polymer with $LiCF_3SO_3$, were prepared by the solution casting technique and plasticized with EC or PC. The highest room-temperature conductivity of 4.92×10^{-4} S cm^{-1} was obtained when 10 wt% PC was added. The lithium air batteries containing such polymer gel electrolytes had a midpoint voltage of about 2.7 V for PC and 2.5 V for EC and specific energies of 29.7 and 24.8 Wh kg^{-1} of the entire cell weight for PC and EC, respectively. Zhang et al. [71] studied the polymer composite electrolyte composed of hydrophobic ionic liquid (1,2-dimethyl-3-propylimidazolium *bis*(trifluoromethanesulfonyl) imide) (PMMITFSI), silica, and PVdF-HFP for lithium air batteries. The composite based on PVdF-HFP, LiTFSI-PMMITFSI, and hydrophobic silica at a weight ratio of 27:70:3 had a room-temperature ionic conductivity as high as 1.83×10^{-3} S cm^{-1}. Lithium air batteries using this composite electrolyte membrane in ambient atmosphere had an initial capacity of 2,800 mAh g^{-1} of carbon in the absence of O_2 catalyst but 4,080 mAh g^{-1} of carbon when the catalyst α-MnO_2 was added; the latter discharge capacity of the cell was calculated as 2,040 mAh g^{-1} for the total mass of the cathode. A flat discharge plateau and large discharge capacity indicated that the hydrophobic ionic liquid–silica–PVdF-HFP polymer composite electrolyte membrane could effectively protect the Li metal anode from moisture invasion.

A solid-state Li–O_2 battery has been reported by Kumar et al. [20]. The battery consists of a highly Li-ion-conductive solid electrolyte membrane and a solid-state

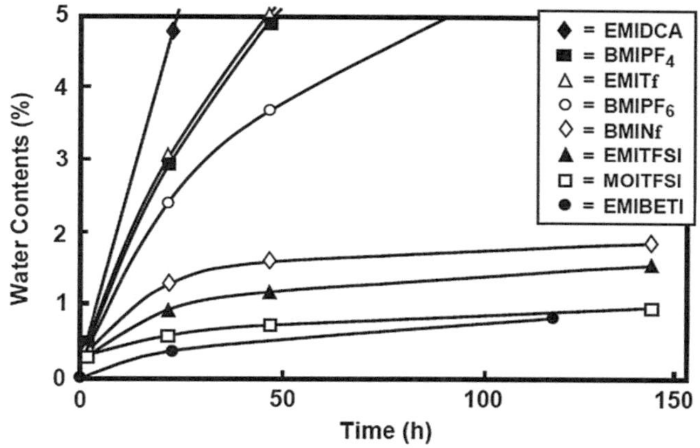

Fig. 10.11 Water contents of ionic liquids in high humidity condition (20 °C, 90 % relative humidity). *Filled circles*, EMIBETI; *open squares*, MOITFSI; *Filled triangles*, EMITFSI; *open diamonds*, BMINf; *open circles*, $BMIPF_6$; *open triangles*, EMITf; *filled squares*, $EMIBF_4$; *filled diamonds*, EMIDCA (Reproduced from Kuboki et al. [4])

composite air cathode. This solid electrolyte membrane laminate was prepared from glass-ceramic and polymer–ceramic materials in which dried poly(ethylene oxide) (PEO), $LiN(SO_2C_2F_5)_2$ (LiBETI), Li_2O, and boron nitride were employed, and the solid-state composite air cathode was prepared from high surface area carbon and ionically conducting glass-ceramic powder. These solid-state $Li–O_2$ batteries exhibited excellent thermal stability and rechargeability at temperatures up to 105 °C. Forty charge–discharge cycles at current densities ranging from 0.05 to 0.25 mA cm^{-2} were obtained. However, their room-temperature performance is limited due to the use of PEO in the composite electrode.

10.3.3 Ionic Liquid Electrolyte

Electrolyte evaporation in metal–air batteries is a serious concern for their long-term operation. Although some liquid electrolytes are relatively stable in the ambient environment [29, 30], eventual loss of solvent will still shorten the lifetime of batteries. Kuboki et al. [4] investigated the use of hydrophobic ionic liquid electrolytes in lithium air batteries. The changes in the water contents of the ionic liquids in the temperature and humidity chamber (20 °C, 90 % relative humidity) are shown in Fig. 10.11. $BMIPF_6$, 1-ethyl-3-methylimidazolium *bis*(trifluoromethylsulfonyl)amide (EMITFSI), 1-ethyl-3-methylimidazolium *bis*(pentafluoroethylsulfonyl)imide (EMIBETI), and 1-methyl-3-octylimidazolium *bis* (trifluoromethylsulfonyl)amide (MOITFSI) were particularly immiscible with moisture, and the water contents after 100 h were 1.7, 1.4, 0.8, and 0.7 wt%, respectively. Among several 1-alkyl-3-methyl imidazolium cations and perfluoroalkylsulfonyl imide anions investigated, 1-ethyl-3-methylimidazolium *bis* (trifluoromethylsulfonyl)amide, which has high conductivity, showed the best electrolyte performance. The cell operated for 56 days in air, and the cathode carbon materials showed a high discharge capacity of 5,360 mAh g^{-1}.

Ye et al. [72] reported Li-ion-conducting polymer gel electrolytes based on ionic liquid/PVdF-HFP blends for Li–O_2 battery application, in which the ionic liquids were thermodynamically compatible with Li metal. A Li–O_2 battery using a polymer gel electrolyte based on 1 M LiTFSI/1-methyl-3-propyl-pyrrolidinium *bis*(trifluoromethanesulfonyl)imide (Pyr13TFSI)/PVdF-HFP could deliver a capacity of about 900 mAh g^{-1} of carbon and catalyst and a discharge voltage plateau around 2.5 V during the first discharge process. Although the cost of ionic liquids is still high compared to that of other electrolytes, it is expected that it can be significantly reduced in the near future through production scale-up. Because of their low volatility and low water solubility, the use of ionic liquids in metal–air batteries may help increase operating time.

10.3.4 Aqueous and Hybrid Electrolytes

It is well known that water in aqueous electrolytes will be decomposed at 1.23 V to release H_2 and O_2. Therefore, aqueous electrolyte itself cannot be used in lithium air batteries that have an open-circuit voltage of more than 2.9 V. In 1980, a Japanese group claimed a lithium air battery with an aqueous alkali electrolyte by placing a Li_3N solid electrolyte between the Li metal anode and the carbon-air electrode. In 1993, Mizumoto and Horiba claimed lithium air batteries with aqueous LiOH electrolytes. An H-permeable and liquid-nonpermeable electron-conductive separator was installed parallel to the electrodes, where the separator could be an H-absorbing alloy or a hydrophobically treated porous carbonaceous material.

One of the most important advances in the development of aqueous lithium air batteries is to use a glass electrolyte (Li_{1+x+y} Al_xTi_{2-x} $Si_yP_{3-y}O_{12}$ (LATP), Ohara Corp, Japan) to separate Li and aqueous electrolyte in lithium air batteries. In the combined solid glass/aqueous electrolyte system, transfer of H^+ ions from the aqueous electrolyte to the Li anode is blocked by the glass electrolyte so the decomposition of water can be effectively prevented. In addition, the instability of glass in contact with a Li metal anode can be mitigated by addition of an interfacial layer that is stable against Li metal (it can be a solid electrolyte, nonaqueous electrolyte, or polymer electrolyte) between the Li and the glass electrolyte. Therefore, most of the aqueous lithium air batteries in fact use a three-layer hybrid electrolyte. Two of these layers (the interfacial layer and the solid electrolyte layer) are used to block the unwanted chemical reaction, and only the third layer (the aqueous electrolyte) was used to facilitate Li–O_2 reaction. These aqueous electrolytes also participate in Li–O_2 reactions and will be consumed in the discharge process as shown in Table 10.1.

In a hybrid electrolyte system, various interfacial layers have been investigated. Visco et al. combined a solid-state interfacial layer (Cu_3N, LiPON, etc.) with a LISICON glass, thus forming a protected Li electrode (PLE) [3, 73]. Similar protection of the Li metal anode from LATP glass with an interfacial layer has also been reported by Yamamoto and coworkers [74, 75]. Affinito et al. [76] also claimed rechargeable Li–water and lithium air batteries by using a multilayered composite to protect the Li metal anode from aqueous electrolytes. The multilayered structure consisted of a single-ion conductive material layer and a polymeric layer positioned between the Li anode and the single-ion conductive material. This assembly formed a stabilized Li metal anode that was compatible with the ambient environment. Recently, several groups also reported using an organic electrolyte or polymer electrolyte as the interfacial layer between the Li metal anode and the LISICON glass [31–33, 77–83]. A high discharge capacity per gram of carbon has been reported using this kind of structure, and the long-term operation of lithium air batteries has also been achieved. Other inorganic ionic conductive materials have also been reported to replace LISICON, such as $Li_{0.33}La_{0.57}TiO_3$ [84], and lithium–aluminum–germanium–phosphorus glass-ceramics (LAGP) [85]. One disadvantage of lithium air batteries using aqueous-based electrolytes is that water will be consumed during the discharge process. Addition of excessive water in the electrolyte will lead to reduced specific energy of the total batteries. Another concern for aqueous-based lithium air batteries is the violent reaction between Li and water in the case of battery abuse. Development of new additives and passivation layers may help alleviate these concerns.

As we discussed above, the electrolyte plays a critical role in the performance of lithium air batteries. An air-stable electrolyte is not only important for rechargeable lithium air batteries but also crucial for primary lithium air batteries. The selection of the electrolytes also needs to be matched with the selection of air electrode materials to get the best results for the complete batteries.

10.4 Design of Primary Lithium Air Batteries

Because the air electrode is usually based on carbon, most of the previous research on lithium air batteries has reported the specific capacity (mAh g^{-1}) of the battery based on the weight of carbon alone in the air electrode instead of on the weight of the complete electrode or battery [22]. Some of the high specific capacity values were reported with very low carbon loadings when the batteries were tested in a pure oxygen environment (where the oxygen pressure is higher than 1 atm). These studies have established a solid foundation concerning the effects of various components on the performance of the lithium air batteries. However, the integration of these components into a high-performance lithium air battery is still a significant challenge. As with many new technologies, a combination of the individually optimized components may not necessarily lead to an optimized device configuration. To fully realize the potential of lithium air batteries, the

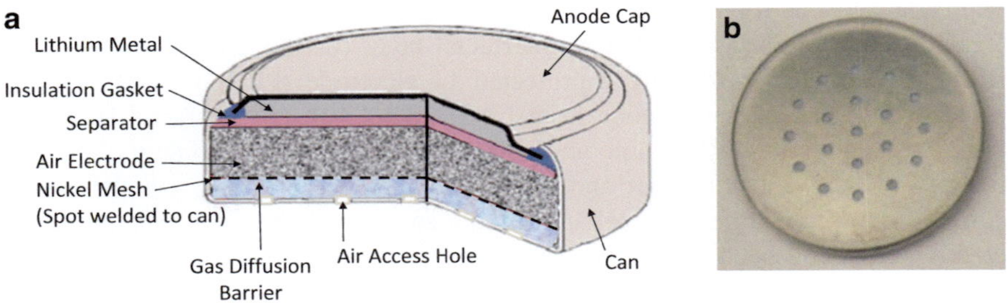

Fig. 10.12 (**a**) Structure of coin-cell-type lithium air batteries. (**b**) Photograph of a coin-cell-type lithium air battery (2.3 cm cell diameter) (Reproduced from Xiao [10])

structures of the air electrode (i.e., its material selection, porosity, and thickness) and the complete battery (including the air electrode, electrolyte, separator, current collector, lithium metal anode, and package) need to be further optimized.

In Sect. 10.2, the effect of the mesopore volume of the carbon material [4, 10] and the thickness of the carbon electrode were discussed [10, 22]. The capacity of the carbon electrode increases with increasing mesopore volume of the carbon material and there is an optimal carbon loading (or thickness) for the air electrode. In Sect. 10.3, we also described the selection of electrolyte and we notice that the selection of the electrolytes needs to be matched with the selection of air electrode materials to get the best results for the complete batteries. These examples are clear evidence of strong interrelation between the performances of different components of lithium air batteries. The full lithium air batteries need to be designed as an integrated system to realize the full potential of this promising energy-storage system.

Recently, Zhang and colleagues [27–29] analyzed the performance of lithium air batteries operated in ambient conditions in both coin cells (type 2325 kits from National Research Council, Canada) and pouch cells. Figure 10.12a, b shows the schematic structure and a photograph of a lithium air coin cell, respectively. The cells are 23 mm in diameter and 2.5 mm thick. The positive pans of the kits have 19 × ∅1.0-mm holes that are evenly spaced for air distribution. Lithium foil was used as the anode, and Whatman® GF/D glass microfiber filter paper was used as the separator because it holds more electrolyte than most polymer-based separators. These coin cells are easy to prepare and also can be reproduced easily. Therefore, they have been used in the selection and optimization process for various components (electrolyte, carbon, separator, oxygen-selective diffusion membrane, etc.) used for lithium air batteries when the battery was tested in a dry air environment (oxygen partial pressure, 0.21 atm; relative humidity, ~20 %). The performance of lithium air batteries is closely related to the microstructure and loading of carbon used in the air electrode [10, 22]. For the selected carbon source, there is an optimized carbon loading (or thickness) for the air electrode. Further increasing or decreasing the carbon loading led to a reduced area-specific capacity. Xiao et al. [28] found that a hybrid air electrode can significantly improve the energy

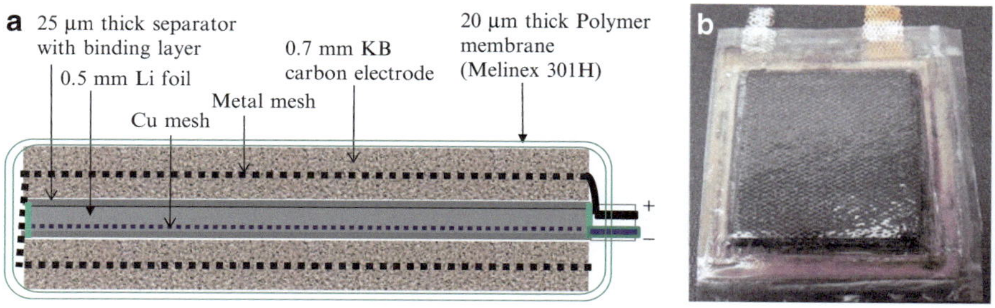

Fig. 10.13 (**a**) Schematic of a typical pouch cell with polymer membranes. (**b**) Photograph of a pouch cell lithium air battery (4.6 cm × 4.6 cm) operated in the ambient environment (Reproduced from Zhang et al. [9])

density of lithium air batteries. When compared with pure carbon-based lithium air batteries, the batteries using CF_x-based hybrid air electrodes demonstrate significantly improved specific capacities and energies.

To extend the lifetime of lithium air batteries operated in ambient air, protecting the porous electrode with an oxygen-diffusion membrane is desired. For long-term (or low-rate) operation of a primary lithium air battery (such as for 30 days), only a small amount of oxygen flow is required to maintain its small discharge current. Therefore, it is possible to use a low-oxygen-permeability membrane as the packaging for the battery. The appropriate membrane may also act as a good barrier to minimize diffusion of moisture into the battery and of electrolyte out of the battery. Experimental data demonstrated the feasibility of this approach. A polymer membrane has been used as both an oxygen-diffusion membrane and as a moisture barrier for lithium air batteries [75]. The membrane also can minimize evaporation of the electrolyte from the batteries. Figure 10.13a is a schematic of a typical pouch cell encased within a polymer membrane, and Fig. 10.13b shows a prototype pouch cell (4.6 cm × 4.6 cm) operated in air. These cells used a heat-sealable polymer (20-μm thick Melinex 301H, DuPont Teijin Films, Wilmington, Delaware) as both the packaging and the oxygen-diffusion membrane. Melinex 301H (referred to as ML hereafter) is a bilayer membrane with a biaxially oriented polyethylene terephthalate (PET) layer and a terephthalate/isophthalate copolyester of ethylene glycol thermal bonding layer. The cell consists of a 0.5-mm-thick Li foil, fully wrapped by a separator (glass fiber filter paper GF/C or Celgard® 5550). A copper mesh was pressed into the Li foil and extended to the outside of the cell as the anode current collector. Two carbon-based air electrodes were stacked on the outside of the separators to form an air electrode/separator/lithium/separator/air-electrode stack (or dry-cell stack). Nickel mesh was pre-laminated onto the carbon electrode as a current collector. The dry-cell stacks were soaked in the selected electrolytes for more than 2 h and then heat-sealed in ML or other selected membranes.

Figure 10.14 shows the voltage profile of pouch-type lithium air batteries as a function of capacity. The specific energy of the complete cell (including the packaging) was 362 Wh kg^{-1} based on the total weight of the battery (including

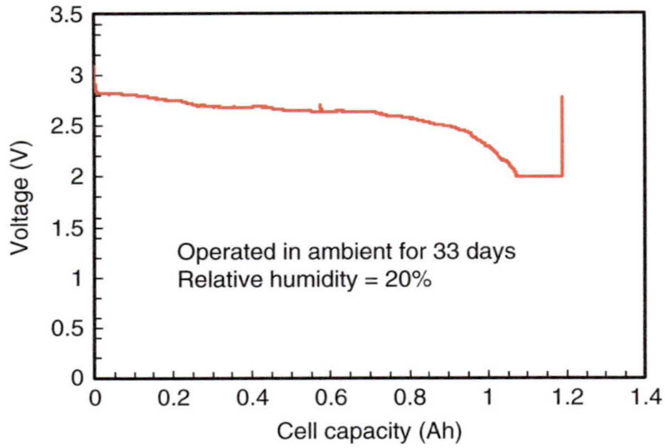

Fig. 10.14 Voltage profile as a function of capacity for a lithium air cell. The specific energy of the complete cell (including the package) was 362 Wh kg^{-1} (Reproduced from Zhang et al. [9])

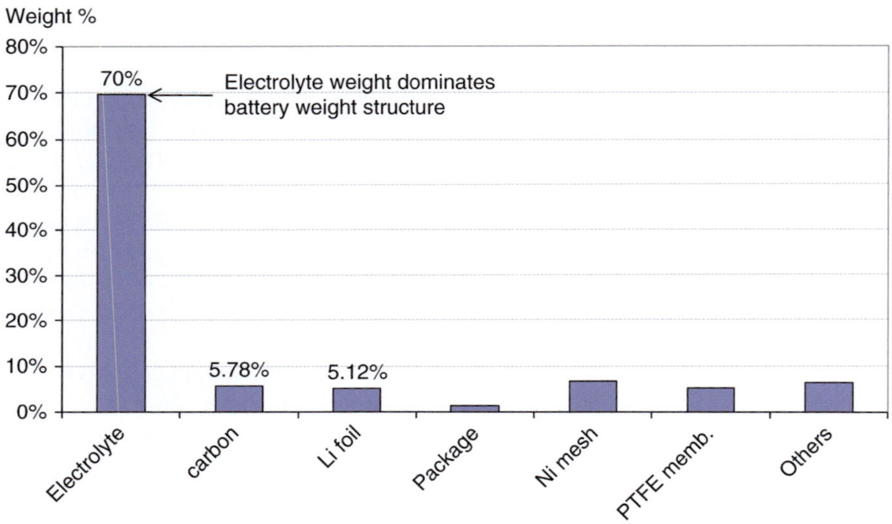

Fig. 10.15 Weight distribution in a practical lithium air battery (Reproduced from Zhang et al. [9])

its packaging). It has been operated in ambient conditions for more than 1 month at a current density of ~0.05 mA cm^{-2}. The specific capacity based on the carbon weight alone is 2,340 mAh g^{-1} when operated in an ambient environment with an oxygen partial pressure of only 0.21 atm and a relative humidity of 20 %.

To further improve the performance of lithium air batteries, Zhang et al. [9] established a model to analyze the effect of various components on the performance of practical lithium air batteries. Figure 10.15 shows the weight distribution of all of the components in a typical lithium air pouch cell with ML membrane as shown in Fig. 10.13b. In previously published papers, most theoretical work predicted the performance of lithium air batteries based on the weight of only the Li metal or only

Table 10.2 Key parameters in a practical lithium air battery

Dry air electrode porosity (%)	88.7
Carbon mesopore volume (cc g^{-1})	4.95
Mesopore expansion efficiency (%)	87.0
Percent of pore volume occupied by Li_2O_2/L_2O (%)	20.3
Lithium utilization (%)	68.5
Capacity (Ah)	1.12
Carbon-based capacity (mAh g^{-1})	2,340
Average voltage (V)	2.67
Specific energy (Wh kg^{-1})	362

the weight of the carbon (a few recent reports also included the binder and catalyst). However, the weight distribution of practical lithium air batteries given in Fig. 10.15 shows that the Li anode and the carbon each occupy less than 6 % of the total battery weight. The weight of the nickel mesh (cathode current collector) alone is greater (6.7 %) than those of either the Li or the carbon. Therefore, it is important to find a better cathode current collector to increase the specific energy of lithium air batteries. In the given battery design, the package/oxygen-diffusion membrane contributes less than 2 % of the total weight of the battery. The most important finding shown in Fig. 10.15 is that the electrolyte contributes nearly 70 % of the battery weight. As a comparison, the electrolyte normally occupies only 10–20 % of the weight in lithium-ion batteries. Although the weight distribution in the lithium air batteries depends on the specific design of the battery, the general pattern of weight distribution should be similar. Therefore, to increase the specific energy of practical lithium air batteries, one of the main challenges is to reduce the weight of the electrolyte while maintaining high capacities.

Several reports [10, 86] have shown that the capacity of carbon-based air electrodes increases with the mesopore volume of the carbon, but it is not very sensitive to the bulk porosity of the carbon electrode. In other words, expansion of mesopores during electrolyte soaking generated the majority of the tri-phase regions required for the Li/O_2 reaction. In practice, inter-particle spaces (bulk porosity) and mesopore spaces in an air electrode are similar to the arteries and capillaries in a living organism. Oxygen and Li ions are transported through inter-particle spaces (i.e., rely on the bulk porosity of the electrode), but the final Li/O_2 reaction occurs mainly in the mesopore spaces within the carbon particles. To reflect this phenomenon, carbon mesopore volume has been introduced into this model. After soaking in liquid electrolyte, these mesopores will fully expand to form a tri-phase region to facilitate the Li/O_2 reaction. Consequently, the reaction by-product will partially occupy these spaces after reaction. In fact, only ~20 % of the mesopore space was occupied by reaction by-products as shown in Table 10.2. Although KB has a large pore volume (4.95 cc g^{-1}), not all of the volume can be utilized for the Li/O_2 reaction. If the pore size is too small, reactant entry into the pore will be blocked quickly by initial reaction products. If the pore size is too large, most of the volume will become inactive after the internal active surface of the pore is fully covered by a critical thickness of reaction products. These large pores also will absorb a large amount of electrolyte, which further increases the weight of lithium air batteries. A nanostructured carbon with a large mesopore volume and an optimized pore size may lead to the better use of carbons, less electrolyte absorption, and a significant improvement in the specific energy of lithium air batteries.

The main structure and performance parameters of the above battery were analyzed further as shown in Table 10.2. Lithium utilization in the given sample is near 70 %, which is very close to its practical limit. This is also an indication that the selected oxygen-diffusion membrane (20-μm-thick ML) provided good protection for the battery, and corrosion of the Li metal by moisture was minimal during the discharge process. Because KB carbon has a very high mesopore volume of 4.95 cc g^{-1}, the dry cell has a very large porosity (88.7 %). Assuming that all the porous structure of a dry carbon electrode was occupied by electrode, the weight ratio of the electrolyte can be calculated to be about 30 % of the full battery. However, after soaking with electrolyte, the thickness of the air electrode increased 87 %. The large increase in the thickness of the electrode after it was soaked with electrolyte is one of the main reasons that KB-based air electrodes absorbed more than twice as much electrolyte (~70 % of the battery weight) as those estimated values based on the initial thickness and porosity of the electrode. This large deviation between simulated and practical batteries also leads to a greatly reduced practical specific energy of lithium air batteries. This is good evidence that individually optimized components may not be the best choices for the complete devices. To further improve the performance of lithium air batteries, a better carbon source that has a good balance between mesopore volume and electrolyte absorption needs to be identified. The full battery needs to be designed to have an optimized mesopore volume and the optimized amount of electrolyte absorption (minimum required to facilitate Li–O_2 reaction).

10.5 Challenges and Future Work on Primary Lithium Air Batteries

Lithium air batteries exhibit much higher energy density than most other conventional energy-storage systems, but they also pose significant challenges. Although it is necessary to optimize the various parameters in these batteries, the key to realizing the potential of lithium air batteries is a balanced design. The most important criterion in these balanced designs is maximum energy density (Wh kg^{-1} or Wh L^{-1}) and/or power density (W kg^{-1} or W L^{-1}) of the full batteries. For primary lithium air batteries, the following challenges need to be overcome before their practical application:

1. A balance between the mesopore and the macropore structures of the air electrode. A carbon electrode needs to have well-defined mesopore structure to provide required space to store the reaction products, but carbon with very high mesopore volume may absorb too much electrolyte and lead to overall decrease

in the energy density of the battery. On the other hand, the electrode also needs to have enough macropore channels to facilitate gas transport. Ideally, an electrode with a channel opening that increases gradually from inside to outside will minimize electrode blocking during discharge.

2. A balance between oxygen solubility in electrolyte and polarity of the electrolyte. An electrolyte needs to have a high oxygen solubility to allow high rate discharge, but an electrolyte with good oxygen solubility often has very low polarity and is easy to lead to electrolyte flooding. It is also critical for an electrolyte to have a suitable polarity when in contact with the given carbon electrode. An optimized polarity will generate a maximum amount of tri-phase regions for $Li–O_2$ reaction, but still prevent electrode flooding and allow easy transport of oxygen in the electrode.
3. A balance between oxygen-diffusion rate and water penetration. Ideally, an oxygen-selective membrane (with an oxygen-to-water selectivity >100) should be used in lithium air batteries, which allows only oxygen to pass and blocks water vapor and other gases. This membrane is very critical for the long-term operation of rechargeable lithium air batteries. For primary lithium air batteries, the required selectivity ratio can be relaxed. For example, a membrane with an oxygen-to-water selectivity of ~10 and a high oxygen permeability will be suitable for primary lithium air batteries.
4. A balance between the battery energy density and power density. It will be difficult to design a lithium air battery with both high-energy density and high-power density, but it is easier to design a lithium air battery with either high-energy density or high-power density. The combination of these two configurations or a combination of a lithium air battery with other energy-storage devices (such as supercapacitors and lithium-ion batteries) may open an easy path to practical applications for lithium air batteries.

Although there are still significant barriers that need to be overcome before the commercialization of primary lithium air batteries, these barriers are less difficult to be solved than those in the rechargeable lithium air batteries. Therefore, we expect that primary lithium air batteries will find their commercial applications in the next decade. Because of many similarities between primary and rechargeable lithium air batteries, success in primary lithium air batteries can also largely help the development of rechargeable lithium air batteries, which have the great potential to be used in the next-generation energy-storage systems beyond lithium-ion batteries.

References

1. Abraham KM, Jiang Z (1996) A polymer electrolyte-based rechargeable lithium/oxygen battery. J Electrochem Soc 143:1–5
2. Abraham KM, Jiang Z (1996) Solid polymer electrolyte-based oxygen batteries. US Patent 5,510,209, April 1996

3. Visco SJ et al (2004) Lithium fuel cells. In: Proceedings of the 12th international meeting on lithium batteries. The Electrochemical Society, Nara, Japan, 27 June to 2 July 2004, Abstract No. 396
4. Kuboki T et al (2005) Lithium-air batteries using hydrophobic room temperature ionic liquid electrolyte. J Power Sources 146:766–769
5. Kinoshita K (1992) Electrochemical oxygen technology. The Electrochemical Society series. Wiley, New York, p 431
6. Read J (2006) Ether-based electrolytes for the lithium/oxygen organic electrolyte battery. J Electrochem Soc 153:A96–A100
7. Read J (2002) Characterization of the lithium/oxygen organic electrolyte battery. J Electrochem Soc 149:A1190–A1195
8. Zheng JP et al (2008) Theoretical energy density of Li-air batteries. J Electrochem Soc 155: A432–A437
9. Zhang JG et al (2010) Ambient operation of Li-air batteries. J Power Sources 195:4332–4337
10. Xiao J (2010) Optimization of air electrode for Li-air batteries. J Electrochem Soc 157: A487–A492
11. Korovin N (2006) Metal-air batteries with carbonaceous air electrodes and nonmetallic catalysts. In: Barsukov IV et al (eds) New carbon based materials for electrochemical energy storage systems: batteries, supercapacitors and fuel cells. IOS Press, Amsterdam, pp 137–149
12. Lan CJ, Chi YF, Chin TS (2008) Bamboo-carbon as a novel carbon-support/electrocatalyst of air cathodes. ECS Trans 3:51–65
13. Zhang SS, Foster D, Read J (2010) Discharge characteristic of a non-aqueous electrolyte Li/O-2 battery. J Power Sources 195:1235–1240
14. Energizer (2004) Application manual: zinc air (ZN/O_2). http://data.energizer.com/PDFs/zincair_appman.pdf. Accessed 10 May 2012
15. Zhang XG (2008) Novel anode for high power zinc-air batteries. ECS Trans 3:1–11
16. Electric Fuel Battery Corporation (EFB) (2010) About zinc air. http://www.efbpower.com/aboutza.html. Accessed 10 May 2012
17. Reynolds TA, Brose DJ, Golovin MN (1999) Membrane for selective transport of oxygen over water vapor and metal-air electrochemical cell including said membrane. AER Energy Resources, US Patent 5,985,475, 16 Nov 1999
18. Littauer EL, Tsai KC (1997) Corrosion of lithium in alkaline-solution. J Electrochem Soc 124:850–855
19. Ye H, Xu JJ (2008) Polymer electrolytes based on ionic liquids and their application to solid-state thin-film Li-oxygen batteries. ECS Trans 3:73–81
20. Kumar B et al (2010) A solid-state, rechargeable, long cycle life lithium-air battery. J Electrochem Soc 157:A50–A54
21. Visco SJ et al (2007) Protected active metal electrode and battery cell structures with non-aqueous interlayer architecture. PolyPlus Battery Company. US Patent 7,282,295, 16 Oct 2007
22. Beattie SD, Manolescu DM, Blair SL (2009) High-capacity lithium-air cathodes. J Electrochem Soc 156:A44–A47
23. Abraham KM, Jiang Z, Carroll B (1997) Highly conductive PEO-like polymer electrolytes. Chem Mater 9:1978–1988
24. Yang XH, Xia YY (2010) The effect of oxygen pressures on the electrochemical profile of lithium/oxygen battery. J Solid State Electrochem 14:109–114
25. Dobley A, Morein C, Abraham KM (2005) Cathode optimization for lithium-air batteries. In: Abstracts of the 208th meeting of the Electrochemical Society: energy technology and battery joint general session, Los Angeles, 16–21 Oct 2005, Abstract No. 823
26. Shiga T, Nakano H, Imagawa H (2008) Non-aqueous air battery and catalyst therefor. US Patent 7,887,957B2, 15 Feb 2011
27. Williford RE, Zhang JG (2009) Air electrode design for sustained high power operation of Li-air batteries. J Power Sources 194:1164–1170

28. Xiao J et al (2010) Hybrid air-electrode for Li-air batteries. J Electrochem Soc 157: A294–A297
29. Xu W et al (2009) Optimization of nonaqueous electrolytes for primary lithium/air batteries operated in ambient environment. J Electrochem Soc 156:A773–A779
30. Xu W et al (2010) Effects of nonaqueous electrolytes on the performance of lithium/air batteries. J Electrochem Soc 157:A219–A224
31. Kowalczk I, Read J, Salomon M (2007) Li-air batteries: a classic example of limitations owing to solubilities. Pure Appl Chem 79:851–860
32. Wang YG, Zhou HS (2010) A lithium-air battery with a potential to continuously reduce O-2 from air for delivering energy. J Power Sources 195:358–361
33. Shimonishi Y et al (2010) A study on lithium/air secondary batteries—stability of NASICON-type glass ceramics in acid solutions. J Power Sources 195:6187–6191
34. Debart A et al (2007) Effect of catalyst on the performance of rechargeable lithium/air batteries. ECS Trans 3:225–232
35. Debart A et al (2008) Alpha-MnO_2 nanowires: a catalyst for the O-2 electrode in rechargeable lithium batteries. Angew Chem Int Ed 47:4521–4524
36. Debart A et al (2007) An O_2 cathode for rechargeable lithium batteries: the effect of a catalyst. J Power Sources 147:1177–1182
37. Giordani V et al (2010) H_2O_2 decomposition reaction as selecting tool for catalysts in Li-O_2 cells. Electrochem Solid State Lett 13:A180–A183
38. Laoire CO et al (2009) Elucidating the mechanism of oxygen reduction for lithium-air battery applications. J Phys Chem C 113:20127–20134
39. Hamlen RP, Atwater TB (2001) Metal/air batteries. In: Linden D, Reddy T (eds) Handbook of batteries, 3rd edn. McGraw-Hill, New York, pp 38.1–38.53
40. Visco SJ, Nimon E, De Jonghe LC (2009) Secondary batteries—metal-air systems: lithium-air. In: Garche J, Dyer C, Moseley P, Ogumi Z, Rand D, Scrosati B (eds) Encyclopedia of electrochemical power sources, vol 5. Elsevier, Amsterdam, pp 376–383
41. Jöerissen L (2009) Secondary batteries—metal-air systems: bifunctional oxygen electrodes. In: Garche J, Dyer C, Moseley P, Ogumi Z, Rand D, Scrosati B (eds) Encyclopedia of electrochemical power sources, vol 5. Elsevier, Amsterdam, pp 356–371
42. Luo J-Y, Cui W-J, He P, Xia YY (2010) Raising the cycling stability of aqueous lithium-ion batteries by eliminating oxygen in the electrolyte. Nat Chem 2:760–765
43. Zheng JP, Andrei P, Hendrickson M, Plichta EJ (2011) The theoretical energy densities of dual-electrolytes rechargeable Li-air and Li-air flow batteries. J Electrochem Soc 158: A43–A46
44. Read J et al (2003) Oxygen transport properties of organic electrolytes and performance of lithium/oxygen battery. J Electrochem Soc 150:A1351–A1356
45. Cheng H, Scott K (2010) Carbon-supported manganese oxide nanocatalysts for rechargeable lithium–air batteries. J Power Sources 195:1370–1374
46. Lee JS et al (2011) Metal–air batteries with high energy density: Li-air versus Zn-air. Adv Energy Mater 1:34–50
47. Wang G et al (2008) Improving the DMFC performance with Ketjen Black EC 300J as the additive in the cathode catalyst layer. J Power Sources 180:176–180
48. Mirzaeian M, Hall PJ (2009) Preparation of controlled porosity carbon aerogels for energy storage in rechargeable lithium oxygen batteries. Electrochim Acta 54:7444–7451
49. Tran C, Yang X-Q, Qu D (2010) Investigation of the gas-diffusion-electrode used as lithium/air cathode in non-aqueous electrolyte and the importance of carbon material porosity. J Power Sources 195:2057–2063
50. Zhang GQ et al (2010) Lithium–air batteries using SWNT/CNF buckypapers as air electrodes. J Electrochem Soc 157:A953–A956
51. Mitchell RR, Gallant BM, Thompson CV, Shao-Horn Y (2011) All-carbon-nanofiber electrodes for high-energy rechargeable Li-O_2 batteries. Energy Environ Sci 4:2952–2958

52. Liu C, Yu Z, Neff D, Zhamu A, Jang B (2010) Graphene-based supercapacitor with an ultrahigh energy density. Nano Lett 10:4863–4868
53. Wang D et al (2009) Self-assembled TiO_2-graphene hybrid nanostructures for enhanced Li-ion insertion. ACS Nano 3:907–914
54. Yoo E, Zhou H (2011) Li-air rechargeable battery based on metal-free graphene nanosheet catalysts. ACS Nano 5:3020–3026
55. Li Y, Wang J, Li X, Geng D, Li R, Sun X (2011) Superior energy capacity of graphene nanosheets for a nonaqueous lithium-oxygen battery. Chem Commun 47:9438–9440
56. Xiao J et al (2011) Hierarchically porous graphene as a lithium-air battery electrode. Nano Lett 11:5071–5078
57. Xu Y, Shelton WA (2011) Oxygen reduction by lithium on model carbon and oxidized carbon structures. J Electrochem Soc 158:A1177–A1184
58. Zhang S, Foster D, Read J (2010) A high energy density lithium/sulfur–oxygen hybrid battery. J Power Sources 195:3684–3688
59. Xiao J et al (2011) Investigation of the rechargeability of Li-O_2 batteries in non-aqueous electrolyte. J Power Sources 196:5674–5678
60. Lu Y-C et al (2011) The discharge rate capability of rechargeable Li-O2 batteries. Energy Environ Sci 4:2999–3007
61. Lu Y-C et al (2010) Platinum-gold nanoparticles: a highly active bifunctional electrocatalyst for rechargeable lithium-air batteries. J Am Chem Soc 132:12170–12171
62. Shiga T, Nakano H, Imagawa H (2008) US 2008/0299456 A1, 4 Dec 2008
63. Xu W, Xiao J, Wang D, Zhang J, Zhang JG (2010) Crown ethers in nonaqueous electrolytes for lithium/air batteries. Electrochem Solid State Lett 13:A48–A51
64. Zhang SS, Read J (2011) Partially fluorinated solvent as a co-solvent for the non-aqueous electrolyte of Li-air battery. J Power Sources 196:2867–2870
65. Zhang S, Xu K, Read J (2011) A non-aqueous electrolyte for the operation of Li-air battery in ambient environment. J Power Sources 196:3906–3910
66. Crowther O, Meyer B, Salomon M (2011) Methoxybenzene as an electrolyte solvent for the primary lithium metal air battery. Electrochem Solid State Lett 14:A113–A115
67. Zhang S, Foster D, Read J (2010) The effect of quaternary ammonium on discharge characteristic of a non-aqueous electrolyte Li/O_2 battery. Electrochim Acta 56:1283–1287
68. Wang Y, Zheng D, Yang X-Q, Yang D (2011) High rate oxygen reduction in non-aqueous electrolytes with the addition of perfluorinated additives. Energy Environ Sci 4:3697–3702
69. Zhang Z et al (2011) Increased stability toward oxygen reduction products for lithium-air batteries with oligoether-functionalized silane electrolytes. J Phys Chem C 115:25535–25542
70. Mohamed SN et al (2008) Electrochemical studies on epoxidised natural rubber-based gel polymer electrolytes for lithium-air cells. J Power Sources 183:351–354
71. Zhang D, Li R, Huang T, Yu A (2010) Novel composite polymer electrolyte for lithium air batteries. J Power Sources 195:1202–1206
72. Ye H, Huang J, Xu JJ, Khalfan A, Greenbaum SG (2007) Li ion conducting polymer gel electrolytes based on ionic liquid/PVDF-HFP blends. J Electrochem Soc 154:A1048–A1057
73. Visco SJ, Katz BD, Nimon YS, Dejonghe LD (2007) Protected active metal electrode and battery cell structures with non-aqueous interlayer architecture. US Patent 7,282,295, 16 Oct 2007
74. Imanishi N et al (2008) Lithium anode for lithium-air secondary batteries. J Power Sources 185:1392–1397
75. Hasegawa S et al (2009) Study on lithium/air secondary batteries—stability of NASICON-type lithium ion conducting glass-ceramics with water. J Power Sources 189:371–377
76. Affinito JD et al (2007) Rechargeable lithium/water, lithium/air batteries. US Patent 20070221265 A1
77. Zhang T et al (2008) Li/polymer electrolyte/water stable lithium-conducting glass ceramics composite for lithium-air secondary batteries with an aqueous electrolyte. J Electrochem Soc 155:A965–A969

78. Zhang T et al (2009) Water-stable lithium anode with the three-layer construction for aqueous lithium–air secondary batteries. Electrochem Solid State Lett 12:A132–A135
79. Zhang T et al (2010) Stability of a water-stable lithium metal anode for a lithium-air battery with acetic acid-water solutions. J Electrochem Soc 157:A214–A218
80. Zhou H, Wang Y, Li H, He P (2010) The development of a new type of rechargeable batteries based on hybrid electrolytes. ChemSusChem 3:1009–1019
81. Shimonishi Y et al (2011) A study on lithium/air secondary batteries—stability of the NASICON-type lithium ion conducting solid electrolyte in alkaline aqueous solutions. J Power Sources 196:5128–5132
82. He P, Wang Y, Zhou H (2011) The effect of alkalinity and temperature on the performance of lithium-air fuel cell with hybrid electrolytes. J Power Sources 196:5611–5616
83. Li L, Zhao X, Manthiram A (2012) A dual-electrolyte rechargeable Li-air battery with phosphate buffer catholyte. Electrochem Commun 14:78–81
84. Wolfenstine J, Allen JL (2008) Reaction of $Li_{0.33}La_{0.57}TiO_3$ with water. J Mater Sci 43:7247–7249
85. Aleshin GY et al (2011) Protected anodes for lithium-air batteries. Solid State Ion 184:62–64
86. Yang XH, He P, Xia YY (2009) Preparation of mesocellular carbon foam and its application for lithium/oxygen battery. Electrochem Commun 11:1127–1130

Chapter 11
Overview of LiO_2 Battery Systems, with a Focus on Oxygen Handling Requirements and Technologies

Paul Albertus, Timm Lohmann, and Jake Christensen

11.1 Introduction

The preceding chapters of this book have focused on some of the specific components required to make a rechargeable battery that reacts Li and O_2 (and possibly other species), especially the electrodes and the electrolyte. In this chapter we consider the design of a Li/O_2 battery system, with a particular focus on the supply of O_2 to the reaction. We find that, because the Li/O_2 reactions with the highest energies (and receiving the most attention) yield intermediates and products that react with components in air, the supply of uncontaminated O_2 is critical to providing a complete Li/O_2 *system* with high energy content. The reactions that we consider in this chapter are shown in Table 11.1; reactions (1) and (2) include Li and O_2 as reactants, while reactions (3) and (4) are included for the sake of comparison. Reaction (1) is the Li/O_2 reaction taking place in a nonaqueous electrolyte; essentially all experiments report the formation of Li_2O_2 rather than Li_2O, consistent with the difficulty of fully breaking the O–O bond that is required for the formation of Li_2O [1–3]. Reaction (2) is one of a larger set of reactions involving O_2 and H_2O that can be carried out in an aqueous medium [3]. Reaction (2) actually consists of the formation of dissolved LiOH in the reaction $2Li + \frac{1}{2}O_2 + H_2O \leftrightarrow 2LiOH\ (aq)$, followed by the precipitation of LiOH as a monohydrate when its solubility limit is reached (~5.25 M at 25 °C) [4]. $LiOH \cdot H_2O$ must be precipitated out of the solution in order to exceed the specific energy of ~430 Wh/kg corresponding to the saturated solution. Note that for reaction (2) there is a 6:1 ratio between the number of moles of O_2 and H_2O consumed during discharge (and produced on charge), indicating the need to consider H_2O as well as O_2 handling for this couple. Theory and experiments suggest a significantly larger kinetic overpotential for reaction (2) than reaction (1) [5], consistent with the need to

P. Albertus • T. Lohmann • J. Christensen (✉)
Bosch Research and Technology Center, 4005 Miranda Avenue, Palo Alto, CA 94304, USA
e-mail: jake.christensen@us.bosch.com

N. Imanishi et al. (eds.), *The Lithium Air Battery: Fundamentals*,
DOI 10.1007/978-1-4899-8062-5_11,

Table 11.1 Reactions considered in this chapter as well as select reaction or discharge product properties

Reaction or discharge product property	Reactions and discharge products			
	(1) $2Li + O_2 \leftrightarrow Li_2O_2$	(2) $2Li + \frac{1}{2}O_2 + 3H_2O \leftrightarrow 2LiOH{\cdot}H_2O$	(3) $Li + MO_2 \leftrightarrow LiMO_2$	(4) $H_2 + \frac{1}{2}O_2 \leftrightarrow H_2O$
Specific capacity (mAh/g)	1,168	639	275	3,150
Density (g/cm^3)	2.31	1.51	4.25	1.0
Capacity density (mAh/cm^3)	2,698	965	1,169	3,150
Equilibrium potential (V)	2.96	3.45[a]	3.75	1.23
Approx. practical cell discharge potential (V)	2.65	2.95	3.50	0.70

[a]This is the standard potential for the reactions $Li \leftrightarrow Li^+ + e^-$ and $2H_2O + 4e^- + O_2 \leftrightarrow 4OH^-$

break the O–O bond completely in reaction (2). Numerous other reactions involving Li and O_2 may be envisioned, as we discussed in a previous publication, but here we focus on the two that have received the most attention so far (other reactions have a lower specific energy) [3]. Although we include reactions (1) and (2) throughout our analysis, we focus on reaction (1) because of its higher theoretical specific energy.

Reactions (3) and (4) are included for comparison with the Li/O_2 couples. Reaction (3) involves a Li metal anode and an advanced metal oxide cathode into which Li is inserted on discharge and extracted on charge. We assume the $LiMO_2$ material has a specific capacity of 275 mAh/g, a density of 4.25 g/cm^3, and an average potential of 3.75 V vs. Li; while such a material is not in widespread use today, there are research efforts on materials with similar properties [6]. We also include the H_2/O_2 reaction taking place in a proton-exchange membrane fuel cell (PEMFC) because it is also an electrochemical system in which air is a reactant and which is under consideration to replace combustion engines [6].

Table 11.1 also lists some of the properties of the reaction products, which are instructive for assessing the energy storage potential of the different systems. For example, we see that while the specific capacity of Li_2O_2 is more than a factor of 4 higher than that of $LiMO_2$, the capacity densities are closer (a factor of 2.3), and the capacity density of $LiMO_2$ is slightly higher than that of $LiOH{\cdot}H_2O$. Interestingly, the capacity density of Li metal (2,062 mAh/cm^3) is actually lower than that of Li_2O_2. All of the couples with Li metal will have significant cell volume change take place during cycling if the cathode thickness is fixed, as we discuss in more detail in a later section. We can also see major differences in ratios among practical and equilibrium potentials, though one should keep in mind the practical potentials for reactions (1) and (2) remain somewhat uncertain given the early stage of research. In particular, while the Li-based reactions (1)–(3) are typically embodied in cells with a discharge energy efficiency above 80 %, the H_2/O_2 reaction is often run in cells with a relatively high current density (>1 A/cm^2) and an energy efficiency of ~55 %.

Now that we have covered the reactions and discharge products for which we are building full battery systems, in the remaining sections of this chapter we cover Li/O_2 battery system design and targets (Sect. 11.2), the additional mass and volume beyond the active materials in going to a full system (Sect. 11.3), storing O_2 in tanks in a closed-system configuration (Sect. 11.4), removing contaminants from air in an open-system configuration (Sect. 11.5), and conclusions (Sect. 11.6).

11.2 Li/O_2 Battery System Design and Targets

While there has been significant research on materials and reaction processes in Li/O_2 cells, no material set and cell design have yet been identified that meet the full set of performance criteria for a secondary cell, especially with regard to the cycle life, power, and energy content. Thus, much remains uncertain about how a Li/O_2 system may ultimately look, and our comments in this section on system design

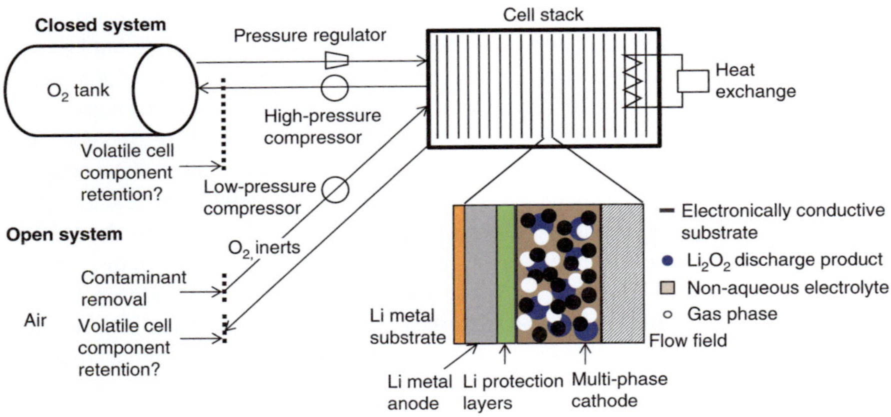

Fig. 11.1 Schematic of the two basic configurations (open and closed) for a couple reacting Li and O_2. A nonaqueous cell is shown here

should be taken in the context of the early stage of research. Figure 11.1 shows a simplified schematic of two versions of a hypothetical Li/O_2 system. Taking the components related to oxygen handling first, a critical point to realize is that air cannot be fed directly to the reaction without the removal of species that result in deleterious side reactions. In particular, for reaction (1) Li_2O_2 reacts with H_2O and CO_2 to form LiOH or Li_2CO_3, while for reaction (2) it is well known that carbonates form when CO_2 is exposed to a basic medium such as LiOH (aq) [2, 3, 8]. It is possible that side reaction products may be removed either continually or periodically (e.g., by charging to high potentials), but generally side reactions lead to capacity fade as reactants such as Li may be consumed irreversibly. While the issue of O_2 supply has received little treatment so far, there are now reports specifically addressing the effect of H_2O on the reaction chemistry [9]. In addition to side reactions in the cathode or separator with components of air, the Li metal contained in the cell must be protected because it is unstable with components in air such as H_2O, O_2, CO_2, and N_2. Ceramic, glass, polymer, or composite layers may be used to isolate the Li metal from species with which it would react [3].

Given the need to supply O_2 but exclude CO_2 (and H_2O for the nonaqueous case), there are at least two basic pathways for achieving this. One is a closed system in which the O_2 is stored in tanks, while the other is an open system in which there is a separation process that removes deleterious components such as CO_2 (and H_2O in the nonaqueous case); both schemes are shown in Fig. 11.1. Note that for both schemes, there may be a need to keep any volatile components (e.g., electrolyte solvents) within the cell, including H_2O for reaction (2). The use of ionic liquids for reaction (1) may be a promising route to limit loss of electrolyte solvent, as ionic liquids have a low vapor pressure [10]. If volatile components cannot be retained in the cells, for the closed system the volatile components should be nonreactive with high-pressure O_2, while for the open system they should be suitable for periodic (or continuous) solvent replenishment. Both designs also require a compressor. In the case of the closed system, the O_2 would need to be pressurized to achieve a reasonable system energy density (later we show results for 350 bar), requiring a high-pressure, multistage, mechanical compressor. Using an external compressor may work, and then home charging would require the installation of a compressor. A compressor could also be carried on board a vehicle such that only an electrical outlet would be needed for charging but that would add to the system size, cost, and complexity. Full or partial electrochemical compression may be considered to eliminate or reduce the pump size, but its use would depend on a design optimization that took into account additional stack costs associated with exposure to high pressures. For the closed-system configuration, a compressor would be needed to drive the air through the separation processes that remove contaminants (we discuss membrane and adsorption units below) and to increase the pressure of air within the stack (for PEMFCs a pressure of 1.5–3.0 bar is typical) to reduce O_2 mass-transfer limitations; such compressors have already been designed for PEMFC and other applications [11, 12].

Turning next to the design of the cell sandwich and stack, it is clear that in addition to the Li metal, Li metal protection, and cathode layers that comprise a Li metal/insertion-cathode cell, a Li/O_2 battery would also need to have a flow field to distribute the O_2 over the surface of the cell. This suggests that the cell sandwiches may be stacked and in a bipolar design, as in PEMFCs, rather than wound as is typical of Li-ion cells. While flow fields are often made of carbon or coated stainless steel, the use of Li metal requires a substrate (such as copper) with which Li metal will not interact. In any case, a major difference between a Li/O_2 stack and a PEMFC stack is the area required for a given amount of power. In particular, while a PEMFC is designed to operate at nearly 1 W/cm^2 [13], the much higher impedance of a Li/O_2 cell will likely lead to operation below 0.05 W/cm^2. Thus, for a design with 80 kW of net output, an area of about 200 m^2 would be required, much higher than the ~10 m^2 required for a PEMFCs with an equivalent net output. Flow field designs for a much lower power density than those for PEMFCs will be required. Another major difference between a PEMFC stack and a Li/O_2 stack is the fact that in the latter case one of the reactants (Li metal) is stored within the stack and is in the solid phase. Thus, issues of current distribution and volume change will be a serious issue in the Li/O_2 case.

Finally, there are additional systems-related components beyond those of the oxygen handling equipment and the stack, including a thermal management system (which may interact with more components than just the stack, e.g., the compressor), controls and electronics, and packaging. Many of these items are also required of Li-ion battery systems and are therefore not unique to Li/O_2 battery systems. However, they do have an effect on the system efficiency, as the auxiliary loads reduce the net power output of the battery pack.

What are the performance targets for a Li/O_2 battery system? Perhaps the two most important questions for the design of a battery system are the amount of energy it should store and the amount of power it should produce. Because of the high price of current Li-ion battery systems (~500 \$/kWh or more), there is a strong driving force for packs with <30 kWh, resulting in EVs with a range of <150 km.

However, a major reason for pursuit of the Li/O_2 battery is the potential for a significantly higher range, and for our study here we use the power and energy targets for PEMFC vehicles, another electrochemical system under widespread discussion to replace lithium-ion batteries. The Department of Energy (DOE) PEMFC targets are 80 kW of net power and 5.6 kg of usable H_2 (which works out to ~100 kWh of usable energy, providing a driving range of about 500 km) [13, 14]. In terms of mass and volume, the ultimate PEMFC system targets (for the fuel cell system and the hydrogen storage system combined) are about 200 kg and 175 L. For comparison, the USCAR "Minimum goals for long term commercialization" for a 40 kWh electric vehicle battery pack is a mass of 266 kg and a volume of 174 L, while the "Long Term Goal" is 200 kg and 133 L [15]. We choose targets of 200 kg and 175 L. One should keep in mind that targets are often based on the equivalent systems in an internal combustion vehicle and do not necessarily need to be met to create a successful product.

Recent experimental and theoretical work shows that for reaction (1), the round-trip kinetic efficiency for discharge and charge at a reasonable current density is 85 %, suggesting that a practical discharge potential may be 2.65 V (about 90 % voltaic efficiency) [16]. With a 5 % system power consumption, the discharge and charge efficiency would be 85 % and the total round-trip efficiency would be 72 %. This implies that the total required power from the battery cells should be about 84 kW and the total energy stored in the pack should be about 118 kWh. A pack that produces 80 kW and 100 kWh of usable power and energy with these efficiencies implies a maximum (pulse) required *C*-rate of about 0.7. The maximum required sustained *C*-rate could be lower, as only about 35 kW are required to sustain an efficient passenger vehicle at 130 km/h on flat ground [7]. In other words, a sustained rate of *C*/2 should be adequate, with pulses to 0.7*C*. Here, our choice of a high-energy pack (100 kWh practical) allows us to use a high-energy cell design with a relatively low power/energy ratio. The resulting specific energy and pulse power targets (based on the pack's practical energy and net power) are 500 Wh/kg and 400 W/kg, and the energy density and pulse power targets are 570 Wh/L and 460 W/L. These are substantially higher than the approximately 100 Wh/kg and 150 Wh/L available from today's Li-ion battery packs. Another important performance target is the time to recharge the battery. Charging a 118 kWh battery at a *C*/10 rate (where the efficiency may be 95 %) would require about 12.5 kW, higher than many Level 2 (240 V) charging units. DC fast charging offers significantly higher powers, although the stations are currently too expensive for home installation.

What about other targets? Of course, cost is critical. A large battery pack at the USCAR "Long Term Goal" of 100 $/kWh of usable energy at the system level would result in a pack costing $10,000 for 100 kWh [15]. The cost of a basic combustion engine system with a comparable power output is ~$2,500, and the DOE target for PEMFCs at the system level is 30 $/kW in 2015, or about 24 $/kWh of usable energy. While this leaves a major gap in initial costs, the lower operating costs from the use of electricity rather than oil can eventually offset the difference. Other requirements include the need to operate at a wide range of ambient

Table 11.2 Select Li/O_2 battery full system targets

Usable energy (kWh)	100
Net power (kW)	80
Charge, discharge efficiency (%)	85
Total mass (kg)	200
Total volume (L)	175
Specific energy (Wh/kg)	500
Specific power (W/kg)	400
Energy density (Wh/L)	570
Power density (W/L)	460
Ambient temperature range (°C)	−40 to +50
Cumulative driving distance (km)	240,000
Calendar life (years)	10
Cost ($/kWh of practical energy)	100

temperatures (from −40 to +50 °C), provide a cumulative driving distance of around 240,000 km and a calendar life of 10 years, and meet rigorous safety requirements. A short summary of our system design and resulting system targets is given in Table 11.2.

11.3 From Active Materials to a Vehicle System

While estimates of the theoretical energy content of active materials are straightforward, estimates of the energy content of a full system require numerous design assumptions and are therefore subject to change as research and development proceeds. In this section we make some best guess design assumptions based on the early stage of development of Li/O_2 battery chemistry and cells. As an important reference point, today's Li-ion packs have a theoretical specific energy of about 400 Wh/kg and energy density of about 1,400 Wh/L, while 90–120 Wh/kg (about 25 %) and 160–210 Wh/L (about 13 %) may be achieved practically in a pack of ~25 kWh [7, 11].

Figure 11.2 shows the values of the specific energy and energy density for the active materials alone. From this figure we see that including the weight of the O_2 results in a significant reduction of the theoretical specific energy for the Li/O_2 couples, but that the resulting values (3.46 kWh/kg for Li_2O_2, 2.14 kWh/kg for $LiOH{\cdot}H_2O$) are significantly higher than what might be expected from an advanced Li/MO_2 couple (1.03 kWh/kg). For reference, the theoretical specific energy of Li/S is about 2.6 kWh/kg. For $LiOH{\cdot}H_2O$ we also show the impact of including the H_2O in addition to the O_2. Given the need to operate a vehicle in any climate (including dry ones), H_2O will need to be carried on the vehicle. In terms of energy density, in the discharged state the value for Li_2O_2 is nearly a factor of 2 higher than Li/MO_2, while the value for $LiOH{\cdot}H_2O$ is lower than Li/MO_2. We assume the O_2 comes from the atmosphere in the calculations for the theoretical charged energy density and therefore is not counted. With H_2 compressed to 350 bar and O_2

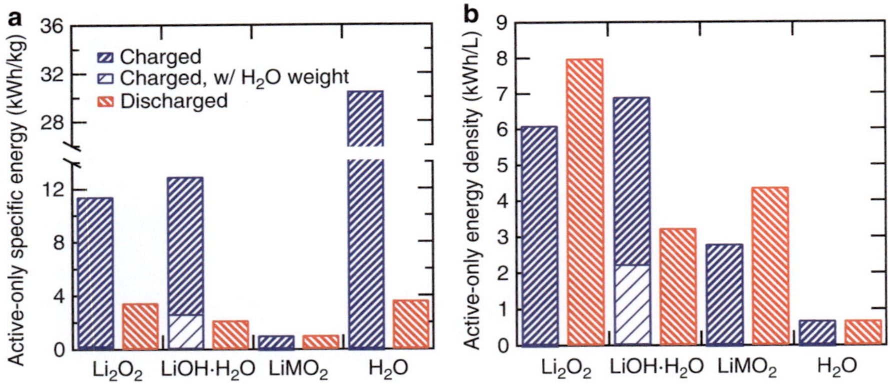

Fig. 11.2 Specific energy (**a**) and energy density (**b**) based on the active materials alone. In (**b**) we exclude the volume of O_2 for the reactions forming Li_2O_2, $LiOH{\cdot}H_2O$, and H_2O (assuming it may be obtained from the air), while for the H_2/O_2 reaction we assume H_2 is stored at 350 bar

from the atmosphere, the theoretical energy density of the H_2/O_2 couple is significantly lower than the others. Storing H_2 at 700 bar would increase the energy density by about 68 %. If O_2 is stored in a tank for a Li/O_2 system then the theoretical energy density falls significantly.

Now that we have looked at the energy content of the active materials alone, we turn to what might be achieved in a practical cell sandwich and a collection of cells or a stack. Here we base our values for the nonaqueous cell forming Li_2O_2 on a Li-ion cell, assuming the electrolyte transport properties will be similar, and for the aqueous Li/O_2 cell forming $LiOH{\cdot}H_2O$, we assume the cathode (or product reservoir [17]) can be significantly thicker due to the better transport properties of aqueous electrolytes (e.g., the conductivity of 5.25 M (saturated) LiOH is ~0.4 S/cm at 25 °C, compared with ~0.008 S/cm for 1 M $LiPF_6$ in EC:DEC). In both cases we assume that O_2 transport is not limiting, which will likely require a continuous gas network within the electrode in addition to the electrolyte network, reducing the volume fraction of active material compared to a Li-ion cell. For a similar set of calculations, we refer the reader to one of our prior publications [3]. The cell sandwich and other cell parameters are given in Table 11.3, while material properties are given in Table 11.4. We assume a complete lithium protection layer may be made with a thickness of 25 μm, which is thinner than current ceramic layers that are likely to fracture at such small thicknesses [18]. A major advance in ceramic or polymer/ceramic composites will likely be needed to achieve such a thin layer. We assume that the inert material in the electrodes for reactions (1)–(3) has the density of carbon, although recent research shows that for reaction (1) carbon probably cannot be used because it reacts with Li_2O_2 to form Li_2CO_3 [21]. Coating carbon with an inert material may be possible. Reaction (2) needs a catalyst for both oxygen reduction and evolution, but we ignore those weights here. The thicknesses of Li in the fully charged state (ranging between 61 and 140 μm) indicate that in these cell designs

Table 11.3 Properties related to the cell sandwiches for reactions (1)–(3)

		Values		
Cell-sandwich property	Units	Reaction (1)	Reaction (2)	Reaction (3)
$L_{positive}$	μm	150	500	150
$L_{Li\ protection\ layer}$	μm	25	25	25
L_{Li}	μm	118	140	61
$L_{neg,\ current\ collector\ or\ Li\ substrate}$	μm	7.5	7.5	7.5
$L_{pos,\ current\ collector}$	μm	0	0	10
$L_{flow\ field}$	μm	50	50	0
ε_{active} at end of discharge		0.5	0.5	0.6
$\varepsilon_{electrolyte}$ throughout discharge		0.25	0.25	0.25
$\varepsilon_{gas\ phase}$ at end of discharge		0.10	0.10	0
ε_{inerts}		0.15	0.15	0.15
ε_{gas} in flow field		0.7	0.7	0

Table 11.4 Properties of materials used in cells embodying reactions (1)–(3)

Property	Value	Units
$\rho_{LiOH\ (aq)\ elyte.\ (saturated)}$	1.11	g/cm^3
$\rho_{nonaqueous\ electrolyte}$	1.2	g/cm^3
ρ_{inert}	2.2	g/cm^3
ρ_{LPSL}	3.0	g/cm^3
$\rho_{Li_2O_2}$	2.31	g/cm^3
$\rho_{LiOH \cdot H_2O}$	1.51	g/cm^3
ρ_{LiMO_2}	4.25	g/cm^3
$\rho_{separator}$	1.3	g/cm^3

thicker layers of Li will need to cycle while avoiding dendrites and other morphology changes than has been demonstrated to date [19]. We assume 20 % excess Li; we consider larger quantities unrealistic as the consumption of significantly more Li is likely to seriously degrade cell performance (e.g., reduce power). The areal capacities for the cell sandwiches embodying reactions (1)–(3) are shown in Fig. 11.3. Note that the cell for reaction (1), with a capacity of around 20 mAh/cm^2, requires about 10 mA/cm^2 for a *C*/2 rate, a potentially challenging current density to sustain for today's nonaqueous electrolytes and the positive electrode thickness and porosity given in Table 11.3. Also, in order to achieve a potential drop across the cell corresponding to 90 % cell efficiency (the 2.65 V discharge potential specified above), for a 10 mA/cm^2 current density, an average area-specific impedance over the course of the discharge needs to be ~30 $\Omega\ cm^2$, also a challenging target given the need to have a Li metal protection layer in the cell. Also on the issue of cell impedance, we stress the need to have nearly uniform filling of the positive electrode with product during discharge in order to maintain a continuous gas and electrolyte network. Nonuniform current distributions may lead to nonuniform filling (especially for the case of reaction (1), where thus far the Li_2O_2 discharge product has appeared insoluble and therefore

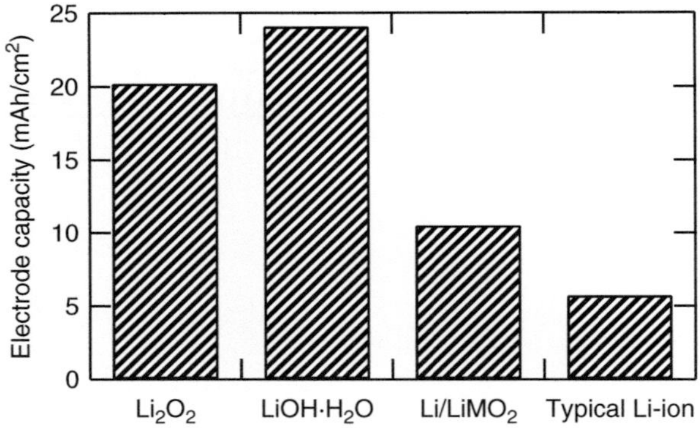

Fig. 11.3 Areal capacity for reactions (1)–(3) embodied in the cell-sandwich designs described in Table 11.3, as well as the value for a Li-ion cell for comparison (assuming 150 μm cathode, 60 % active material, a density of 4.25 g/cm^3, and a capacity of 150 mAh/g)

Table 11.5 Packing factors for mass and volume used in our analysis

Property	Value (%)
Cell-sandwich mass/stack mass	80
Stack mass/system mass (excluding oxygen handling)	70
Cell-sandwich volume/stack volume	80
Stack volume/system volume (excluding oxygen handling)	70

remains where it is formed) and drive up the impedance by introducing mass-transfer limitations, especially near the end of discharge.

From the cell-sandwich design, it is necessary to estimate the added mass and volume in going to a cell (or stack of cells) and then in going from a cell (or stack) to a full system. Our estimates are given in Table 11.5. We assume that the cell sandwich is 80 % of the cell or stack mass and volume. The components included in the stack include end plates and tie rods, seals, thermal management, and other minor components. Also, there will be significant volume change in the stack during cycling, which may require additional components to manage. Next, we need to estimate a ratio between the mass and volume of the cells or stack and the mass and volume of a full system. We will treat the mass and volume of the oxygen handling components in the following sections, so here we only include components such as the thermal management system, controls, and additional packaging. Keeping in mind that we are looking at a pack where the balance of system components contribute to a system with a large amount of energy (100 kWh practical), we assume 70 % of the pack mass and volume are composed of the cells (or stack), exclusive of the oxygen handling equipment.

Another matter of significant importance that is related to the design of the cell sandwich is volume change. In particular, unlike in a Li-ion battery with two

Table 11.6 Amount of volume change in the cell sandwich that takes place during a full discharge for the cell-sandwich properties given in Table 11.3

Reaction	Discharged/charged cell-sandwich thickness
(1) $2Li + O_2 \leftrightarrow Li_2O_2$	0.72
(2) $2Li + ½O_2 + 3H_2O \leftrightarrow 2LiOH{\cdot}H_2O$	0.84
(3) $Li + MO_2 \leftrightarrow LiMO_2$	0.80

insertion electrodes or in a PEMFC, in a Li/O_2 battery there is a significant amount of volume change in the anode (i.e., the Li metal) and possibly in the cathode region. Assuming a design in which the volume fraction of electrolyte within the cathode remains constant throughout cycling and only the volume fractions of gas and discharge product change (i.e., that gas rather than electrolyte is pushed out of the cathode during discharge), in Table 11.6 we show the amount of volume change in the cell sandwich over the course of a full discharge for the cell-sandwich properties given in Table 11.3. Please see Christensen et al. [3] for a more detailed discussion of volume changes. The value shown for reaction (2) assumes that all of the H_2O consumed is stored within the cathode and the cathode thickness is fixed. Reaction (1) has a volume change of nearly 30 %, which will likely present a significant challenge.

Now that we have looked at the issue of cell sandwich, cell (or stack), and full system energy content values exclusive of oxygen handling, we turn to the issue of how oxygen handling may be implemented in a Li/O_2 system.

11.4 Storing O_2 in Tanks in a Closed Li/O_2 Battery System Configuration

As discussed before in the context of Fig. 11.1, one option for oxygen handling for a Li/O_2 battery is to store the oxygen in tanks carried on board a vehicle so the system is totally closed. While this has the disadvantage that the mass and volume of O_2 are added to the vehicle, it has the significant advantage that it creates a completely closed system such that the contents of the cell are never exposed to any components of the atmosphere and no cell components may escape. When considering the suitability of storing O_2 in tanks, it is useful to compare the results with the storage of H_2 on a vehicle, where tanks with pressures of about 350 and 700 bar have been explored for commercialization [20]. Table 11.1 provided the reaction stoichiometry as well as the theoretical and practical potentials for reactions (1), (2), and (4). Of particular note is that the practical potential of the reaction forming Li_2O_2 may be nearly four times higher (2.65 V vs. 0.7 V) than the practical potential of the reaction forming H_2O, with an equal number of electrons consumed per mole of stored gas (O_2 is stored for reaction (1), H_2 is stored for reaction (4)). This means

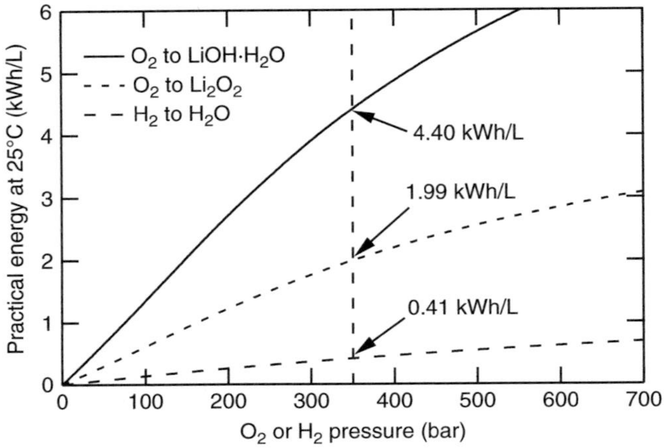

Fig. 11.4 Energy contained in the gases consumed in reactions (1), (2), and (4) as a function of pressure, based on the practical potential of the reactions given in Table 11.1. The van der Waals equation was used to calculate the compressibilities

that a significantly smaller number of moles of O_2 may be stored than moles of H_2 for a given unit of practical energy delivered, and hence the volume can be smaller, as quantified in Fig. 11.4. Because reaction (2) consumes only ½O_2 per $2e^-$, the energy content of stored O_2 is even higher than for reaction (2) than for reaction (1).

Besides a tank, additional equipment is needed to implement a closed Li/O_2 battery system storing O_2, including a compression unit, valves, piping, regulators, and other components. Small, multistage, mechanical air compressors with output pressures near 350 bar are available commercially, and a unit (not optimized for automotive applications) that offers an output of 100 L/min weighs about 40 kg, has a volume of about 125 L, and has a price of about $3,000 [21]. For a *C*/10 charge of a 118 kWh Li/O_2 battery pack, we would need to have a flow rate of ~31 L/min of O_2 at 1 bar and 25 °C, so a compressor of this size would provide a maximum charging rate of ~*C*/3, which would be more than sufficient for Level 2 charging. However, compared with the target system volume and mass given in Table 11.2, such a compressor would be too large. Significant optimization would be required for an onboard automotive application. It should also be noted that a multistage adiabatic compression unit typically has interstage cooling in order to increase the efficiency; integration of the compressor with the vehicle's thermal management system would therefore be needed.

Additional equipment to carry out a separation process to contain volatiles within the cells or stack and prevent them from entering the stack may also be required. Until working cell chemistry is identified, it is not possible to identify potential separation processes. Indeed, if all volatiles are safe in the presence of high-pressure O_2, then it may be possible to simply allow any volatiles to enter the O_2 tanks.

Table 11.7 Tank system design specifications, based on Hua et al. [20]. We assume the tank is a cylinder capped by hemispherical ends and that the system includes components such as a pressure regulator, pressure transducer, solenoid control valves, fill tube/port, and others

Tank design and materials	Type IV tank with carbon fiber and a polymer lining	
Maximum working pressure	350	bar
Tank safety factor	2.25	
Tensile strength of carbon fibers	2,550	MPa
Carbon fiber density	1.8	g/cm^3
Length/diameter ratio	3.0	
Gas/system volume	0.81	
Carbon fiber/system mass	0.53	

Finally, it is important to point out that some of the equipment required for a closed system that stores and supplies oxygen does not scale linearly with the system size. In particular, we would not expect a compressor to scale linearly, as the same maximum pressure may be used regardless of system size, and while the output rate may fall (if charge occurs at a fixed *C*-rate), the size can only be reduced somewhat. In addition, parts like pressure regulators and piping will also likely not scale in a linear manner. While we do not carry out a detailed analysis here, we stress that there are advantages, in terms of the fraction of the total system size occupied by the balance of plant, for large systems. The same argument also applies to the open system.

What, then, is the impact on the system specific energy and energy density of storing gas in a tank? It depends, of course, on the tank design, and our assumptions are shown in Table 11.7 [20]. Figure 11.5 shows the estimated practical system specific energy with and without O_2 tank systems. The energy density, in particular, depends on the assumption of the pressure of O_2 storage, and here we assume 350 bar. Note that the calculations in Fig. 11.5 *exclude* the mass and volume of a compressor and a separation unit for volatiles, given the current uncertainty in the design. The values in Fig. 11.5 should therefore be considered an upper bound. The tank system masses (including O_2 weight) are 42.1 and 18.1 for reactions (1) and (2), respectively, (18.3 and 7.9 kg without O_2), while the tank system volumes are 66.2 and 28.4 L.

There are numerous points of note that can be drawn from Fig. 11.5. First, Li_2O_2 can meet the specific energy and energy density targets listed in Table 11.2 both with and without a tank. The specific energy and energy density of $LiOH{\cdot}H_2O$ both fall below the targets, even in the absence of the tank system. The significantly lower capacity density of $LiOH{\cdot}H_2O$ compared with Li_2O_2, as shown in Table 11.1, is the principal reason for the difference. $LiMO_2$ has a specific energy that is lower than Li_2O_2 and $LiOH{\cdot}H_2O$ and slightly lower than H_2O, but the energy density is nearly the same as Li_2O_2 (without a tank) and significantly higher than $LiOH{\cdot}H_2O$ and H_2O. Also note that the values for $LiMO_2$ shown in Fig. 11.5 are more certain than those for Li_2O_2 and $LiOH{\cdot}H_2O$ because there is no need for oxygen handling

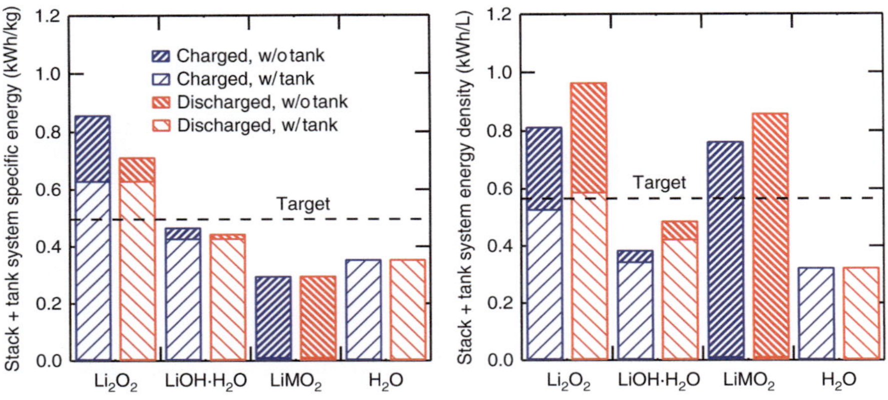

Fig. 11.5 Estimated practical stack and tank system specific energy and energy density with and without O_2 tanks. H_2/O_2 fuel cell values are from Wagner et al. (2010). For Li_2O_2 and $LiOH{\cdot}H_2O$, components such as a compressor or volatiles handling are excluded due to a lack of information. For reference, today's Li-ion packs achieve about 0.09–0.12 kWh/kg and 0.16–0.21 kWh/L

equipment. The significant volume occupied by an O_2 tank—even when pressurized to 350 bar—reduces the energy density of the Li_2O_2 system, justifying a search for technologies that avoid onboard gas storage.

As a simple check on the results presented in Fig. 11.5, as mentioned above, today's Li-ion packs achieve about 25 % of the specific energy and 13 % of the energy density of the active materials alone. The values for $LiMO_2$ in Fig. 11.5 are about 29 % (for mass) and 18 % (for volume); thus our packaging assumptions represent moderate but not dramatic advances.

The differences in energy density in the charged and discharged state shown for Li_2O_2, $LiOH{\cdot}H_2O$, and $LiMO_2$ result from the change in the thickness of Li metal. As we don't expect the actual system boundaries to move in practice, it is best to use the charged values rather than the discharged values. While smaller than the cell-sandwich ratios given in Table 11.6, the differences give an indication of the significance of the issue of the volume change within the cell sandwich manifesting even at the full system level.

In summary, compared to a $LiMO_2$ cell, Li_2O_2 may offer a significantly higher specific energy and a slightly higher energy density, but the advantage for energy density will depend on the volume of oxygen handling components. While the specific energy of the $LiOH{\cdot}H_2O$ cell is somewhat higher than the $LiMO_2$ cell, the energy density is significantly lower, and that is without accounting for the oxygen handling components. Overall, we expect that the need for oxygen handling equipment will result in lower pack/active material ratios for Li/O_2 compared to Li-ion packs.

Besides the issue of the mass and volume of storing O_2 on a vehicle, another important issue for storing compressed gas on board a vehicle is the amount of energy required to compress the gas. In particular, while the compression energies

Table 11.8 Comparison of the compression energy to the reaction energy for reactions (1), (2), and (4) at 25 °C. Note that the cell efficiency rather than the system efficiency is used here to calculate the discharge energy

Reaction	Stored gas	Isothermal compression energy (1–350 bar)/practical potential discharge energy
$2Li + O_2 \leftrightarrow Li_2O_2$	O_2	0.028
$2Li + ½O_2 + 3H_2O \leftrightarrow 2LiOH{\cdot}H_2O$	O_2	0.013
$H_2 + ½O_2 \leftrightarrow H_2O$	H_2	0.11

per mole of gas are similar for O_2 and H_2 in the pressure range considered here, there are significant differences in the compression energy compared with the reaction energy. In particular, Table 11.8 shows the energy for isothermal compression for reactions (1), (2), and (4) (assuming ideal gas behavior, which gives very similar results to real gases up to 350 bar) divided by the discharge energy based on the practical potential. As the table shows, the ratio of the energy for compression to the practical reaction energy is significantly lower for the formation of Li_2O_2 than the formation of H_2O, again a result of the higher potential of the reaction forming Li_2O_2. Note that for multistage mechanical compression, the required work is about twice the theoretical work [20]. While the energy of compression is not taken from the energy contained within the battery itself at full charge (or a "charged" PEMFC), it does enter into a technology's overall efficiency. For charging of an EV from a socket with a fixed power output, the need to run a compressor during charge will reduce the power entering the cells, adding to the charging time.

What about the other system requirements shown in Table 11.2? In terms of the ambient temperature operating range, in addition to the question of whether a Li/O_2 system could operate with sufficient power at very low temperatures, there is also a potential issue associated with the maximum pressure design point for the tank. In particular, a fixed number of moles of a gas in a fixed volume has a pressure that increases linearly with temperature. For the range of required ambient operating temperatures shown in Table 11.2, −40 to +50 °C, there would be a 28 % difference in the tank pressure. Hence, unlike in a PEMFC where the stored gas (H_2) is refilled and can therefore be vented if needed, in a completely closed Li/O_2 battery, the tank may need to be designed for the pressure at the maximum expected temperature, potentially resulting in a temperature-dependent range and/or an increase in the tank cost and size. Second, considering that many of the solvents being explored thus far for Li/O_2 cells have some volatility, there is a major question of whether a volatile and combustible solvent (or other cell component) would get into a high-pressure oxygen tank and potentially cause a fire or explosion. This should be kept in mind by those developing electrolytes as well as other cell components. Finally, in terms of cost, tanks for storing H_2 on vehicles are again a relevant point of comparison. H_2 tank systems designed for pressures of 350 bar cost about 11 $/L of gas volume (or $15/kWh of theoretical energy) and have a gas volume of about 258 L in order to meet the target of storing 5.6 kg of usable H_2 [20]. While the tank systems for storing the O_2 required for a battery with 100 kWh of practical energy running reaction (2) would be much smaller, the significantly higher practical specific energy of the stored gas (see Fig. 11.4) in the case of the Li/O_2 reaction (1) should result in a cost of about 5 $/kWh of theoretical energy, a potentially acceptable value.

In summary, the major advantage of a Li/O_2 system with a tank is that a closed system is created that can exclude any atmospheric contaminants from interfering with the reaction chemistry. However, the additional volume (and, to a lesser extent, mass) and the need for oxygen compression are significant challenges.

11.5 Separating O_2 from Air in an Open Li/O_2 System Configuration

While a plausible design for a closed Li/O_2 battery system that uses an O_2 tank is fairly clear, there are more questions when it comes to the design of an open system that removes contaminants from air and sends O_2 and inerts to the cell stack. In particular, the required purity of the gas stream is uncertain, and it will be an important point of investigation once a cell chemistry that works adequately with pure O_2 has been identified. The required purity also has, of course, a major influence on the nature of the separation technologies that can be used and the mass, volume, cost, and other features.

On the issue of required purity, Table 11.9 quantifies the approximate amount of air passing through a Li/O_2 system carrying out reaction (1) over the course of 240,000 km of driving. A separation process that reduced the amounts of H_2O and CO_2 to 1 ppmv would pass about 0.10 kg of CO_2 and 0.040 kg of H_2O into the cell where it would interact with the reaction processes. These values are a small fraction of the ~34.1 kg of Li_2O_2 stored in the battery at full discharge, but it remains too early to assess whether air purification to above or below 1 ppmv will be necessary.

Regarding the expected separation energy, the theoretical energy required for the separation of the five highest-concentration species in air (N_2, O_2, H_2O, Ar, and

Table 11.9 Amount of air, CO_2, and H_2O that would pass through a Li/O_2 battery system forming Li_2O_2 over the course of 240,000 km of driving. We assume air with a moisture content of 0.01 kg of H_2O/kg of dry air, an O_2 stoichiometry of 1.5 (1/1.5 of the O_2 that enters the cell is consumed), and that 175 Wh/km are required by the vehicle from the battery pack

Species	Volume (L)	Mass (kg)
Air	5.5×10^7	7.1×10^4
CO_2	2.2×10^4	39
H_2O	9.8×10^5	710

CO_2), which is a worst-case scenario, is <0.5 Wh/mol, much less than the energy of compression. While in theory a separation process that removes contaminants and provides O_2 to the cells should have a minimal impact on the system efficiency, in practice specific designs must be evaluated because achieving separations at a desired rate will result in higher energy requirements.

What technologies might be used to remove CO_2 and (for the nonaqueous system) H_2O from air? There are numerous air separation technologies, and the ones that are most suitable for use on a vehicle include membranes, pressure-swing adsorption, and temperature-swing adsorption. The best analysis thus far (from Søren Dahl of Haldor Topsoe) suggests that a membrane separation process (to remove a significant amount of H_2O and some CO_2) could produce an output stream with H_2O below 4,000 ppm depending on the H_2O content of the incoming air and the required flow rate, and CO_2 to below 200 ppm [22]. An Air Products dehydration membrane made of hollow polymeric fibers (the Prism Cactus membrane, Model PC4030-D2-6A-20) has a volume of 13 L and a mass of 7 kg and can produce air with water content below 4,000 ppm at a flow rate of about 600 L/min for a pressure drop of 3.5 bar (for supply air with a dewpoint of ~40 °C at 7 bar) [23]. Two of these units would be nearly sufficient to meet the peak power requirement (about 1,500 L/min of air). While a membrane can achieve a significant reduction in the amount of contaminants, an additional step would be necessary to reduce the amount of H_2O and CO_2 to the ppm level. A promising approach for further reduction of H_2O and CO_2 is the use of adsorption onto solids [22]. A zeolite may be used for H_2O adsorption [24] and a diamine-functionalized silica gel for CO_2 adsorption [25]. For our assumptions of an O_2 stoichiometry of 1.5, about 2.6 kg of zeolite (achieving a weight fraction of 16.5 % H_2O at full loading) and 4.7 kg of diamine-functionalized silica gel (achieving a weight fraction of 1.25 % at full loading) would be required; with packing the adsorption units may require about 15 kg and 15 L. The adsorption unit could be regenerated thermally (e. g., by heating to 150 °C), with an initial estimate suggesting that 5–6 % of the battery's energy would be required for a thermal regeneration that would take 1 h [22]. During such a regeneration cycle, the vehicle could not be driven unless a backup adsorption bed was available, and the adsorption processes would be temperature dependent, making it challenging to operate over a wide range of ambient temperatures. Another issue with a thermal regeneration process is that the adsorption unit may need to be carefully insulated so that other battery pack (and possibly vehicle) components are not exposed to high temperatures. While pressure-swing adsorption could be used in place of temperature-swing adsorption, that would introduce a load on the battery pack during operation that temperature-swing adsorption avoids.

In addition to a membrane unit and an adsorption unit, an open system would also need a compressor to increase the partial pressure of O_2 moving through the stack. A unit for a PEMFC designed for a ~3 bar pressure drop would add an additional 12 kg and 10 L [11]. Pressures higher than 3 bar would likely be required to drive air through the separation units in addition to the stack.

Adding up the mass and volume of the membrane, adsorption units, and low-pressure compressor and adding 20 % for the packaging and controls associated with these units, we come out with about 55 kg and 65 L. We stress that these are preliminary values, but it is interesting to note that they are similar to the values for a tank system for Li_2O_2 of 42.1 kg and 66.2 L. In other words, with the important proviso that we aren't including the mass and volume of a high-pressure compressor for the closed-system option, the mass and volume of the oxygen handling equipment for the closed and open systems may be comparable.

How does an open system fit to other performance targets? An open system avoids some of the challenges faced by storing O_2 in high-pressure tanks, including compression, safety issues, and the issue of tank sizing for the maximum expected ambient temperature. The cost is unclear, but will likely be higher than tanks. A safety factor not associated with a closed system is that, during a full charge of a Li/O_2 battery, a significant amount of O_2 will be released. If charging is carried out in a closed space (as is the case with a closed garage), this could significantly increase the partial pressure of O_2, potentially creating safety issues. A 118 kWh Li/O_2 battery back forming Li_2O_2 on discharge would produce 18.5 m^3 of O_2 at 1 bar and 25 °C during a full charge. Also, if the cell chemistry is very sensitive to contaminants from the atmosphere, it would be necessary to ensure that when O_2 is being ejected from the cells no contaminants could enter.

Finally, an open system would also need to limit the escape of volatile components of the cell chemistry. With significant amounts of O_2 being ejected from the cell during charge (and a significant amount of air circulating through the cell during discharge), it may be necessary to either have a trap for volatiles or to be able to replenish them. Given the present uncertainty about the identity and volatility of the cell components, we refrain from speculation about potential processes.

11.6 Conclusions

In this chapter we considered some of the factors relevant to the oxygen handling required by a Li/O_2 couple, with a discussion of both closed and open systems. Oxygen handling is a critical issue because for both the nonaqueous (discharge product: Li_2O_2) and the aqueous (discharge product: $LiOH{\cdot}H_2O$) chemistries, there are components of air (CO_2, H_2O) that react with the discharge products and lead to rapid cell degradation. Our analysis should be seen as preliminary because a chemistry has not yet been identified that results in stable cycling and high-energy, high-power cells for either nonaqueous or aqueous Li/O_2 batteries. We found that the specific energy of a battery system that forms Li_2O_2 or $LiOH{\cdot}H_2O$ significantly exceeds that of an advanced $LiMO_2$ cell (when the oxygen handling components are excluded), although the energy densities of the Li_2O_2 and $LiMO_2$ cells are similar and the energy density of the $LiMO_2$ cell is significantly higher than the $LiOH{\cdot}H_2O$ cell.

While the major advantage of a closed system is that pure oxygen can be used and contaminants from air can be excluded, there is a significant mass and volume associated with carrying a tank on board a vehicle. In addition, a compressor is required during charge (we assume O_2 stored at 350 bar in our calculations here), either requiring it to be stored on board the vehicle and adding to the system mass, volume, and cost or requiring the installation of an external compressor at charging stations. While the tank at 350 bar for the Li/O_2 cell forming Li_2O_2 would occupy a significant fraction of the target system volume (66.2 of 175 L), we note that the energy content of storing O_2 is far higher than storing H_2 (1.99 vs. 0.41 practical kWh/L for O_2 vs. H_2 both at 350 bar).

For an open system, the major advantage is that the oxygen would not need to be carried on board the vehicle, potentially reducing the system mass and volume. The most promising approach identified so far for contaminant removal is the use of a membrane (e.g., a hollow-fiber unit with H_2O as the permeate and a polymeric separation medium) to lower H_2O and CO_2 concentrations, followed by temperature-swing adsorption units that could reduce H_2O and CO_2 concentrations to 1 ppmv (an assumed value; the required purity is still unknown). A compressor would also be required for this approach in order to drive the gases through the separation units and the stack. Our first estimates suggest that the open and closed systems would both require components with a mass of around 50 kg and a volume of around 70 L for the battery design we have chosen. These values exclude the compressor required for the closed system (assuming that an external option may be possible) while including it for the open system; both cases exclude processes that may be required to retain volatiles within the cell stack. Based on our preliminary analysis, it appears that today's oxygen handling technologies will make it difficult to meet the targets for an automotive Li/O_2 system. In order to realize the promise of high-energy Li/O_2 couples therefore, we believe that system design and oxygen handling are critical issues. While advancing gas separation/storage technology itself is one route for advancement, we also encourage those working on cell chemistry to investigate materials that may reduce the deleterious impact of contaminants from air on the cell processes.

References

1. Abraham KM, Jiang Z (1996) A polymer electrolyte-based rechargeable lithium/oxygen battery. J Electrochem Soc 143:1–5
2. Girishkumar G, McCloskey B, Luntz A, Swanson S, Wilcke W (2010) Lithium-air battery: promise and challenges. J Phys Chem Lett 1:2193–2203
3. Christensen J, Albertus P, Sanchez-Carrera RS, Lohmann T, Kozinsky B, Liedtke R et al (2012) A critical review of Li/air batteries. J Electrochem Soc 159:R1–R30
4. Gierszewski P, Finn P, Kirk D (1990) Properties of LiOH and LiNO3 aqueous solutions. Fusion Eng Des 13:59–71
5. McCloskey BD, Scheffler R, Speidel A, Girishkumar G, Luntz AC (2012a) On the mechanism of nonaqueous $Li–O_2$ electrochemistry on C and its kinetic overpotentials: some implications for Li–air batteries. J Phys Chem C 116:23897–23905
6. Kim JS, Johnson CS, Vaughey JT, Thackeray MM, Hackney SA, Yoon W et al (2004) Electrochemical and structural properties of xLi2M′O3 $(1-x)$ LiMn0.5Ni0.5O_2 electrodes for lithium batteries (M′ = Ti, Mn, Zr; 0 x 0.3). Chem Mater 16:1996–2006

7. Wagner FT, Lakshmanan B, Mathias MF (2010) Electrochemistry and the future of the automobile. J Phys Chem Lett 1:2204–2219
8. McCloskey BD, Speidel A, Scheffler R, Miller DC, Viswanathan V, Hummelshøj JS et al (2012b) Twin problems of interfacial carbonate formation in nonaqueous Li/O_2 batteries. J Phys Chem Lett 3(8):997–1001
9. Meini S, Piana M, Tsiouvaras N, Garsuch A, Gasteiger HA (2012) The effect of water on the discharge capacity of a non-catalyzed carbon cathode for Li-O_2 batteries. Electrochem Solid State Lett 15:A45–A48
10. Kuboki T, Okuyama T, Ohsaki T, Takami N (2005) Lithium-air batteries using hydrophobic room temperature ionic liquid electrolyte. J Power Sources 146:766–769
11. Greszler T, Mathias M, Gu W, Goebel S, Masten D, Lakshmanan B (2012) Li-air and Li-sulfur in an automotive system context. Paper presented at beyond lithium ion V, Berkeley, CA, 5–7 June 2012
12. James B, Kalinoski J (2009) Mass-production cost estimation of automotive fuel cell systems. Department of Energy document. Available at http://www.hydrogen.energy.gov/pdfs/review09/fc_30_james.pdf. Accessed 6 Sept 2013
13. Fuel Cell Technologies Office Multi-Year Research, Development and Demonstration Plan (2011b). Section 3.4 Fuel Cells, 2011. Department of Energy document. Available at http://www1.eere.energy.gov/hydrogenandfuelcells/mypp/pdfs/fuel_cells.pdf. Accessed 6 Sept 2013
14. Fuel Cell Technologies Office Multi-Year Research, Development and Demonstration Plan (2011a). Section 3.3 Hydrogen storage, 2011 Interim Update. Department of Energy document. Available at http://www1.eere.energy.gov/hydrogenandfuelcells/mypp/pdfs/storage.pdf. Accessed 6 Sept 2013
15. Goals for Advanced Batteries for EVs (2013) United States Council for Automotive Research document. Available at www.uscar.org/commands/files_download.php?files_id=27. Accessed 6 Sept 2013
16. Viswanathan V, Nørskov J, Speidel A, Scheffler R, Gowda S, Luntz A (2013) Li–O_2 kinetic overpotentials: Tafel plots from experiment and first-principles theory. J Phys Chem Lett 4:556–560
17. Stevens P, Toussaint G, Caillon G, Viaud P, Vinatier P, Cantau C et al (2010) Development of a lithium air rechargeable battery. ECS Trans 28:1–12
18. Nakajima K, Katoh T, Inda Y, Hoffman B (2010) Lithium ion conductive glass ceramics: properties and application in lithium metal batteries. Paper presented at symposium on energy storage beyond lithium ion: materials perspective, Oak Ridge National Laboratory, TN, 7–8 Oct 2010
19. Dudney NJ (2005) Solid-state thin-film rechargeable batteries. Mater Sci Eng B 116:245–249
20. Hua T, Ahluwalia R, Peng J-K, Kromer M, Lasher S, McKenney K et al (2011) Technical assessment of compressed hydrogen storage tank systems for automotive applications. Int J Hydrogen Energy 36:3037–3049
21. Couper JR, Penney WR, Fair JR (2010) Chemical process equipment: selection and design, 2nd edn. Gulf Professional, Houston
22. Dahl S (2012) Air cleaning—perspectives from catalytic processes. Paper presented at beyond lithium ion V, Berkeley, CA, 5–7 June 2012
23. Air Products Cactus® Dryer Performance at 100 psig (2011) Available at http://www.airproducts.com/~/media/Files/PDF/products/prism-membrane-low-pressure-cactus-chart.pdf. Accessed
24. Sircar S, Myers AL (2003) Gas separation by zeolites. In: Auerbarch SM, Carrado KA, Dutta PK (eds) Handbook of zeolite science and technology. Marcel Dekker, Inc., New York (22)
25. Wurzbacher JA, Gebald C, Steinfeld A (2011) Separation of CO_2 from air by temperature-vacuum swing adsorption using diamine-functionalized silica gel. Energy Environ Sci 4:3584–3592

Index

N. Imanishi et al. (eds.), *The Lithium Air Battery: Fundamentals*,
DOI 10.1007/978-1-4899-8062-5, © Springer Science+Business Media New York 2014

D

S

T

U

V

W

X

MIX
Papier aus verantwortungsvollen Quellen
Paper from responsible sources
FSC® C105338

If you have any concerns about our products,
you can contact us on
ProductSafety@springernature.com

In case Publisher is established outside the EU,
the EU authorized representative is:
Springer Nature Customer Service Center GmbH
Europaplatz 3, 69115 Heidelberg, Germany

Printed by Libri Plureos GmbH
in Hamburg, Germany